312887
20/11/79

STRATHCLYDE UNIVERSITY LIBRARY

**MODELING AND CONTROL OF
RIVER QUALITY**

McGraw-Hill Series in Water Resources and Environmental Engineering

Ven Te Chow, Rolf Eliassen, and Ray K. Linsley
Consulting Editors

Bailey and Ollis Biochemical Engineering Fundamentals
Bear Hydraulics of Groundwater
Bockrath Environmental Law for Engineers, Scientists, and Managers
Canter Environmental Impact Assessment
Chanlett Environmental Protection
Graf Hydraulics of Sediment Transport
Haimes Hierarchical Analyses of Water Resources Systems
Hall and Dracup Water Resources Systems Engineering
James and Lee Economics of Water Resources Planning
Linsley and Franzini Water Resources Engineering
Linsley, Kohler, and Paulhus Hydrology for Engineers
Metcalf and Eddy, Inc. Wastewater Engineering: Collection, Treatment, Disposal
Nemerow Scientific Stream Pollution Analysis
Rich Environmental Systems Engineering
Rinaldi, Soncini-Sessa, Stehfest, and Tamura Modeling and Control of River Quality
Schroeder Water and Wastewater Treatment
Tchobanoglous, Theisen, and Eliassen Solid Wastes: Engineering Principles and Management Issues
Walton Groundwater Resources Evaluation
Wiener The Role of Water in Development: An Analysis of Principles of Comprehensive Planning

**McGRAW-HILL
INTERNATIONAL
BOOK COMPANY**

New York
St. Louis
San Francisco
Auckland
Beirut
Bogotá
Düsseldorf
Johannesburg
Lisbon
London
Lucerne
Madrid
Mexico
New Delhi
Panama
Paris
San Juan
São Paulo
Singapore
Sydney
Tokyo
Toronto

S. RINALDI

*Centro Teoria dei Sistemi,
C.N.R.,
Politecnico di Milano,
Italy*

R. SONCINI-SESSA

*Centro Teoria dei Sistemi,
C.N.R.,
Politecnico di Milano,
Italy*

H. STEHFEST

*Kernforschungszentrum, Karlsruhe,
Abteilung für angewandte Systemanalyse,
Karlsruhe,
West Germany*

H. TAMURA

*Department of Precision Engineering,
Osaka University,
Japan*

Modeling and Control of River Quality

This book was set in Times Roman, Series 327

British Library Cataloging in Publication Data

Modeling and control of river quality.
  (McGraw-Hill advanced book program)
  1. Water—Pollution
  I. Rinaldi, S.
  628.1′688′1693   TD425   77-30475

ISBN 0-07-052925-6

**MODELING AND CONTROL OF
RIVER QUALITY**

Copyright © 1979 McGraw-Hill Inc. All rights reserved.
No part of this publication may be reproduced,
stored in a retrieval system, or transmitted, in any form or by any means,
electronics, mechanical, photocopying, or otherwise,
without the prior permission of the publisher.

1 2 3 4   W.&J.M.   8 0 7 9 8

Printed and bound in Great Britain

CONTENTS

Preface ix

List of Symbols xi

1   General Remarks on Modeling 1
   1-1   Internal and external description of dynamical systems 1
   1-2   Different types of models 3
   1-3   The model building process 14
        References 16

2   Water Pollution Processes and Quality Indicators 17
   2-1   Hydrologic phenomena 17
   2-2   Thermal phenomena 20
   2-3   Biochemical and self-purification phenomena 24
   2-4   Other phenomena 41
   2-5   Interrelationships between the phenomena 43
        References 44

3   Structure of the Models 47
   3-1   Balance equations 47
   3-2   Attributes of the models 58
   3-3   The hydrologic submodel 59
   3-4   The thermal submodel 68
   3-5   The biochemical submodel 73
        References 90

4 Some Particular Self-Purification Models — 94
- 4-1 Streeter-Phelps model — 94
- 4-2 Other chemical models — 106
- 4-3 Approximated Streeter-Phelps dispersion models — 117
- 4-4 An ecological model — 129
- 4-5 Other ecological models — 134
- References — 137

5 State and Parameter Estimation — 140
- 5-1 General remarks — 140
- 5-2 Nonlinear parameter estimation of Streeter-Phelps models — 147
- 5-3 Quasilinearization technique with application to the Rhine river — 158
- 5-4 Kalman filtering for parameter and state estimation — 174
- 5-5 Extended Kalman filtering technique and model discrimination — 193
- 5-6 Other applications — 198
- References — 201

6 General Remarks on Control — 204
- 6-1 Control problems — 204
- 6-2 Mathematical programming — 208
- References — 220

7 A Short Survey of Water Pollution Control Facilities — 221
- 7-1 Wastewater treatment — 221
- 7-2 Temperature control — 230
- 7-3 Artificial aeration — 235
- 7-4 Other solutions — 238
- References — 242

8 Steady State Control — 244
- 8-1 General remarks — 244
- 8-2 Linear programming — 246
- 8-3 Nonlinear programming — 254
- 8-4 Dynamic programming — 260
- References — 281

9 Unsteady State Control — 283
- 9-1 General remarks — 283
- 9-2 Second variation and artificial aeration control — 286
- 9-3 Pole assignment and effluent discharge control — 292
- 9-4 Feedforward emergency control — 300
- References — 307

10	**Management of the River Basin**	**309**
	10-1 Tasks of river quality management	309
	10-2 Multiobjective programming in water quality management	318
	10-3 A sequencing problem	323
	10-4 Taxation schemes: static analysis	334
	10-5 Taxation schemes: dynamic analysis	353
	References	365

Index   **368**

PREFACE

The overall aim of this book is to show how the basic principles and techniques of systems analysis can be applied in the description and solution of river pollution problems. Therefore, the book is relatively complete so far as the methodologies are concerned, while some really important river pollution problems have been dealt with only briefly. For instance, relatively little is said about eutrophication phenomena in rivers and about measurement techniques. The real applications shown have been selected with regard to how well they illustrate the methodologies described, rather than with regard to the severity of the pollution problem or the accuracy of the data.

Emphasis is placed on the analytical treatment of the problems. Consequently, simple models and techniques are analyzed fairly thoroughly, while very little attention is paid to large scale simulation studies. For the same reason we decided not to discuss in detail the techniques for numerical integration of distributed parameter models. As a consequence of this, problems of estuarial pollution are only mentioned incidentally, because they can be analyzed in a realistic manner only by means of those integration techniques.

The book is, hopefully, of interest to all people working in the water quality area: hydrologists, hydrobiologists, ecologists, chemical engineers, civil and environmental engineers, resource economists, mathematicians, and systems analysts. Although the level of abstraction is relatively high, the mathematical formalisms are gradually introduced and explained by means of long introductory sections.

The book is organized into two parts. The *first part* of the book deals with the mathematical modeling of river quality phenomena and ends with Chapter 5 (State and Parameter Estimation) where the principles of the estimation techniques are described, and four major applications are presented. A summary of this first part will be given in Sec. 1-3, after having made clear what the notion of modeling means. The *second part* of the book shows how river quality models can be used to achieve or maintain a desired level of river quality in a rational way. Both steady and unsteady state problems are discussed and illustrated by examples. A brief survey of this second part appears in Sec. 6-1.

We would like to thank our home institutions, namely the *Department of Electrical and Electronics Engineering of the Politecnico*, Milano, Italy, the *Division for Applied Systems Analysis of the Nuclear Research Center*, Karlsruhe, W. Germany, and the *Department of Precision Engineering, Osaka University*, Osaka, Japan, for their interest and support in the development of this work. Moreover, a particular acknowledgement is due to the *International Institute for Applied Systems Analysis*, Laxenburg, Austria, and to the *Centro per lo Studio della Teoria dei Sistemi*, C.N.R., Milano, Italy, which jointly sponsored this international cooperative effort. Our warmest thanks are due to Elizabeth Ampt of the International Institute for Applied Systems Analysis, Laxenburg, Austria. She not only typed the uncounted versions of the manuscript, at the same time removing the roughest linguistic mistakes, but she also did much of the coordinating work. Her persistence and enthusiasm were really essential for the completion of the book.

1977

S.R.
R.S.
H.S.
H.T.

LIST OF SYMBOLS[†]

Latin Lower Case

- b biochemical oxygen demand concentration (BOD)
- $\bar{b}$ BOD standard
- c dissolved oxygen concentration (DO)
- $\underline{c}$ DO standard
- c_s oxygen saturation concentration
- $\text{cov}(\cdot)$ covariance
- d dissolved oxygen deficit
- $\bar{d}$ DO deficit standard
- d_c critical DO deficit
- $\det(\mathbf{A})$ determinant of matrix $\mathbf{A}$
- $\text{div}(\cdot)$ divergence operator
- $e^{\mathbf{F}(t-t_0)}$ transition matrix
- $\mathbf{g}$ input-state vector
- $\text{grad}(\cdot)$ gradient operator
- h elevation of the river bottom from a reference level
- $\mathbf{h}^T$ state-output row
- k_1 deoxygenation (degradation) coefficient
- k_2 reaeration coefficient
- l spatial coordinate along the river

[†] This list contains only the symbols which are used in at least two sections of the book. Boldface letters indicate vectors, matrices, or sets. If a vector or matrix degenerates to a scalar, the corresponding italic letter is used.

l_c location of critical DO deficit
m mass
$m(t)$ impulse response
p concentration
$p(A)$ probability density function
$p(A|B)$ probability density function of A conditioned to B
q lateral inflow per unit of length
$r(c)$ reaeration rate
s complex variable (in Laplace transform), substrate concentration
$\mathbf{s}(t)$ sensitivity vector
t time
$\mathbf{u}(\cdot)$ input function
$\mathbf{u}(t)$ input value at time t
v velocity
$\mathbf{v}(t)$ process noise
$\mathrm{var}(\cdot)$ variance
w propagation velocity
$\mathbf{w}(t)$ output (measurement) noise
w_1 easily degradable pollutant concentration
w_2 slowly degradable pollutant concentration
w_3 nondegradable pollutant concentration
$\mathbf{x}(t)$ state at time t
$\mathbf{y}(\cdot)$ output function
$\mathbf{y}(t)$ output value at time t
z vertical spatial coordinate, decision vector
$\mathbf{z}(t)$ extended state at time t

Greek Lower Case

δ variation, perturbation
$\delta(\cdot)$ impulse function
$\varepsilon(t)$ estimation error
$\boldsymbol{\eta}(t, \mathbf{x})$ output transformation
$\boldsymbol{\theta}$ parameter vector
λ eigenvalue, Lagrange multiplier
μ static gain
ξ elevation of the river surface from a reference level
ρ mass density
σ standard deviation
τ flow time
$\tau(l, u)$ tax, charge
ϕ phase
$\phi(\omega)$ phase of frequency response
$\boldsymbol{\phi}(t_0, t, \mathbf{x}_0, \mathbf{u}_{[t_0, t)})$ transition function
$\boldsymbol{\psi}_{t_0, \mathbf{x}_0}$ input–output relationship
ω frequency

Latin Upper Case

- A cross-sectional area of the river
- BOD   biochemical oxygen demand
- B bacterial biomass concentration
- $\mathscr{C}$ cost
- $\mathbf{C}$ controllability matrix
- COD   chemical oxygen demand
- $\mathscr{D}$ dispersion term
- $\mathbf{D}$ dispersion matrix (coefficient)
- DO   dissolved oxygen
- E environmental damage
- $E[\cdot]$ expectation operator
- $\mathbf{F}(t)$ "state-state" matrix
- $\mathbf{G}(t)$ "input-state" matrix
- H river depth
- $\mathbf{H}(t)$ "state-output" matrix
- $\mathbf{I}$ identity matrix
- J performance index
- $\mathbf{K}(t)$ control law matrix
- K_L laboratory BOD decay rate
- $\mathscr{L}[\cdot]$ Laplace transform
- L length of river stretch
- L_c limit value of l_c
- $L(l, t)$ distributed BOD load
- $\mathbf{L}(t)$ Kalman filter gain matrix
- $L(z, \lambda)$ Lagrangian
- $\mathbf{M}(s)$ transfer function
- $\mathbf{O}$ observability matrix
- $\mathscr{P}$ pollution index
- P protozoa biomass concentration, oxygen production
- Q flow rate
- R oxygen consumption due to respiration
- $R(\omega)$ module of frequency response
- S source term in balance equation
- S_Q rate of water inflow
- T time, temperature
- U input set
- $\mathbf{U}(s)$ Laplace transform of $\mathbf{u}(t)$
- V volume
- $\mathbf{V}_\varepsilon$ covariance matrix of the estimation error
- X state set
- $\mathbf{X}(s)$ Laplace transform of $\mathbf{x}(t)$
- Y river width, output set
- $\mathbf{Y}(s)$ Laplace transform of $\mathbf{y}(t)$
- Z feasibility set

Greek Upper Case

Γ	set of output functions $\mathbf{y}(\cdot)$
Δ	variation
$\Delta_{\mathbf{F}}(s)$	characteristic polynomial of the matrix $\mathbf{F}$
$\Theta(t, t_0)$	"input-state" linear transformation
Λ	linear output transformation
$\Phi(t, t_0)$	transition matrix
Ω	set of input functions $\mathbf{u}(\cdot)$
∇	gradient operator

Superscripts

i	ordering index
o	optimal
T	transpose
$\hat{}$	estimate
$*$	particular value
$-$	equilibrium, mean, nominal, particular value, upperbound, standard
$\cdot$	derivative

Subscripts

b	boundary
f	final
h	ordering index
i	initial, ordering index
j	ordering index
k	ordering index
n	ordering index
0	initial, particular value
$-$	lower bound, standard

CHAPTER
ONE

GENERAL REMARKS ON MODELING

1-1 INTERNAL AND EXTERNAL DESCRIPTION OF DYNAMICAL SYSTEMS

Since the first part of this book is devoted to the problem of modeling water quality in rivers it is worth while for the reader to be acquainted with the terminology and with the main concepts and results of what is known today as "system theory," which is nothing but the theoretical background of the "art" of modeling. This chapter is not an equilibrate survey of such a theory, since only the concepts needed in the book are presented; thus, the reader interested in more details should refer to some of the books which are listed in the references to this chapter.

There are many alternative ways of introducing the notion of those abstract objects called systems (or models), but certainly the simplest one (although crude and incomplete) is that of thinking of a system as an object with two groups of variables, called causes and effects, or *inputs* and *outputs*, with some particular relationship between them.

These variables are usually denoted by **u** and **y**, and the system is represented as in Fig. 1-1-1. In general, **u** and **y** are vector-valued functions of time t so that the input and output values at time t are indicated by $\mathbf{u}(t)$ and $\mathbf{y}(t)$ and the corresponding input and output functions by $\mathbf{u}(\cdot)$ and $\mathbf{y}(\cdot)$. The interactions of the system

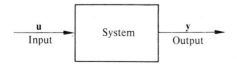

Figure 1-1-1 General representation of a system.

2 GENERAL REMARKS ON MODELING

with other systems (the "rest of the world") are usually reflected in some suitable constraints on t (i.e., $t \in T$), on the values $\mathbf{u}(t)$ and $\mathbf{y}(t)$ and on the functions $\mathbf{u}(\cdot)$ and $\mathbf{y}(\cdot)$, so that we can write

$$\mathbf{u}(t) \in U \qquad \mathbf{u}(\cdot) \in \Omega$$
$$\mathbf{y}(t) \in Y \qquad \mathbf{y}(\cdot) \in \Gamma$$

In some special systems the output $\mathbf{y}(t)$ can be computed from the knowledge of $\mathbf{u}(t)$ simply by means of a function $\mathbf{y}(t) = \mathbf{f}(\mathbf{u}(t))$, but this is not often possible. For example, the rate $u(t)$ at which a biodegradable compound is discharged into a perfectly mixed pool is the ultimate cause of the reactions taking place in the pool, and it is reasonable to say that the concentration $y(t)$ of the dissolved oxygen in the pool is the corresponding effect. Nevertheless, it cannot be assumed that $y(t) = f(u(t))$, since the oxygen concentration at time t is the result of all the past "history" of the pool, and actually does not depend at all upon the value of u at time t. The concentration $y(t)$ can be determined only if the "internal condition" of the pool is given at a specified initial time $t_0 (t_0 < t)$ together with the input segment $u_{[t_0, t)}(\cdot)$ ($u_{[t_0, t)}(\cdot)$ means the input function $u(\cdot)$ restricted to the time interval $[t_0, t)$). This internal condition at time t_0 is represented by the concentrations of all living species, such as bacteria and protozoa, by the concentrations of all chemical compounds, and by the oxygen concentration itself $y(t_0)$. It is therefore apparent that the output of the system can be determined at any time from the internal condition of the system at that time, while the converse is, in general, not true.

In conclusion, it may be said that the output of the system at time t depends upon some initial condition and upon the preceding values of the input; hence these types of systems are called *dynamical systems*.

The internal condition, denoted by $\mathbf{x}(t)$, is called *state*, so that dynamical systems are defined (see, for instance, Kalman et al., 1969, and Zadeh and Polak, 1969) as abstract objects characterized by three dependent variables, namely, input $\mathbf{u}(t) \in U$, state $\mathbf{x}(t) \in X$ and output $\mathbf{y}(t) \in Y$, satisfying the following two properties

$$\mathbf{x}(t) = \boldsymbol{\phi}(t_0, t, \mathbf{x}(t_0), \mathbf{u}_{[t_0, t)}(\cdot)) \qquad t > t_0$$
$$\mathbf{y}(t) = \boldsymbol{\eta}(t, \mathbf{x}(t))$$

The first property says that the state can be updated on any time interval $[t_0, t)$ provided the input function $\mathbf{u}(\cdot)$ is known on that interval (with the exclusion

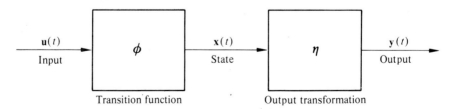

Figure 1-1-2 Internal description of a dynamical system.

of the value of **u** at time t), while the second property says that the output can be determined from the state at the same instant of time. The two functions ϕ and η characterizing any dynamical system are called *transition function* and *output transformation*. Systems described by these two functions are represented as in Fig. 1-1-2 where the internal variable (the state) is explicitly shown. This description of dynamical systems is called, for obvious reasons, *internal description*, while the corresponding models are called input-state-output models or, more briefly, *state models* (sometimes *mechanistic models*, since the knowledge of the state implies that the internal mechanisms of the process have been understood).

In an alternative description of dynamical systems, called *external description*, only input and output variables are mentioned. Of course, from the preceding discussion it is understood that this can only be done by implicitly assuming that the initial conditions are fixed. In fact, if the initial time t_0 and the initial state $\mathbf{x}(t_0) = \mathbf{x}_0$ are fixed the output becomes a function of t and $\mathbf{u}_{[t_0,t)}(\cdot)$ only, i.e.,

$$\mathbf{y}(t) = \psi_{t_0,\mathbf{x}_0}(t, \mathbf{u}_{[t_0,t)}(\cdot))$$

where $\psi_{t_0,\mathbf{x}_0}$ is called *input–output relationship*. Of course, in a dynamical system, there are as many input–output relationships as pairs $(t_0, \mathbf{x}_0)$, although quite often only one of these functions is of practical interest. The determination of an *input–output model* from a state model, is, in principle, always possible in one and only one way since

$$\psi_{t_0,\mathbf{x}_0}(t, \mathbf{u}_{[t_0,t)}(\cdot)) = \eta(t, \phi(t_0, t, \mathbf{x}_0, \mathbf{u}_{[t_0,t)}(\cdot)))$$

while the inverse problem (called *realization* problem) has a unique solution only if all input–output relationships $\psi_{t_0,\mathbf{x}_0}$ are given.

It is clear that from input–output observations over an interval $[t_1, t_2]$ with initial state $\mathbf{x}(t_1) = \mathbf{x}^*$ one will, in general, obtain information about the input–output relationship $\psi_{t_1,\mathbf{x}^*}$, while, only under particular circumstances, can input–output observations be expected to allow the association of a state model with the observed physical system.

1-2 DIFFERENT TYPES OF MODELS

In the preceding section it was shown that two mathematical models can be associated with a physical system, namely a state model (characterized by the transition function ϕ and by the output transformation η) and an input–output model (characterized by the input–output relationship $\psi_{t_0,\mathbf{x}_0}$). But it was not possible to specify these functions, since the analysis must be restricted to particular subclasses of dynamical systems if something about the mathematical tools suitable for their description is to be known. For this reason, the main attributes of the models are now qualified and some results about the required mathematical formalisms given. While discussing these attributes reference will be made to the internal description of dynamical systems

$$\mathbf{x}(t) = \phi(t_0, t, \mathbf{x}(t_0), \mathbf{u}_{[t_0,t)}(\cdot)) \qquad (1\text{-}2\text{-}1a)$$

$$\mathbf{y}(t) = \eta(t, \mathbf{x}(t)) \qquad (1\text{-}2\text{-}1b)$$

since this is the description most frequently used in the book.

The first important attribute refers to the independent variable t(time): if t varies continuously (i.e., if **T** is the set of the real numbers) the model is said to be *continuous* in time, while if t is discrete (i.e., if **T** is the set of the integers) the model is said to be *discrete* in time. Water pollution phenomena are naturally described by continuous models since the variables involved are defined at any instant of time. Nevertheless, if input and output variables are sampled over time, a discrete model in which t represents the sampling time would possibly be more appropriate. Discrete models are, in general, more simple to deal with especially as far as their simulation on a digital computer is concerned, and this is why discretization over time has recently become quite a popular topic.

A second attribute of a system is the dependence of its characteristics upon time. It could be that the same "experiment," i.e., the same initial state and the same input function, gives rise to different output functions when starting from a different initial time. This would be the case in our pool if the photosynthetic oxygen production, which is a function of solar radiation, is not negligible. Systems of this kind are said to be *time-varying* while, on the other hand, systems in which experiments can be shifted over time are called *time-invariant*.

In time-invariant models, the output transformation η does not explicitly depend upon t and the transition function ϕ depends upon the length $(t - t_0)$ of the time interval only, so that any initial time can be, for simplicity, called "zero" and Eq. (1-2-1) becomes

$$\mathbf{x}(t) = \phi(t, \mathbf{x}(0), \mathbf{u}_{[0,t)}(\cdot))$$

$$\mathbf{y}(t) = \eta(\mathbf{x}(t))$$

Another distinction which will be encountered in this book is between *lumped parameter* and *distributed parameter models*. Lumped parameter models are those in which input, state, and output values $\mathbf{u}(t)$, $\mathbf{x}(t)$, and $\mathbf{y}(t)$ can be represented as points of suitable finite dimensional vector spaces. Our pool for example, is a system of this kind under the assumption that the number of living species and chemical compounds is finite. On the other hand, distributed parameter models are those in which at least one of the three variables $\mathbf{u}(t)$, $\mathbf{x}(t)$, $\mathbf{y}(t)$ belongs to an infinite dimensional vector space. Systems of this kind are frequently used in river pollution modeling. In fact, the determination of the concentration of some particular compound at some particular point of the river requires, in general, the knowledge of the initial conditions of the river in all other points, which is equivalent to saying that the state is a function of space. Of course, distributed parameter models can be approximated by lumped parameter models by means of a suitable spatial discretization (e.g., a river can be thought of as a finite sequence of pools, see Sec. 4-3). Since distributed parameter systems are usually described by partial differential equations while lumped parameter systems are described by ordinary differential equations, this discretization can often be interpreted as a particular scheme for the numerical integration of the associated partial differential equation.

A somewhat similar distinction (based on dimensionality) is made between *single input–single output systems* and *multivariable systems*, the former often being much more simple to deal with. In single input–single output systems input $\mathbf{u}(t)$ and output $\mathbf{y}(t)$ are one-dimensional vectors while multivariable systems are those in which input and output are proper vectors. It must be noticed that this attribute of the model is independent from the preceding one, since it is possible to have single input–single output systems whose state space X is infinite-dimensional (an example is a river in which the input is the rate of discharge of a biodegradable compound at a certain point and the output is the dissolved oxygen concentration at a downstream point).

Finally, the distinction between *linear* and *nonlinear systems* is also very important because of the practical implications that this fact can have. Linear systems are defined as those systems in which the transition function is linear in the pair $[\mathbf{x}(t_0), \mathbf{u}(\cdot)]$ and the output transformation is linear in $\mathbf{x}(t)$. Thus, Eq. (1-2-1) can be written as

$$\mathbf{x}(t) = \mathbf{\Phi}(t_0, t)\mathbf{x}(t_0) + \mathbf{\Theta}(t_0, t)\mathbf{u}_{[t_0, t)}(\cdot) \qquad (1\text{-}2\text{-}2a)$$

$$\mathbf{y}(t) = \mathbf{\Lambda}(t)\mathbf{x}(t) \qquad (1\text{-}2\text{-}2b)$$

where $\mathbf{\Phi}(t_0, t)$, $\mathbf{\Theta}(t_0, t)$, and $\mathbf{\Lambda}(t)$ are suitable linear transformations. Equation (1-2-2) is the formal statement of the so-called *superposition principle*, which says that if $\mathbf{y}'(t)$ and $\mathbf{y}''(t)$ are the output values corresponding to the pairs $[\mathbf{x}'(t_0), \mathbf{u}'_{[t_0, t)}(\cdot)]$ and $[\mathbf{x}''(t_0), \mathbf{u}''_{[t_0, t)}(\cdot)]$, respectively, then the output $\mathbf{y}(t)$ corresponding to the pair $[\alpha\mathbf{x}'(t_0) + \beta\mathbf{x}''(t_0), \alpha\mathbf{u}'_{[t_0, t)}(\cdot) + \beta\mathbf{u}''_{[t_0, t)}(\cdot)]$ is $[\alpha\mathbf{y}'(t) + \beta\mathbf{y}''(t)]$. Equation (1-2-2a) also says that the *motion* of the system, i.e., the function $\mathbf{x}(\cdot)$, is the sum of the so-called *free motion* $\mathbf{\Phi}(t_0, t)\mathbf{x}(t_0)$ depending upon the "internal cause" $\mathbf{x}(t_0)$ and the so-called *forced motion* $\mathbf{\Theta}(t_0, t)\mathbf{u}_{[t_0, t)}(\cdot)$ depending upon the "external cause" $\mathbf{u}_{[t_0, t)}(\cdot)$. In other words, in a linear system state and output vectors can be decomposed at any instant of time into two vectors which represent the contributions to the evolution of the system due to the initial state and to the input function.

Particular classes of systems which enjoy more than one of the properties mentioned above will now be considered.

Of particular interest is the class of *continuous, lumped parameter, linear systems* since it can be shown that under very general assumptions (see, for example, Kalman et al., 1969) the evolution of their state is described by n linear ordinary differential equations of the form

$$\frac{dx_1(t)}{dt} = f_{11}(t)x_1(t) + \cdots + f_{1n}(t)x_n(t) + g_{11}(t)u_1(t) + \cdots + g_{1m}(t)u_m(t)$$

$$\vdots$$

$$\frac{dx_n(t)}{dt} = f_{n1}(t)x_1(t) + \cdots + f_{nn}(t)x_n(t) + g_{n1}(t)u_1(t) + \cdots + g_{nm}(t)u_m(t)$$

6 GENERAL REMARKS ON MODELING

where n and m are the dimensions of the state and input vectors, $\mathbf{x}(t) = [x_1(t) \cdots x_n(t)]^T$ and $\mathbf{u}(t) = [u_1(t) \cdots u_m(t)]^T$ (T means transpose) and $f_{ij}(t)$ and $g_{ij}(t)$ are suitable functions of time. The output transformation is described by p linear equations of the form

$$y_1(t) = h_{11}(t)x_1(t) + \cdots + h_{1n}(t)x_n(t)$$
$$\vdots$$
$$y_p(t) = h_{p1}(t)x_1(t) + \cdots + h_{pn}(t)x_n(t)$$

where p is the dimension of the output vector, $\mathbf{y}(t) = [y_1(t) \cdots y_p(t)]^T$ and $h_{ij}(t)$ are again suitable functions of time. These equations can be written in the compact form

$$\dot{\mathbf{x}}(t) = \mathbf{F}(t)\mathbf{x}(t) + \mathbf{G}(t)\mathbf{u}(t) \tag{1-2-3a}$$

$$\mathbf{y}(t) = \mathbf{H}(t)\mathbf{x}(t) \tag{1-2-3b}$$

where $\dot{\mathbf{x}}(t)$ stands for $d\mathbf{x}(t)/dt$ and $\mathbf{F}(t)$, $\mathbf{G}(t)$, and $\mathbf{H}(t)$ are the three matrices $[f_{ij}(t)]$, $[g_{ij}(t)]$, and $[h_{ij}(t)]$ of dimension $n \times n$, $n \times m$, and $p \times n$ respectively. A self-explanatory block-diagram representation of Eq. (1-2-3) is given in Fig. 1-2-1. Equation (1-2-3a) relates the state and the input vector at the same instant of time and can be seen as an indirect way of defining Eq. (1-2-2a) (actually Eq. (1-2-2a) represents the solution of Eq. (1-2-3a)) while Eq. (1-2-3b) directly corresponds to Eq. (1-2-2b).

From the general theory of linear differential equations it follows that the solution of Eq. (1-2-3a) can be given the form

$$\mathbf{x}(t) = \mathbf{\Phi}(t_0, t)\mathbf{x}(t_0) + \int_{t_0}^{t} \mathbf{\Phi}(\xi, t)\mathbf{G}(\xi)\mathbf{u}(\xi)\, d\xi \tag{1-2-4}$$

where the matrix $\mathbf{\Phi}(t_0, t)$, called *transition matrix*, is the solution of the (matrix) differential equation

$$\frac{d\mathbf{\Phi}(t_0, t)}{dt} = \mathbf{F}(t)\mathbf{\Phi}(t_0, t) \tag{1-2-5}$$

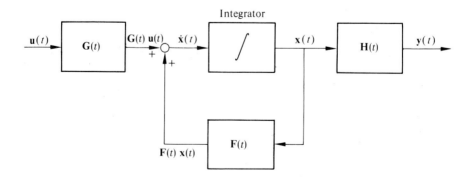

Figure 1-2-1 Block diagram representation of a linear system (internal description).

with initial conditions

$$\Phi(t_0, t_0) = \mathbf{I} \text{ (identity matrix)} \tag{1-2-6}$$

If the system is also *time-invariant* the three matrices $(\mathbf{F}, \mathbf{G}, \mathbf{H})$ appearing in Eq. (1-2-3) are constant in time and, as a consequence, Eq. (1-2-4) can be simplified. In fact, if $\mathbf{F}$ is constant Eq. (1-2-5), with the initial condition (1-2-6), can be solved in closed form:

$$\Phi(t_0, t) = e^{\mathbf{F}(t-t_0)} = \mathbf{I} + \mathbf{F}(t - t_0) + \mathbf{F}^2 \frac{(t - t_0)^2}{2!} + \cdots$$

Moreover, in this case $t_0 = 0$ can be assumed without loss of generality, so that Eq. (1-2-4) can be given the well-known form

$$\mathbf{x}(t) = e^{\mathbf{F}t} \mathbf{x}(0) + \int_0^t e^{\mathbf{F}(t-\xi)} \mathbf{G} \mathbf{u}(\xi) \, d\xi \tag{1-2-7}$$

From this expression it follows that a perturbation $\delta \mathbf{x}(0)$ of the initial state gives rise to a perturbation $\delta \mathbf{x}(t)$ of the state at time t given by

$$\delta \mathbf{x}(t) = e^{\mathbf{F}t} \delta \mathbf{x}(0)$$

which means that any perturbation of the initial state is asymptotically "absorbed" by the system if the transition matrix $e^{\mathbf{F}t}$ tends to the zero matrix for t going to infinity. Systems in which this happens are called *asymptotically stable systems* and it can be shown that the transition matrix tends to zero if and only if all the *eigenvalues* of the matrix $\mathbf{F}$ have a strictly negative real part (recall that the eigenvalues of a matrix $\mathbf{F}$ are the solutions of the *characteristic equation* $\det(\lambda \mathbf{I} - \mathbf{F}) = 0$).

Finally, for single input–single output systems the matrices $\mathbf{G}$ and $\mathbf{H}$ degenerate to a column vector $\mathbf{g}$ and to a row vector $\mathbf{h}^T$ so that Eq. (1-2-7) together with Eq. (1-2-3b) gives

$$y(t) = \mathbf{h}^T e^{\mathbf{F}t} \mathbf{x}(0) + \int_0^t \mathbf{h}^T e^{\mathbf{F}(t-\xi)} g u(\xi) \, d\xi \tag{1-2-8}$$

This is, obviously, the general form of the input–output relationship $\psi_{t_0, \mathbf{x}_0}$ with $t_0 = 0$ and $\mathbf{x}_0 = \mathbf{x}(0)$. A particular input–output relationship is that corresponding to zero initial state $(\mathbf{x}(0) = 0)$, which is often referred to as the input–output relationship of systems initially "at rest." From Eq. (1-2-8) we obtain

$$y(t) = \psi_{0,0}(t, u_{[0,t)}(\cdot)) = \int_0^t \mathbf{h}^T e^{\mathbf{F}(t-\xi)} g u(\xi) \, d\xi \tag{1-2-9}$$

from which it can be concluded that the input–output relationship $\psi_{0,0}$ is known if the function

$$m(t) = \mathbf{h}^T e^{\mathbf{F}t} \mathbf{g} \tag{1-2-10}$$

is known. The function $m(t)$ given by Eq. (1-2-10) is called the *impulse response*

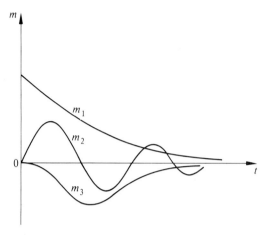

Figure 1-2-2 Three examples of impulse responses of asymptotically stable systems.

of the system since if $u(t)$ is substituted by an impulse function $\delta(t)$ in Eq. (1-2-9) $y(t) = m(t)$ is obtained. The impulse response $m(t)$ turns out to be (see, for example, Zadeh and Desoer, 1963) a linear combination of terms of the kind $t^k e^{\lambda_i t}$ where k is a non-negative integer and λ_i is an eigenvalue of $\mathbf{F}$. Thus, the impulse responses of asymptotically stable systems vanish for t going to infinity as shown in Fig. 1-2-2 for some particular cases.

It is worth while noting that impulse responses different from the one given by Eq. (1-2-10) are obtained for distributed parameter systems. For example, the system described by an input–output relationship of the kind $y(t) = u(t - T)$, which represents a pure time delay T, is obviously a linear distributed parameter system since the state $\mathbf{x}(t)$ is represented by the function $u_{[t-T, t]}(\cdot)$. Thus, this system cannot be described by a triplet $(\mathbf{F}, \mathbf{g}, \mathbf{h}^T)$ since the state is infinite dimensional. Nevertheless, the impulse response can still be defined and is given by the delayed impulse

$$m(t) = \delta(t - T)$$

which is obviously not of the form (1-2-10).

Another important and widely used notion for time invariant, single input–single output, linear systems is the *transfer function* of the system which is defined as the *Laplace transform* $\mathscr{L}$ of its impulse response, i.e.,

$$M(s) = \mathscr{L}[m(t)] \tag{1-2-11}$$

The Laplace transformation $\mathscr{L}[\cdot]$ is the linear operator given by

$$F(s) = \mathscr{L}[f(t)] = \int_0^\infty e^{-st} f(t)\, dt \tag{1-2-12}$$

where s is the complex variable, so that the transfer function of a system turns out to be a complex function of the complex variable s. The Laplace transformation enjoys some remarkable properties including

$$\mathscr{L}[\dot{f}(t)] = sF(s) - f(0)$$

$$\mathscr{L}\left[\int_0^t f(\xi)\,d\xi\right] = \frac{1}{s} F(s)$$

$$\mathscr{L}[f(t-T)] = e^{-Ts} F(s)$$

from which it follows that "derivation" and "integration" in the time domain are equivalent to "multiplication" and "division" by s in the so-called *frequency domain* (the reason for this terminology will shortly become clear), while the shift operation of T units of time applied to the function $f(t)$ is equivalent to the multiplication of its transform $F(s)$ by e^{-Ts}.

From Eqs. (1-2-10)–(1-2-12) it follows that the transfer function of a system described by a triplet $(\mathbf{F}, \mathbf{g}, \mathbf{h}^T)$ is given by

$$M(s) = \mathbf{h}^T \left[\int_0^\infty e^{(-s\mathbf{I}+\mathbf{F})t} \mathbf{g}\,dt\right] = \mathbf{h}^T(s\mathbf{I} - \mathbf{F})^{-1}\mathbf{g} \qquad (1\text{-}2\text{-}13)$$

which is a proper rational function of the kind

$$M(s) = \frac{b_1 s^{n-1} + b_2 s^{n-2} + \cdots + b_n}{s^n + a_1 s^{n-1} + \cdots + a_n} \qquad (1\text{-}2\text{-}14)$$

where n is the order of the system (the dimension of the matrix $\mathbf{F}$). The transfer function (1-2-14) is often used in the alternative form

$$M(s) = \mu \frac{\prod_{i=1}^{n-1}(1 + st_i)}{\prod_{i=1}^{n}(1 + sT_i)} \qquad (1\text{-}2\text{-}15)$$

where μ is the *static gain* and t_i and T_i are the so-called *time constants* of the system. In particular, the time constants T_i are related to the eigenvalues λ_i of the matrix $\mathbf{F}$ by the simple relationship $\lambda_i T_i = -1$ since the polynomial at the denominator of the transfer function (1-2-14) is obviously the *characteristic polynomial*

$$\Delta_\mathbf{F}(s) = \det(s\mathbf{I} - \mathbf{F})$$

Thus, asymptotically stable systems have positive time constants in the denominator of their transfer function.

Of course, transfer functions can be defined in the same way for distributed parameter systems. For example, the following is obtained for the pure time delay $y(t) = u(t - T)$

$$M(s) = \mathscr{L}[m(t)] = \mathscr{L}[\delta(t - T)] = \int_0^\infty e^{-st} \delta(t - T) \, dt = e^{-sT}$$

which is a transcendental transfer function, as is usually the case in distributed parameter systems.

If the equations

$$\dot{\mathbf{x}}(t) = \mathbf{F}\mathbf{x}(t) + \mathbf{g}u(t)$$

$$y(t) = \mathbf{h}^T\mathbf{x}(t)$$

are transformed assuming that the initial state is zero

$$s\mathbf{X}(s) = \mathbf{F}\mathbf{X}(s) + \mathbf{g}U(s)$$

$$Y(s) = \mathbf{h}^T\mathbf{X}(s)$$

is obtained, where $U(s)$, $\mathbf{X}(s)$, and $Y(s)$ are the Laplace transforms of $u(t)$, $\mathbf{x}(t)$, and $y(t)$, and from these equations we get

$$Y(s) = [\mathbf{h}^T(s\mathbf{I} - \mathbf{F})^{-1}\mathbf{g}]U(s)$$

which, taking Eq. (1-2-13) into account, can be rewritten as

$$Y(s) = M(s)U(s) \qquad (1\text{-}2\text{-}16)$$

Equation (1-2-16) is one of the reasons why transfer functions have been and, to some extent, still are so popular: the output (transform) of the system can in fact be immediately obtained by multiplying the transfer function of the system by the input (transform). This property is particularly useful when dealing with systems constituted by suitably connected subsystems: the transfer function of a system constituted by two systems connected in *series (parallel)* is the product (sum) of the single transfer functions as shown in Fig. 1-2-3.

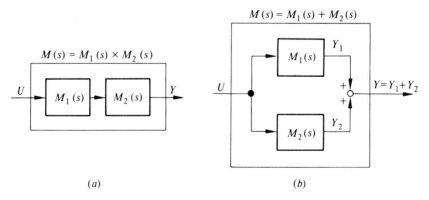

(a) (b)

Figure 1-2-3 Simple connections of two linear systems and their transfer functions: (a) serial connection (b) parallel connection.

If we now recall that the complex variable s corresponds to derivation with respect to time, from Eqs. (1-2-14) and (1-2-16) one obtains the following n-th order linear differential equation relating input and output at the same time

$$\frac{d^n y}{dt^n} + a_1 \frac{d^{n-1} y}{dt^{n-1}} + \cdots + a_n y = b_1 \frac{d^{n-1} u}{dt^{n-1}} + b_2 \frac{d^{n-2} u}{dt^{n-2}} + \cdots + b_n u \qquad (1\text{-}2\text{-}17)$$

This way of describing linear systems is widely used and turns out to be particularly useful in parameter estimation (see Chapter 5). Here, it is only necessary to stress that the impulse response, the transfer function, and Eq. (1-2-17) are completely equivalent descriptions of a linear system and that they represent an input–output model since they define the input–output relationship $\psi_{0,0}$ given by Eq. (1-2-9). As a consequence, the problem of realization of linear single input–single output systems consists of finding a triplet $(\mathbf{F}, \mathbf{g}, \mathbf{h}^T)$ which has a given impulse response, or a given transfer function or a given set of parameters $\{a_i, b_i\}$ in Eq. (1-2-17). The theory of realization of linear systems is well established but is not developed here since it is not used in this book (the interested reader can refer to any recent book of linear system theory and, in particular, to Kalman et al., 1969).

This crude survey of linear system theory would be incomplete without mentioning the so-called *frequency response* which also justifies the interest in the notion of transfer function. Under very general conditions (usually satisfied in applications) linear systems are such that one and only one periodic output function $\tilde{y}(\cdot)$ can be associated with a periodic input function $\tilde{u}(\cdot)$ and that the periods of $\tilde{u}(\cdot)$ and $\tilde{y}(\cdot)$ are the same. Moreover, if the system is asymptotically stable and the input $\tilde{u}(t)$ is applied to the system the output $y(t)$ tends to $\tilde{y}(t)$ for t going to infinity, no matter what the initial state. In particular, if $\tilde{u}(\cdot)$ is a sine wave, i.e.,

$$\tilde{u}(t) = \bar{u} \sin(\omega t)$$

the corresponding output $\tilde{y}(\cdot)$ is a sine wave of the same frequency, i.e.,

$$\tilde{y}(t) = \bar{y} \sin(\omega t + \phi)$$

The output amplitude $\bar{y}$ depends upon $\bar{u}$ and ω and is linear in $\bar{u}$, while the phase ϕ depends only upon ω, i.e.,

$$\bar{y} = R(\omega)\bar{u} \qquad \phi = \phi(\omega)$$

The pair of functions $R(\cdot)$ and $\phi(\cdot)$ is called frequency response and the function $R(\cdot)$ is particularly important since it allows one to specify the filtering properties of a system. In fact if, for example, $R(\cdot)$ is as shown in Fig. 1-2-4 we can say that the system is a low pass filter since low frequency ($\omega < \bar{\omega}$) variations of the input are practically unattenuated while high frequency ($\omega > \bar{\omega}$) components cannot be transferred from the input to the output. Physical systems often have this property so that it becomes very important to be able to compute their frequency response and, in particular, to derive the frequency $\bar{\omega}$ at which the system starts to drastically attenuate. This task turns out to be particularly

12 GENERAL REMARKS ON MODELING

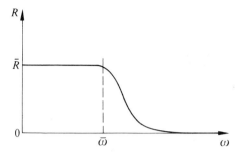

Figure 1-2-4 A low pass filter.

simple to accomplish since it is possible to prove (see, for example, Zadeh and Desoer, 1963) that

$$M(i\omega) = R(\omega) \, e^{i\phi(\omega)} \qquad (1\text{-}2\text{-}18)$$

Thus, $R(\omega)$ and $\phi(\omega)$ can be computed as the module and the argument of the complex number $M(i\omega)$. Moreover, starting from Eq. (1-2-18), it can be shown that, if the eigenvalues λ_i are real, the limit frequency $\bar{\omega}$ is given by the inverse of the maximum time constant of the system

$$\bar{\omega} = 1/\max_i T_i$$

or, which is the same, by

$$\bar{\omega} = -\min_i \lambda_i$$

where λ_i are the eigenvalues of the matrix **F**.

Many of the results just described for single input–single output systems can be generalized to the case of multivariable systems. The only difference is that the vectors **g** and $\mathbf{h}^T$ are substituted by two matrices **G** and **H** and the system is characterized by a matrix of impulse responses given by $\mathbf{H} \, e^{\mathbf{F}t} \, \mathbf{G}$ and by a matrix of transfer functions $\mathbf{H}(s\mathbf{I} - \mathbf{F})^{-1}\mathbf{G}$ so that in Eq. (1-2-16) U and Y are now vectors (**U** and **Y**) and M is a matrix (**M**).

All the results mentioned earlier, with the exception of frequency response, can easily be extended to linear, lumped, *discrete* time systems. As far as the internal description is concerned, these systems are represented by linear difference equations of the form

$$\mathbf{x}(t+1) = \mathbf{F}(t)\mathbf{x}(t) + \mathbf{G}(t)\mathbf{u}(t)$$

$$\mathbf{y}(t) = \mathbf{H}(t)\mathbf{x}(t)$$

where t is now an integer number, and where the matrices $(\mathbf{F}, \mathbf{G}, \mathbf{H})$ are constant if the system is time-invariant. The matrices **G** and **H** are substituted by two vectors **g** and $\mathbf{h}^T$ when dealing with single input–single output systems. The impulse response of these systems can still be defined and is given by

$$m(t) = \begin{cases} 0 & t = 0 \\ \mathbf{h}^T \mathbf{F}^{t-1} \mathbf{g} & t \geq 1 \end{cases}$$

while the transfer function is defined as the so-called Z-transform of the impulse response and turns out to be given, as previously, by

$$M(z) = \mathbf{h}^T(z\mathbf{I} - \mathbf{F})^{-1}\mathbf{g}$$

The functional meaning of the new complex variable z is that of one step anticipation (z^{-1} means one step delay) so that it can be verified immediately that such kinds of systems can also be described by an n-th order difference equation of the kind

$$y(t+n) + a_1 y(t+n-1) + \cdots + a_n y(t) = b_1 u(t+n-1) + \cdots + b_n u(t)$$

which is a form that will frequently be used in this book.

If the system is nonlinear the only thing we can state (see, for instance, Kalman et al., 1969) is that, in general, continuous, lumped parameter systems are described by differential equations of the kind

$$\dot{\mathbf{x}}(t) = \mathbf{f}(\mathbf{x}(t), \mathbf{u}(t), t) \qquad (1\text{-}2\text{-}19a)$$

$$\mathbf{y}(t) = \boldsymbol{\eta}(t, \mathbf{x}(t)) \qquad (1\text{-}2\text{-}19b)$$

while discrete, lumped parameter systems are described by analogous difference equations, i.e.,

$$\mathbf{x}(t+1) = \mathbf{f}(\mathbf{x}(t), \mathbf{u}(t), t) \qquad (1\text{-}2\text{-}20a)$$

$$\mathbf{y}(t) = \boldsymbol{\eta}(t, \mathbf{x}(t)) \qquad (1\text{-}2\text{-}20b)$$

In common practice, linear systems are often used to approximate nonlinear systems. The procedure to be followed is known as *linearization* and is now briefly described for continuous, nonlinear systems of the kind (1-2-19). Suppose that the variables $\mathbf{u}(t)$, $\mathbf{x}(t)$, and $\mathbf{y}(t)$ vary in the vicinity of a particular motion $\mathbf{u}^*(t)$, $\mathbf{x}^*(t)$, and $\mathbf{y}^*(t)$ of the system. Thus, the variations

$$\delta \mathbf{u}(t) = \mathbf{u}(t) - \mathbf{u}^*(t)$$

$$\delta \mathbf{x}(t) = \mathbf{x}(t) - \mathbf{x}^*(t)$$

$$\delta \mathbf{y}(t) = \mathbf{y}(t) - \mathbf{y}^*(t)$$

satisfy the following equations

$$\delta \dot{\mathbf{x}}(t) = -\dot{\mathbf{x}}^* + \mathbf{f}(\mathbf{x}^*(t) + \delta \mathbf{x}(t), \mathbf{u}^*(t) + \delta \mathbf{u}(t), t)$$

$$\delta \mathbf{y}(t) = -\mathbf{y}^*(t) + \boldsymbol{\eta}(t, \mathbf{x}^*(t) + \delta \mathbf{x}(t))$$

The functions $\mathbf{f}$ and $\boldsymbol{\eta}$ can be developed in a Taylor series which can reasonably be truncated after the first order term if $\delta\mathbf{u}$, $\delta\mathbf{x}$, and $\delta\mathbf{y}$ are sufficiently small. Thus, recalling that

$$\dot{\mathbf{x}}^*(t) = \mathbf{f}(\mathbf{x}^*(t), \mathbf{u}^*(t), t)$$

$$\mathbf{y}^*(t) = \boldsymbol{\eta}(t, \mathbf{x}^*(t))$$

14 GENERAL REMARKS ON MODELING

we obtain

$$\delta\dot{\mathbf{x}}(t) = \left[\frac{\partial \mathbf{f}}{\partial \mathbf{x}}\right]_{\mathbf{x}^*,\mathbf{u}^*} \delta\mathbf{x}(t) + \left[\frac{\partial \mathbf{f}}{\partial \mathbf{u}}\right]_{\mathbf{x}^*,\mathbf{u}^*} \delta\mathbf{u}(t) \qquad (1\text{-}2\text{-}21a)$$

$$\delta\mathbf{y}(t) = \left[\frac{\partial \boldsymbol{\eta}}{\partial \mathbf{x}}\right]_{\mathbf{x}^*,\mathbf{u}^*} \delta\mathbf{x}(t) \qquad (1\text{-}2\text{-}21b)$$

which is a linear system of the form (1-2-3) with

$$\mathbf{F}(t) = \left[\frac{\partial \mathbf{f}}{\partial \mathbf{x}}\right]_{\mathbf{x}^*,\mathbf{u}^*} \qquad \mathbf{G}(t) = \left[\frac{\partial \mathbf{f}}{\partial \mathbf{u}}\right]_{\mathbf{x}^*,\mathbf{u}^*} \qquad \mathbf{H}(t) = \left[\frac{\partial \boldsymbol{\eta}}{\partial \mathbf{x}}\right]_{\mathbf{x}^*,\mathbf{u}^*}$$

It is worth while noting that system (1-2-21), called *linearized system*, is in general time-varying even if the nonlinear system (1-2-19) is time-invariant (i.e., if $\mathbf{f}$ and $\boldsymbol{\eta}$ in Eq. (1-2-19) do not explicitly depend upon time). The only significant case in which the linearized system (1-2-21) is time-invariant is when we start from a time-invariant nonlinear system and linearize around an *equilibrium* $(\bar{\mathbf{u}}, \bar{\mathbf{x}}, \bar{\mathbf{y}})$ $(\mathbf{f}(\bar{\mathbf{x}}, \bar{\mathbf{u}}) = \mathbf{0}, \bar{\mathbf{y}} = \boldsymbol{\eta}(\bar{\mathbf{x}}))$, since the three matrices $\mathbf{F}$, $\mathbf{G}$ and $\mathbf{H}$ are then constant.

Finally, it should be mentioned that one way of generalizing the notions introduced in this section is to deal with stochastic dynamical systems. General definitions of such objects are available in the technical literature but reference will not be made to them since the approach used will be mainly deterministic. In the following, and in particular in Chapter 5, some stochastic elements will be introduced into the description, but this will be done by simply modifying what is presented in this section. For example, a deterministic nonlinear system of the form (1-2-19) will be modified into the stochastic system

$$\dot{\mathbf{x}}(t) = \mathbf{f}(\mathbf{x}(t), \mathbf{u}(t), t) + \mathbf{v}(t)$$

$$\mathbf{y}(t) = \boldsymbol{\eta}(t, \mathbf{x}(t)) + \mathbf{w}(t)$$

where $\mathbf{v}(t)$ and $\mathbf{w}(t)$ are called *process* and *output* (*measurement*) *noise* respectively and are assumed to be suitable stochastic processes. Thus, even if $\mathbf{u}(t)$ is a known deterministic function of time, the state $\mathbf{x}(t)$ and the output $\mathbf{y}(t)$ of the system become stochastic processes.

1-3 THE MODEL BUILDING PROCESS

Having presented the basic mathematics of "modeling," the different phases that one usually goes through in associating a model with a real system can now be briefly described. The process of building a mathematical model can be divided into several steps, as shown in Fig. 1-3-1.

The first step, called *conceptualization*, includes selection of relevant variables, formation of ideas on how these variables change and interact, and the establishment of the *structure* of the model. Chapters 2 and 3 are devoted to these problems both from a qualitative and a quantitative point of view. Exploration

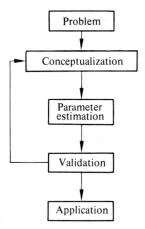

Figure 1-3-1 The process of model building.

of general properties of the model, such as existence of equilibria, stability, and sensitivity, are also part of the conceptualization step (see Chapter 4). This analysis is very important since it constitutes the only rational basis for validating the structure of the model at this stage of the process. If comparison of this analysis with the available data is not satisfactory, the structure of the model is often modified or a particular survey for collecting new observations is suggested. The final result of conceptualization is a class of models whose mathematical structure is usually quite homogeneous.

In order to specify particular models within the class, it is necessary to assign numerical values to some unknown constants, traditionally called *parameters*. This is why the selection of the model which most accurately reproduces the data is called *parameter estimation*, while conceptualization and parameter estimation as a whole are usually called *system identification*. The models, from among which we have to select the best one by means of some parameter estimation scheme, can be either state models or input–output models. If state models are used, in general, suitable observations of the state of the system are required if we wish to identify all the parameters of the model, although in some particular cases (see, for example, Sec. 5-2) this is not necessary. Otherwise, in general, it is only possible to estimate the parameters defining the input–output relationship of the system. The problem of parameter estimation can be transformed into the problem of *state estimation* of a suitable dynamical system. The latter problem consists of estimating the state $\mathbf{x}(t)$ of the system by means of observations (measurements) of the input and output variables over a certain time interval $[t_0, t]$. For this reason, and because state estimation is an interesting problem per se, Chapter 5 will be devoted to both parameter and state estimation problems.

After identification, the model has to be validated, i.e., it is necessary to check whether the model reproduces observations which were not used for the identification step. The result of the *validation* step may be so unsatisfactory that it is necessary to go back to conceptualization in order to modify the structure of the model (see Fig. 1-3-1). The input, state, and output functions which produce the observations used for identification and validation should cover the range in which inputs, states, and outputs are expected to vary although, in practice, models are often used to simulate situations which have not yet been observed.

If a model has passed the validation step, it is applied, i.e., it is used to envisage possibilities for solving the problem which was the starting point for the modeling effort. Possible applications of river quality models are the subject of Chapters 6 to 10. Although the selection of the model structure is governed to a certain extent by the application one has in view, in principle it is hypothesized that the problem can be solved in two steps, namely system identification and model application (*separation hypothesis*). This hypothesis sounds reasonable, but, in general it is found that the best solution to the problem can be obtained only if the two steps are made simultaneously (see Sec. 5-1). This more general approach, however, leads to enormous computational difficulties, so that only in exceptional cases can practical problems be solved in this way.

REFERENCES

Section 1-1

Kalman, R. E., Falb, P. L., and Arbib, M. A. (1969). *Topics in Mathematical System Theory*. McGraw-Hill, New York.
Zadeh, L. A. and Polak, E. (1969). *System Theory*. McGraw-Hill, New York.

Section 1-2

Kalman, R. E., Falb, P. L., and Arbib, M. A. (1969). *Topics in Mathematical System Theory*. McGraw-Hill, New York.
Zadeh, L. A. and Desoer, C. A. (1963). *Linear System Theory*. McGraw-Hill, New York.
Zadeh, L. A. and Polak, E. (1969). *System Theory*. McGraw-Hill, New York.

Section 1-3

No references.

CHAPTER
TWO
WATER POLLUTION PROCESSES AND QUALITY INDICATORS

2-1 HYDROLOGIC PHENOMENA

The study of the dynamics of river flow constitutes perhaps the most classical chapter of the hydrological sciences (see, for example, Gray, 1970). This is largely due to the importance of these studies in flood control and water management in general, and partly to the possibility of describing the phenomenon of surface runoff in a reasonably simple way, as shown in Sec. 3-3.

In principle one should consider *surface runoff* as a part of the so-called *water cycle* which starts with *evaporation*, continues with *formation of clouds*, *precipitation*, *interception* by vegetation, *infiltration*, and *percolation* and terminates with *overland flow*, *interflow*, and *groundwater flow* generating surface runoff as shown in Fig. 2-1-1. Some of these phenomena, such as groundwater dynamics, are characterized by very smooth variations over time, while some others, such as overland flow, are almost immediate responses to precipitation. Spatial scale is also very diverse and changes of state occur in many points of the water cycle. Nevertheless, hydrologists often simplify the description of such a complex system by isolating the main mechanisms involved and by neglecting some interactions between them. A typical result of this approach is the block diagram of Fig. 2-1-2 which represents a still general conceptual model of the entire water cycle. Simplified conceptual models for describing particular aspects of the water cycle can be obtained from this general model by further neglecting some of the interactions between the blocks. For example, if river flow must be determined from rainfall data, then evaporation from ground and water surfaces together with interception can be neglected provided that the time period of the investigation is sufficiently short. The model obtained in this way is a so-called *rainfall–runoff* model and can be used to estimate the *peak flow rates* and their times of

18 WATER POLLUTION PROCESSES AND QUALITY INDICATORS

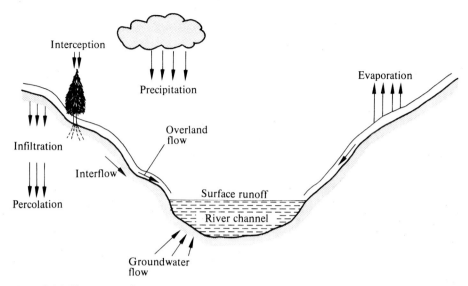

Figure 2-1-1 The water cycle.

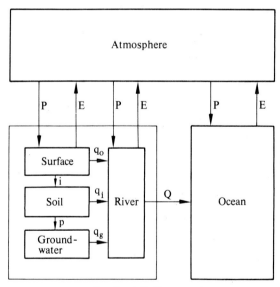

P = Precipitation
E = Evaporation
q_o = Overland flow
q_i = Interflow
q_g = Groundwater flow
Q = River flow
i = Infiltration
p = Percolation

Figure 2-1-2 A general conceptual model of the water cycle.

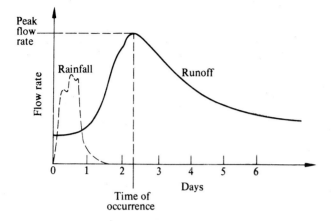

Figure 2-1-3 The runoff hydrograph corresponding to a given rainfall event.

occurrence corresponding to given rainfall events as shown in Fig. 2-1-3 for a particular runoff *hydrograph*. The main properties of this phenomenon are the delay between the rainfall and runoff peaks due to transport time, the smoothing of the rainfall variations due to the diversity among the paths of water particles, and the long tail of the runoff curve due to groundwater flow.

Another problem extensively studied by hydrologists is the propagation of the flood wave in a river stretch. One way of looking at the problem is as follows: suppose a very high variation of flow rate occurs at time t_0 at the upstream end of the river stretch and assume that this variation takes place only during a very short period of time. Subsequent observations at downstream stations show that the wave propagation along the river can be described by a sequence of bell-shaped curves as shown in Fig. 2-1-4. The time of occurrence of the peaks increases with the distance of the station from the upstream end and the bell curves are smoother for downstream stations. The area under each curve is the

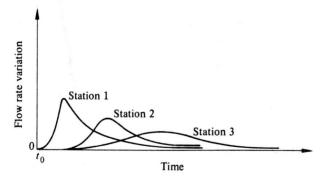

Figure 2-1-4 Time variations of flow rate in three given stations due to an impulse of flow rate at the upstream end of the river.

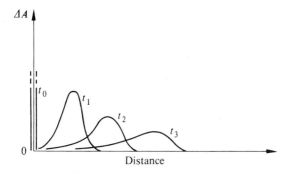

Figure 2-1-5 Spatial variation of the cross-sectional area at different times.

variation of the total volume (or mass) of the water in the river and must therefore be constant if there are no losses in the stretch; thus, the spreading out of the curves implies that the peaks decrease downstream. Another way of looking at the problem is to sample over time the variations of flow rate, or better, of cross-sectional area along the stretch. Again a sequence of bell curves is obtained, as shown in Fig. 2-1-5, and the area underlying the curves is constant since it represents, as before, the total volume of water characterizing the input disturbance. Moreover, quantitative observations show that the bell curves are skewed (steeper at the front of the wave), that the wave propagation velocity is greater than stream velocity, and that the difference between the two velocities increases with the depth of the river.

The relevance of these phenomena in river pollution modeling and control is still relatively limited since for most studies the hydrologic variables can be assumed to be constant in time (this is not the case in estuaries, lakes, and seas). Of course, some relationships between hydrological variables are needed. For example, it is often important to know how the stream velocity is related to the flow rate in steady state conditions. This function, which is increasing and concave, allows one to determine the traveling time of the pollutants between different points on the river when the flow rate is known. Moreover, a hydrologic phenomenon which is very important for river pollution studies is *dispersion*, which is due to the random variations of the velocity in the river (turbulence). Suspended particles and/or dissolved compounds are transferred downstream along different paths so that, on average, pollutants are dispersed in all directions. Although this effect could correctly be explained only by higher dimensional models (models in which there are two or three spatial coordinates), the following will show how dispersion along the axis of the river can be taken into account, together with *molecular diffusion*, by using simple one-dimensional models.

2-2 THERMAL PHENOMENA

An important variable for all river quality considerations is water temperature. Its importance is mainly due to the temperature dependence of many processes through which other quality indicators, like oxygen concentration or pollutant concentration, are determined (these influences will be discussed in the next section). On the other hand, temperature has also to be considered as a quality indicator per se, which affects the kind of fish in the river, the temperature of the drinking water produced from the river water, fog frequency along the river, flow rate, etc. Hence, modeling and control of river quality comprises modeling and control of river temperature, and a prerequisite for this is the understanding of the processes which affect river temperature (see, for example, Krenkel and Parker, 1969; Jobson and Yotsukura, 1972; Heidt, 1975).

These processes are shown symbolically in Fig. 2-2-1, and can be divided into four groups: energy exchange with the atmosphere, energy exchange with the river-bed, internal heat production, and anthropogenic heat addition or subtraction. The radiative energy transport through the water surface consists of three components: short-wave solar radiation, long-wave atmospheric radiation, and long-wave radiation of the river. The *short-wave solar radiation* is that part of the solar radiation which is not absorbed by the atmosphere; it is mainly visible light. The incident energy depends on the atmospheric conditions (e.g., cloudiness) and on the orography of the valley. Part of the incident visible light is reflected, the rest is converted into heat within the river. The proportion reflected depends on the incidence angle, and is usually less than 10 percent. The *long-*

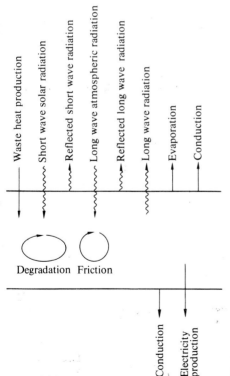

Figure 2-2-1 Heat transfer mechanisms of a river.

wave atmospheric radiation is emitted by the atmosphere above the river (temperature radiation). Its intensity depends on air temperature and humidity. Since the absorption coefficient of air for infrared radiation is low, temperature and humidity in an atmospheric layer of several hundred meters thickness are relevant for this component of the heat balance. About 3 percent of the long-wave atmospheric radiation is reflected. The *long-wave radiation of the water* is emitted only by a very thin surface layer, because the absorption coefficient of water for infrared radiation is high. Hence, the surface temperature is the variable which determines the radiative heat loss of the river.

Energy may also be exchanged between the river and the atmosphere through heat conduction (exchange of *sensible heat*). Pure heat conduction, i.e., heat transport through molecular or atomic collisions, is, however, important only at the air–water interface. Within the two media heat is transported mainly through convection, which occurs in the form of both large scale movement (e.g., advection) and turbulent mixing (*eddy diffusion*, see also Sec. 2-1). The convection in air is primarily forced by the wind. Therefore, the flux of sensible heat through the water surface depends not only on the air and water temperature, but also on wind velocity. Even if no wind is blowing, turbulent mixing is the main transport mechanism for sensible heat in air, because the river flow generates air turbulence through the frictional contact with air. Under particular circumstances *buoyancy forces* may be major sources of convection, for example, if the water temperature is much higher than the air temperature. The convection within the river is, of course, a function of the river flow rate.

The *evaporation* of water from the river also represents a transport of heat, because the change in the state of aggregation requires energy. The heat abstracted from the water through evaporation does not correspond to an increase of air temperature, but occurs as heat only when the water vapor condenses again; that is why one speaks of a flux of *latent heat* from the water to the air. The net evaporation rate is the sum of microscopic evaporation and condensation processes. Thus, the net flux of latent heat obviously depends on the water vapor pressure in the air and on the surface temperature of the water. (The latter determines the saturated vapor pressure of the water.) The dispersion of the evaporated water is governed by the same processes as the dispersion of the sensible heat. Therefore, the dependence of the latent heat flux on wind velocity can be expected to be the same as that of sensible heat. In rare cases the net evaporation is negative, i.e., water vapor condenses at the river surface. The sensible heat carried away by the evaporated water can be neglected. Similarly, the addition of heat through rain, which is not shown in Fig. 2-2-1, is usually negligible.

The heat exchange with the river bed is determined by the heat conductivity of the soil. Since this is very small and also the temperature gradients which occur are moderate, the heat exchange with the river bed can usually also be neglected.

The internal heat production in the river is mainly due to two effects: conversion of potential energy of the water into *frictional heat* and biochemical conversion of chemical energy into heat. The latter plays a minor role even in heavily polluted rivers.

The direct impact of human activities on river temperature consists mainly in the discharge of *waste heat* and in the abstraction of energy, which otherwise would be converted into heat, through *hydropower plants*. Waste heat is a by-product of numerous industrial processes, and the easiest way to dispose of it is by discharging it into rivers (see Sec. 7-2). The main waste heat sources are electric power plants; other sources include chemical plants which have to be cooled, river navigation, or domestic sewage. The abstraction of energy through hydropower plants can often be neglected.

If waste heat is discharged into a river, and one follows the water as it flows downstream, one observes an extra flux of energy from the river to the atmosphere which tends to establish the *natural river temperature*, i.e., the temperature which the river would have if there were no waste heat sources. This extra flux is usually desirable. It can be enhanced by choosing an appropriate outlet structure which influences the way in which the waste heat is admixed to the river. Figure 2-2-2 shows three admixture modes to aim for. The extra heat flux is certainly highest with mode (a), where one is attempting to distribute the heated water quickly over the river surface. But this option has the drawback of hindering the diffusion of oxygen into the water (see Sec. 2-3). The difference between the extra heat fluxes of solution (b) and (c) is small. If rapid mixture is achieved, the heated surface is large but the temperature increase is small; with slow transverse mixing the opposite is the case.

Finally, it should be mentioned that the natural temperature of a river may be changed drastically through hydraulic engineering constructions, which change the heat convection within the river. An extreme example is the construction of a big reservoir with the outlet at the bottom of the dam; then the river temperature immediately downstream of the dam is practically constant throughout the year, while before the construction there may have been large annual and diurnal variations.

Summarizing the discussion of the various heat transfer processes, one can

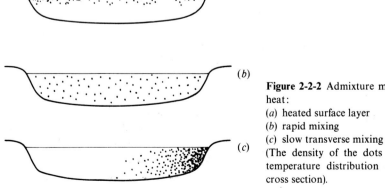

Figure 2-2-2 Admixture modes for waste heat:
(a) heated surface layer
(b) rapid mixing
(c) slow transverse mixing
(The density of the dots represents the temperature distribution over the river cross section).

state that for both the natural river temperature and the additional heat flux due to an artificial temperature increase the most relevant quantities are the convection characteristics of the river and the meteorological parameters which affect the energy transfer through the water surface. The importance of the meteorological parameters has led to the concept of the *equilibrium temperature*, which is defined as the temperature which a completely mixed water column has if the net heat flux through its surface is zero; the column is assumed to be thermally insulated laterally and at the bottom. The equilibrium temperature is determined uniquely by the meteorological conditions; it may be lower or higher than the air temperature. Roughly speaking, one can say that at each moment the energy transfer processes through the river surface tend to establish this temperature level.

2-3 BIOCHEMICAL AND SELF-PURIFICATION PHENOMENA

Most river quality problems are generated by matter which is discharged into the river as a consequence of human activities. Many of these problems are related to the interactions between the discharged matter and river organisms, such as bacteria, rooted plants, and fish; these interactions are now discussed.

A Laboratory Experiment

The impact of a pollutant discharge on river biology can be studied qualitatively through the following laboratory experiment. A sample of river water containing a representative biological community (*biocenosis*) is placed in a reaction tank. A given amount of pollutants at a given instant of time is added, and the resultant processes in the sample are observed. The processes observed correspond to the processes going on in a volume of river water which is flowing downstream; the addition of the pollutant in the experiment corresponds to a single wastewater effluent on the river. While the observations along the river would be disturbed by other effluents and varying hydrology and temperature, in the laboratory experiment described the biochemical processes can be studied under well-defined, constant conditions. That is why observations from such an experiment, which are shown in Fig. 2-3-1 (Münch, 1970), are used as a guide-line for the following discussion. Peptone, which is a mixture of protein fragments, was added at $t = 0$. Hence the "pollutant" was made up of amino acids, which contain considerable amounts of organically bound nitrogen. The development of both the biocenosis and several chemical characteristics was observed. The groups of organisms present other than the bacteria all belong to the subkingdom of *protozoa*, i.e., only unicellular organisms were observed in this experiment. Each group comprises a great variety of species; as an example, Fig. 2-3-2 shows the disaggregation of the class *Ciliata* into single species. Detailed descriptions of the various organisms may be found in textbooks on water biology, e.g., in Liebmann (1962).

Looking at the curves of Figs. 2-3-1 and 2-3-2 it is apparent that peptone

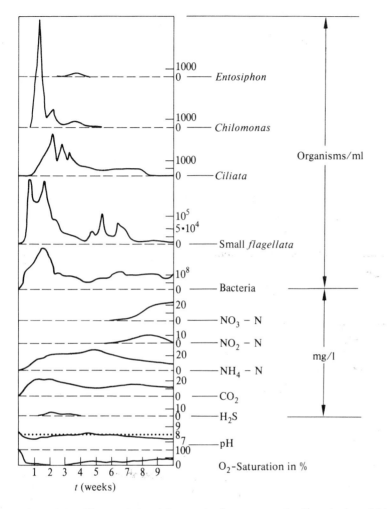

Figure 2-3-1 Changes induced in a natural water sample through the addition of peptone at $t=0$ (after Münch, 1970, $T=10\,°C$. The inorganic nitrogen compounds are represented through their nitrogen content).

stimulates the growth of the species present in the sample. The energy source for this so-called *succession* must be the electrochemical energy of the peptone, since no other energy source is available (the experiment was run in the dark). The appearance of inorganic forms of nitrogen indeed indicates that peptone has been decomposed. The sequential occurrence of ammonium (NH_4^+), nitrite (NO_2^-), and nitrate (NO_3^-), and the subsequent decline of ammonium and nitrite suggest that the primary decomposition product is ammonium, which is then oxidized to nitrate (via nitrite). The concentrations of the dissolved gases carbon dioxide (CO_2) and oxygen (O_2) are at saturation level at the beginning, i.e., the diffusion currents of the gas molecules through the water surface are equal in both directions.

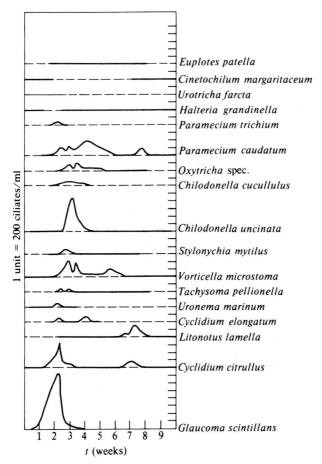

Figure 2-3-2 Development of species of *Ciliata* during the experiment shown in Fig. 2-3-1 (after Münch, 1970).

Hence, the oxygen and carbon dioxide curves in Fig. 2-3-1 indicate that the processes induced by the peptone consume oxygen and liberate carbon dioxide. The curves result from the superposition of these consumption and liberation processes, respectively, and the diffusion, which tends to re-establish the initial equilibrium. Around the third week complete oxygen depletion is observed. The occurrence of hydrogen sulfide (H_2S), which is toxic to most of the organisms, seems to be related to this depletion.

On further inspection of the metabolic activities of the organisms it can be seen that at first the peptone added serves as nutrient for certain bacterial species. The *bacteria* decompose the peptone and use the energy released for growth, multiplication, and maintenance of their life function. They excrete certain organic or inorganic substances, which they cannot use any more. The most relevant of these substances is ammonium. Ammonium is used by some other bacterial species, the *nitrifiers*, as an energy source, the ammonium itself being oxidized (*nitrification*). The organic bacterial excrements are used by *saprozoic flagellates*, to which the groups of small flagellates, *Chilomonas*, and *Entosiphon* in Fig. 2-3-1 belong. These flagellates also decompose the organic matter released after the death of bacteria. The living bacteria and flagellates serve as prey for many ciliates, and these ciliates are eaten by *raptorial ciliates* (e.g., *Litonotus lamella* is a raptorial ciliate, see Fig. 2-3-2). The ciliates also excrete various substances which may be utilized by flagellates or bacteria. Thus, there exists a complex system of nutritional interrelations between the species, which is called *food web* or (less aptly) *food chain*. The population dynamics in Figs. 2-3-1 and 2-3-2 are mainly determined by these interrelations and by competition between species for food.

With all feeding activities chemical energy of the food compounds is set free and used for synthesis of biomass or maintenance of life functions. Hence, part of the released energy is bound again as chemical energy, the rest being converted into heat. The food compounds are thereby oxidized, which means that oxygen is used and organic compounds are converted mainly into carbon dioxide and water. (If no free oxygen is available (*anaerobic conditions*) oxygen from sulfate may be used by some bacteria, resulting in the formation of hydrogen sulfide; see Fig. 2-3-1.) If there is not enough food available, organisms oxidize part of their own matter in order to obtain energy for the maintenance of their life functions until they die of hunger or turn into certain dormant and resistant states, like *bacterial spores*. (The energy consumption for maintenance of life functions, like movement or replenishment of spontaneously degenerated protein molecules, is called *endogenous respiration*.) Hence, after a very long time the biomass in the reaction tank will be as small as at the beginning and the overall result of the experiment will be conversion of peptone into mainly water, carbon dioxide, nitrate, and heat. The water, which may have been fairly turbid during the experiment, becomes as clear as at the beginning. That is why the processes described are designated biochemical *self-purification* (Wuhrmann, 1972). The ability of rivers to purify themselves is an extremely important factor in almost all river quality considerations and plays a major role throughout this book. The importance of self-purification for river quality is illustrated in Fig. 2-3-3 in which the actual organic pollution of the Rhine river, measured as *Chemical Oxygen Demand* (COD) concentration (see Sec. 3-5) is compared with the pollution which would result if the pollutants discharged into the river section shown just accumulated (see Sec. 5-3). The concentration at the downstream end would already be of the same order of magnitude as the concentration in domestic wastewater. Since biochemical self-purification is such an important factor, the most important aspects of it are now discussed in more detail. It will also be necessary to work out the differences between our illustrative experiment and an actual river.

Degradation of Pollutants by Bacteria

The first and most important step in the self-purification process is the degradation

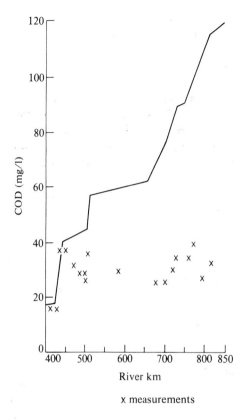

Figure 2-3-3 Organic pollution of the Rhine river: comparison of actual measurements with the pollution which would occur if there were no self-purification.

x measurements

of the originally discharged pollutants by bacteria (and lower order *fungi*). *Degradation* denotes any chemical change of a pollutant which releases electrochemical energy. Bacterial degradation is the most important step because the proportion of the energy of the pollutants which is dissipated is obviously greatest at this level. (Assuming the same efficiency for all energy conversions connected with feeding, and assuming also a strict chain-like structure of the food web, the chemical energy is reduced from link to link of the food chain in a geometric progression.)

The processes related to bacterial degradation are schematically shown in Fig. 2-3-4, in which the energy donors (pollutants) are assumed to be organic molecules. The degradation processes are usually long chains of reactions which are catalyzed by *enzymes*, i.e., degradation takes place practically only if certain highly specific proteins or proteids (enzymes) are present. These enzymes are not changed by the chemical reactions they catalyze. The energy yielding processes during degradation are transfers of electrons such that the potential energy of the electrons is lowered. In the reactions most important for energy deliberation, the electrons are carried by hydrogen (*dehydrogenation* of the energy donor). The energy released is partly bound again through the formation of *adenosine triphosphate* (ATP) from *adenosine di-phosphate* (ADP) and inorganic phosphorus.

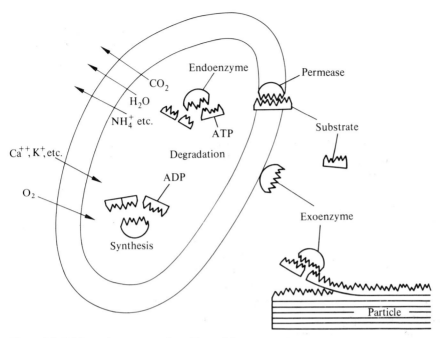

Figure 2-3-4 Schematic representation of bacterial enzyme action.

New bacterial mass is synthesized also by chains of enzymatic reactions. The energy required for this is taken from ATP which is thereby converted into ADP. The building materials for the biomass synthesis (carbon, hydrogen, and small amounts of various other elements) are partly taken from the environment, but also fragments of the pollutants may be integrated directly into the new biomass. The rate at which new bacterial mass is formed is usually limited by the amount of energy which can be gained through degradation rather than by the availability of building materials.

The enzymes mentioned so far are all *endoenzymes*, since they catalyze reactions within the cell. There are particular enzymes, called *permeases*, by which the transport through the cell wall (or, more precisely, through the cytoplasmic membrane) is achieved (diffusion plays no significant role because the transport through the cell wall has to be accomplished against a concentration gradient). If the pollutant molecules are very large (e.g., starch, cellulose, protein), a direct transport into the cell is impossible. In this case, the molecules are decomposed outside the cell into fragments which are small enough. These decompositions are catalyzed by so-called *exoenzymes*. They can be attached to the cell walls as well as be released into the surrounding medium. They differ from the endoenzymes by their small molecular weight (10^4–10^5 as opposed to 10^5–10^6 of the endoenzymes) and by their extremely low cystin and cystein content (Pollock, 1962).

The ability of a bacterium to utilize a pollutant, i.e., to synthesize enzymes which catalyze the degradation of the pollutant, is genetically determined. That is why only those compounds which have been present for a long time in nature are biologically degradable. Compounds which have appeared during the last few decades through the development of chemical technology, cannot often be degraded or can only be partially degraded; among those, for example, are the *chlorinated hydrocarbons* (see, for instance, Eichelberger and Lichtenberg, 1971). Only part of the enzymes, the so-called *constituent enzymes*, is synthesized independently of the available nutrients. The other enzymes are *inducable*, that is, the genetically fixed ability to synthesize them is only realized when the specific substrate (or sometimes other structurally similar compounds) are present.

Four major groups of bacteria, which are defined by the two binary notions *autotrophic–heterotrophic* and *aerobic–anaerobic*, may be distinguished according to the kind of degradation and synthesis processes. A few energy yielding processes which are representative for these groups are shown in Fig. 2-3-5. The distinction between autotrophic and heterotrophic is made according to whether the carbon for synthesis of biomass is of organic or inorganic origin. Autotrophic bacteria get their carbon from carbonic acid or its salts. By and large, this distinction coincides with the distinction as to whether the energy source is inorganic or organic, respectively. The distinction aerobic–anaerobic is based on whether or not oxygen is the final electron acceptor in the degradation process. The two reaction examples given in Fig. 2-3-5 for autotrophic, aerobic organisms are the oxidation of sulfur and ammonium, which can be performed, for example, by the genera *Beggiatoa* and *Nitrosomonas*, respectively. The latter belong to the nitrifying bacteria already mentioned. The autotrophic, anaerobic example is also a sulfur oxidation; since no free oxygen is available nitrate is stripped of oxygen, hence the process is called *denitrification*. *Thiobacillus denitrificans* is an example of a bacterium which is able to denitrify. The two examples in the heterotrophic–aerobic quadrant are the processes performed by baker's yeast and vinegar producing bacteria, respectively. The first example given for heterotrophic–anaerobic organisms is the well-known alcoholic fermentation; the second one, which can be performed by the genus *Desulfovibrio*, reduces sulfate. A reaction similar to this must have occurred around the third week of the experiment

	Autotrophic (C from CO_2 or H_2CO_3)	Heterotrophic (C from organic compounds)
Aerobic	$2S + 2H_2O + 3O_2$ $\rightarrow 2H_2SO_4$ $2NH_4OH + 3O_2$ $\rightarrow 2HNO_2 + 4H_2O$	$C_6H_{12}O_6 + 6O_2$ $\rightarrow 6CO_2 + 6H_2O$ $C_2H_5OH + O_2$ $\rightarrow CH_3COOH + H_2O$
Anaerobic	$5S + 6HNO_3 + 2H_2CO_3$ $\rightarrow 5H_2SO_4 + 2CO_2 + 3N_2$	$C_6H_{12}O_6 \rightarrow 2C_2H_5OH + 2CO_2$ $2C_3H_6O_3 + 3H_2SO_4$ $\rightarrow 3H_2CO_3 + 3H_2O + 3H_2S + 3CO_2$

Figure 2-3-5 Basic types of microbial metabolism and typical energy yielding reactions.

shown in Fig. 2-3-1. Although any bacterial species is adapted to one of the quadrants in Fig. 2-3-5, in many cases bacteria may survive or even reproduce in other quadrants.

The most important type of bacterial metabolism for river quality problems is heterotrophy under aerobic conditions. Anaerobic conditions are avoided whenever possible, because usually unpleasant metabolic by-products, like hydrogen sulfide, occur. And autotrophy is of minor importance because most of the anthropogenic waste discharged into rivers (in terms of electrochemical energy) is organic material. The only important autotrophic bacteria are the *nitrifiers*, because ammonium is a common component of wastewater as well as a common end product of heterotrophic bacterial metabolism. The growth rate of the nitrifiers is, however, quite low, as can be seen from Fig. 2-3-1. Therefore, they play an important part only in slow flowing (e.g., impounded) or overgrown rivers: in overgrown bodies of water the nitrifiers may settle on water plants; if there are no water plants and the river velocity is high, the flow time is too short for the development of a high nitrifier population (Liebmann, 1962; Wezernak and Gannon, 1970; Wolf, 1971; Goering, 1972). Moreover, the growth of the nitrifiers is inhibited by numerous pollutants, so that the influence of nitrification on the self-purification process is often negligible.

The number of different organic compounds which may be found in rivers is immense, and is rapidly increasing as chemical technology develops. Nevertheless, most of these compounds can be decomposed by microorganisms. Because of the great variety of pollutants the bacteria found in rivers are not specialists but show great flexibility in their use of pollutants (see, for example, Hopton, 1970). Most of them belong to the genera *Bacillus, Aerobacter, Pseudomonas, Flavobacterium, Escherichia, Achromobacter, Alcaligenes, Micrococcus, Sphaerotilus*, and *Chromobacterium* (Liebmann, 1962; Hopton, 1970; Frobisher, 1974).

The many degradation pathways which the river bacteria are able to follow are so arranged that with progressing degradation more and more pathways coincide. This is the natural result of evolution, which minimizes the expenditures on enzyme production.

If different bacterial species are able to utilize a certain substance the degradation pathways are in most cases the same. This means, a bacterial community composed of those species behaves similarly to a homogeneous population as far as the degradation of that substance is concerned (see, for example, Gaudy, 1962; Wilderer, 1969). Therefore, the river bacteria, which are so versatile, act to a good approximation like a homogeneous bacterial population against the organic pollutants. This may even be true if just a few species are able to decompose a substance, because often metabolic intermediates are available for the other species. This is especially the case with the end products of reactions catalyzed by *exoenzymes*. Similarly, the small flagellates living on organic bacterial excrements may be looked upon as part of the bacterial population as far as self-purification is concerned, because their population dynamics follow so closely the bacterial population dynamics (see Fig. 2-3-1).

The kinetics of degradation of a certain substance is often specifically influenced by other nutrients or by nondegradable compounds. This influence can consist of the *repression* of the production of an enzyme. Thus, numerous inducible enzymes, especially exoenzymes, are only formed when other more easily degradable nutrients have been used up (Pollock, 1962; Stumm-Zollinger, 1966). Also the activity of enzymes already present may be regulated. This kind of regulation can be achieved through the binding of the regulator molecule to the active site of the enzyme molecule, which is then no longer able to catalyze (*competitive inhibition*, see, for example, Laidler, 1958); in this case the regulator molecule and the substrate molecule are usually structurally similar. [The special case of competitive inhibition in which regulator and substrate molecule are the same occurs if the metabolic pathways of two substrates merge and the slowest (i.e., rate determining) reaction is in the common part of the pathways (Wilderer, 1969).]

In many cases, the regulator molecules are attached to some other part of the enzyme molecule and activate or inhibit it by changing the form of the molecule (*allosteric inhibition*, Laidler, 1958); in these cases there is, in general, no structural similarity between substrate and regulator molecule. Measurement results of a laboratory self-purification experiment in which the degradation of one substance (sorbitol) is inhibited allosterically by another will be shown in Fig. 3-5-7. In competitive inhibition the enzyme activity depends upon the ratio of the concentrations of substrate and regulator; if the substrate concentration is high enough the inhibition can be overcome. On the other hand, in allosteric regulation the enzyme activity depends only on the regulator concentration. Allosteric inhibitions and activations also play an important role in the endogenic regulation of metabolism: the end product of a metabolic pathway acts as an allosteric regulator of the first reaction (feedback). Many substances which occur in wastewaters influence the bacterial metabolism so seriously, even at relatively small concentrations, that bacteria die. *Heavy metals* are an example of such toxic materials.

Great differences in mobility exist within the realm of bacteria. There are attached types as well as various types of flagella. For self-purification considerations, however, the question whether a bacterial species is sessil or motile is less important than the question how many bacteria are actually attached to the river bottom (*benthic*) and how many are suspended in the flowing water (*planctonic*). (The latter category also comprises sessil bacteria, which may be attached to suspended particles or ripped off from the river bottom.) The importance of this distinction, which will be elaborated in the following chapters, can be easily seen. If the conditions for benthic bacteria are favorable downstream of a wastewater inlet (e.g., water weed on which the bacteria may settle) the bacterial degradation activity can be very intense immediately downstream of the inlet. If, however, only planctonic bacteria can grow, and the river upstream of the inlet is quite clean, the maximum of the bacterial degradation may be far downstream because of the time needed to adapt to the wastewater (produce the appropriate enzymes) and to reproduce.

The Role of Higher Order Consumers and of Phototrophs in Self-Purification

Through the bacterial degradation the electrochemical energy of the pollutants has been converted into electrochemical energy of bacterial mass with an efficiency which usually lies between 10 and 60 percent (Servizi and Bogan, 1964; Burkhead and McKinney, 1969; McCarty, 1972). Thus the self-purification cannot be considered finished even if the bacteria have removed all pollutants. The newly created bacterial mass, which must still be considered a kind of pollution (which can be filtered off or sedimented out, however) would decline only very slowly due to *endogenous respiration*; the death rate would become significant only relatively late. But usually self-purification proceeds considerably faster because the bacteria are consumed by protozoa, in particular by ciliates.

The role of *protozoa* in the self-purification process was greatly disputed up to a few years ago, but after several convincing experiments their importance is now an established fact (Javornický and Prokešová, 1963; Bick, 1964; Bhatla and Gaudy, 1965; Straskrabová-Prokešová and Legner, 1966; Münch, 1970; Gaudy, 1972). As an example Fig. 2-3-6 shows the dynamics of bacterial density and oxygen consumption in a laboratory experiment with a river water sample with, and without, the addition of protozoa (Javornický and Prokešová, 1963). One can see that the accumulated oxygen consumption, which can be considered a measure of the pollutants energy which has been converted into heat, is much larger in the first case. The bacterial density is thereby clearly smaller. (After the first day bacterial and protozoan densities are of the order of magnitude of those measured in the river. The smaller bacterial density at the beginning resulted from the fact that in eliminating the natural protozoa many bacteria were inevitably eliminated as well.) Whether the additional oxygen consumption is due solely to the digestion of the bacteria by protozoa, has not yet been clarified. Straskrabová-Prokešová and Legner (1966) suppose, on the basis of their measurements, that the protozoa secrete a substance which enhances the degradation activity of the bacteria. The reason for the high additional oxygen consumption could also be that protozoan grazing, which reduces the competition among the bacterial species, favors the bacteria with high degradation activity more than the others. However, the importance of the protozoa lies mainly in the fact that they control the bacterial density through grazing, and only this aspect is included in the considerations which follow. For example, protozoan grazing should be the main reason for the reduction of the bacterial concentration in the Rhine river between Mainz and Köln which is observed during summer. Figure 2-3-7a shows the bacterial concentration along a section of the Rhine river during summer, calculated as the geometric mean of the measurements taken by the Rhine Water Works during the six summer months of 1967 (ARW, 1969). (Similar values were measured for other years.) Figure 2-3-7b gives the corresponding curves for the six winter months. The opposite behavior between Mainz and Köln in winter (when the self-purification processes are slowed down, see page 39) shows that

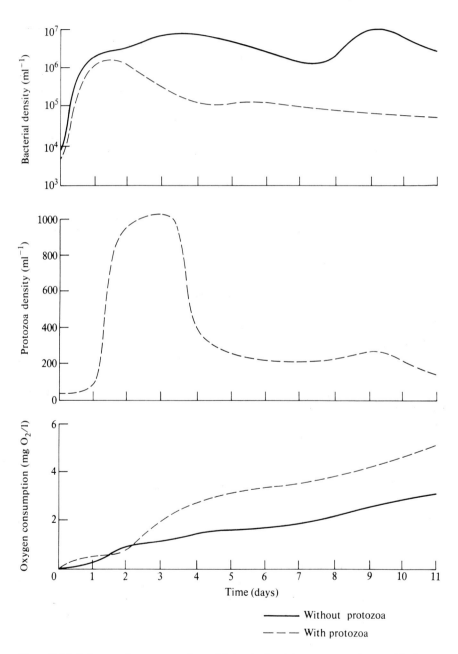

Figure 2-3-6 Influence of protozoa on bacterial density in a laboratory experiment with river water.

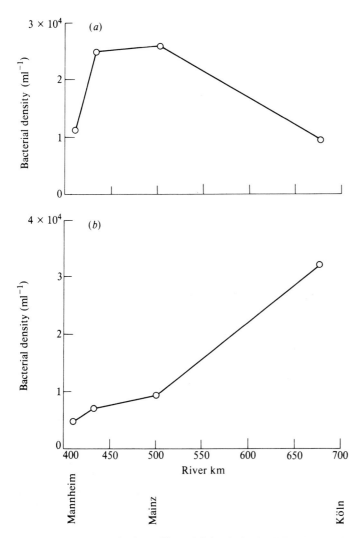

Figure 2-3-7 Measured values of bacterial density in the Rhine river: (a) in summer (b) in winter.

the decline in summer is certainly not caused by differences in the measuring technique. (The plate count technique was used, so one may suspect that the shape of the curve in Fig. 2-3-7a is determined by slight differences in the culture media which are used in the different measurement points.)

The growth rates of bacteria in rivers may vary widely due to the great differences in degradability of their food. For protozoa feeding on bacteria this source of variability of growth rates is much less important, because the protozoan food has essentially always the same composition.

As already mentioned above, the protozoa feeding on bacteria are in turn eaten by higher organisms. Beside the raptorial ciliates, which were involved in the experiment of Fig. 2-3-1, primitive worms (phylum *Nemathelminthes*) and fish are the major predators. Other prey–predator interrelations follow up to the raptorial fish, which lie at the top of the food web.

As with the bacteria, so with higher order consumers *benthic* and *planctonic* organisms must be distinguished, and again among the planctonic organisms there may be sessil forms, which are living on suspended matter or which have been ripped off from the river bottom. Fish can in some cases be assumed to live stationary, i.e., they act like benthic organisms. For other species of fish it might be necessary, however, to take migration into account.

The part of the original electrochemical energy which the higher order consumers dissipate becomes smaller and smaller as one ascends the food web. Nevertheless, the influence of the higher order consumers upon the dynamics of self-purification could be considerable, because they reduce the consumers of lower orders. This, however, is not normally the case, for two reasons. Firstly, the growth rates decrease toward the top of the food web. Therefore those higher consumers which are stream-borne do not have enough time to reach the high population density which could be supported by the nutritional base—the pollutants. Secondly, organisms become, in general, more and more exacting toward the top of the food web, so that many of them cannot survive or breed in heavily polluted water.

Unlike the laboratory experiment described at the beginning, in real rivers the electrochemical energy on which the food web is based, stems not only from the pollutants discharged into the river, but also from *phototrophic* organisms, i.e., organisms which are able to use sunlight as an energy source in building new biomass. The net effect of the phototrophs is practically the inverse of degradation: carbon dioxide, water, and some other substances are combined to energy-rich organic matter, and oxygen is liberated. This organism group contains, beside a few bacteria and many flagellates, algae and higher aquatic plants. Although the phototrophs can use sunlight as an energy source, they often use, either indispensably or facultatively, organic substances. In some cases, growth is possible even in the dark (Round, 1965). The inorganic and organic substances which the phototrophs take up are to a large extent by-products of the degradation processes described above (carbon dioxide, nitrate or ammonium, phosphates, etc.). Thus, the pollutants in fact act as fertilizers for the aquatic flora. This effect, which causes *eutrophication* in lakes, may lead to similar phenomena in slowly flowing rivers, e.g., algal blooms or huge amounts of dying algae in autumn.

In producing new organic matter the phototrophs counteract the self-purification. On the other hand, they may also enhance self-purification. The oxygen released by photosynthesis can prevent anaerobic conditions, which cause the self-purification to proceed much more slowly than under aerobic conditions. However, if the light intensity is not sufficient (during night or in winter) the phototrophs represent an additional oxygen demand because of their endogenous respiration. Figure 2-3-8 gives an example of how the activity of phototrophs influences the oxygen balance of a river. It shows the diurnal variations of oxygen concentration in the Rhine river at Gernsheim during a bright summer day

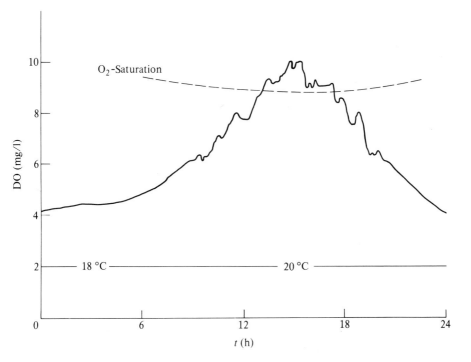

Figure 2-3-8 Diurnal variations of oxygen concentration in the Rhine river at Gernsheim during a bright summer day (after Schulze-Rettmer and Böhnke, 1973).

(Schulze-Rettmer and Böhnke, 1973). It can be seen that the oxygen production through photosynthesis is so intense that oversaturation even occurs. The measurements were taken, however, near the river bank, where the influence of littoral plants and of almost stagnant river branches can be felt. The variations of the cross-sectional mean value are much smaller (see Sec. 5-3). Another beneficial effect on self-purification is that phototrophs provide surfaces for bacteria to attach to, so that the bacterial degradation activity can be very intensive immediately downstream of a wastewater inflow (see page 32). This effect is particularly important for the slowly growing nitrifiers. It should also be mentioned that some species of algae produce compounds which are toxic to higher animals, and which, therefore, disturb drinking water production in particular. These species grow preferably at high temperatures ($> 30\,°C$).

Like the *chemotrophic* organisms, the phototrophic organisms may be benthic or planctonic; the higher plants are without exception stationary. Since the growth rate of most phototrophs is quite small (compared with bacteria, for example), the species which live planctonic are of importance only in very slowly flowing (e.g., impounded) rivers. Phototrophs do not play a major role in deep rivers, because the light intensity in greater depths is not sufficient for their growth; this may be true even for relatively shallow rivers, if they are turbid enough. In general, phototrophs are more exacting than lower order consumers, therefore they cannot grow in heavily polluted rivers. Hence, there are many cases in which the influence of phototrophs on self-purification may be neglected.

Phototrophs serve as food for many animal species, some of which may be able to utilize other kinds of food as well (*omnivores*), while others live exclusively on phototrophs (*herbivores*). The most important group among the latter is the class *Crustacea* (mainly phyllopods and copepods, see Liebmann, 1962). The organisms feeding on phototrophs are subject to the same predatory processes as the organisms mentioned above.

All self-purification processes described up to now are summarized in Fig. 2-3-9 in the form of a food web. The arrows indicate flows of material. The arrows at the compartments "O_2" and "excrements, detritus, etc." which do not end at another compartment symbolize oxygen consumption and waste matter production, respectively, by all living organisms. The slanted lines indicate the environment of the river. It should be appreciated that Fig. 2-3-9 gives only a very approximate picture of reality. The substances and organisms shown are highly aggregated, and only the most important ones are depicted. Nevertheless, it is believed that this picture can be used as a starting point for the second step in conceptualization (see Sec. 1-3), namely formulation of quantitative, mathematical models of self-purification. This will be done in the following chapters. The distinction between benthic and planktonic variables, which has been omitted from Fig. 2-3-9, will also be made.

Each compartment in Fig. 2-3-9 represents one aspect of what is called *water quality*. Too few fish may be perceived as a river quality problem as well as too

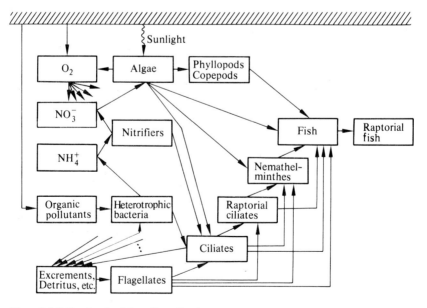

Figure 2-3-9 Food web of the self-purification process.

many bacteria, to give only two examples. Thus, water quality ought to be measured through a quantitative specification of all the compartments in Fig. 2-3-9 (and possibly physical and chemical characteristics, see, for example, Sec. 2-2). This is essentially the way in which Kolkwitz and Marsson (1902) defined water quality when they introduced their *saprobial system* (see also Liebmann, 1962). They defined different categories of water quality through specification of typical *biocenoses*. While this way of measuring water quality is appropriate for descriptive purposes, it is less well suited for anticipatory investigations, because the impact of a pollutant on the whole river biocenosis cannot be evaluated quantitatively. Therefore, water quality is now usually defined through a few elementary variables which characterize the living conditions for the aquatic organisms (Stehfest, 1972). Hence, the quality indicators most frequently used are temperature, oxygen concentration, and pollutant concentration. For the latter, very global measures may be used (see Sec. 3-5) as well as measures which comprise single compounds (e.g., toxins).

The Influence of Physical and Chemical Agents on Self-Purification

The physical river parameters which are most important for self-purification are temperature and flow rate, which have been discussed in the previous sections. If the temperature increases all chemical reactions, and hence all metabolic conversions, are accelerated, unless the temperature is so high that essential compounds of the organisms (e.g., enzymes) denature. If in the experiment of Fig. 2-3-1, for example, the temperature is increased from 10 °C to 20 °C, the curves of Fig. 2-3-10 are observed (Münch, 1970). If one compares Figs. 2-3-1 and 2-3-10 one can see that self-purification proceeds about twice as fast at 20 °C as at 10 °C. At 20 °C the oxygen curve has two clearly distinct minima, the second one obviously being due to nitrification. Some types of organisms occur which have not been observed at 10 °C, even multicellular organisms (e.g., *Rotatoria*). They were probably also present in the 10 °C experiment, but could not reproduce up to an observable population size during the measurement time. In general, not only the growth rates but also the relative abundance of the various species is changed if temperature changes. Each species is adapted to a particular temperature, and, if the actual temperature deviates from this, other species may become superior in the competition for food.

The temperature dependence of the photosynthesis activity of aquatic plants can be much weaker than that of degradation processes if the light intensity is low. The reason is that at low light intensity the limiting reaction is a photochemical one, which should not be influenced by moderate temperature variations. The *endogenous respiration* of the phototrophs, however, depends on temperature as strongly as with the other organisms. This implies that at low light intensities the net growth rate of phototrophs may increase if the temperature is lowered (Round, 1965). The oxygen saturation concentration decreases if the temperature increases, while the decay rate of a given deviation from saturation increases.

It is not clear a priori whether all temperature dependencies mentioned so

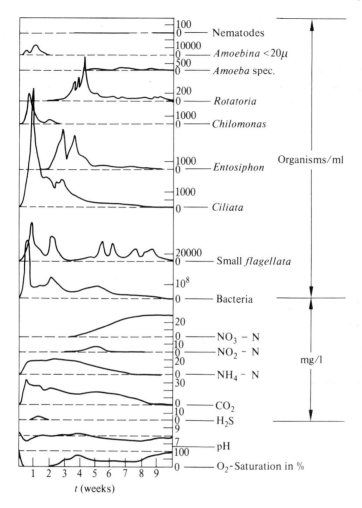

Figure 2-3-10 Laboratory self-purification experiment of Fig. 2-3-1 with increased temperature (after Münch, 1970, $T = 20\,°C$).

far result in an increase or a decrease of oxygen concentration if temperature changes. And, in fact, the river quality models described in the following chapters are such that the oxygen concentration may increase or decrease with decreasing temperature, depending, for example, on the location (see Secs. 4-1 and 5-3).

The main effects of a variation in flow rate are the variations of the flow time between the pollution sources and the variation of the dilution ratio. Again, it is not obvious whether the sum of these effects is a reduction or an increase of the pollutant concentration. Usually a reduction of the pollutant concentration is observed if the flow rate increases; only at very low flow rates, where river velocity increases very quickly with flow rate, may the opposite be observed. A third consequence of flow rate variations, which seems to be less important, is that the ecological interactions within the food web change because the turbulence changes. It is obvious, for example, that the efficiency of bacterial degradation processes which involve exoenzymes drops if turbulence increases.

No general statement can be made about the variation of oxygen concentration with flow rate. Even the influence of flow rate on physical reaeration alone cannot be derived easily, because two consequences have to be considered which compensate each other to a certain extent. If flow rate increases, diffusion of oxygen from air into water is enhanced because of higher turbulence. On the other hand, the oxygen which diffuses through a certain surface area has to be distributed over a greater volume, because the depth has increased.

There are many factors other than temperature and flow rate, which are largely determined externally, and which influence the self-purification processes in rivers. Several of them, such as light intensity and quality of river-bed, have already been mentioned. Another one is the pH value (see Fig. 2-3-1). Its main impact consists in shifting some important chemical equilibria. The equilibrium between ammonia (NH_3) and ammonium (NH_4^+), for example, is shifted toward ammonia if the pH-value is increased; then ammonia may escape and less inorganically bound nitrogen is available (e.g., for phototrophs). But in real rivers this and numerous other possible effects are of minor relevance and are therefore not discussed in detail here.

2-4 OTHER PHENOMENA

There are several processes beside the ones described in Secs. 2-1–2-3 which may be relevant to one or other aspect of river quality. The most important among them are *sedimentation* and *resuspension*, which are obviously closely related to the hydrologic phenomena discussed in Sec. 2-1. If turbulence in a river is weak, which usually happens when river velocity is low, suspended particles may settle. The river water becomes cleaner, so we may speak of *physical self-purification* of the river. Substances which can be degraded biochemically as described in Sec. 2-3 may be degraded within the sediment as well, although the biological species involved are different. Anaerobic conditions are much more likely to occur within the bottom deposits, however, because the oxygen consumed can be replenished only through molecular diffusion, while within the water body turbulent mixing is the dominant oxygen transport mechanism. This implies that biochemical self-purification within the sediments usually proceeds at a slower rate than in the stream above it. Sedimentation of degradable suspended matter has been observed to be relevant only for (cross-sectional mean) velocities smaller than about 0.5 m/s (see, for example, Velz, 1958 and Benoit, 1971). If sedimentation occurred in a river and the flow rate of the river increased considerably, the sediments may be stirred up again. Through this resuspension, large amounts of biodegradable matter can be released into the water instantaneously, which may cause a serious deterioration of the oxygen concentration. The resuspended matter is usually even more easily degradable than the material which has settled out before, due to the slow anaerobic degradation processes within the sediments.

Another phenomenon important for river quality is *sorption*, i.e., the binding of dissolved molecules or ions to solid particles (see, for example, Benoit, 1971). Obviously, this process in conjunction with sedimentation also acts as physical self-purification. The binding forces for sorption are van der Waals' forces in the case of molecules (*adsorption*) and electrostatic forces in the case of ions. When an ion is bound to a particle, another ion, which is less strongly bound, may be displaced into solution (*ion exchange*). Ions of heavy metals, for example, usually displace the common ions of the alkali and alkaline earth metals (like sodium and calcium); this is an important fact for water quality considerations since many heavy metals are extremely toxic (Förstner and Müller, 1974). The particles on which sorption occurs may be either organic or inorganic. The larger the total solid surface, the greater becomes the purification effect, i.e., small particle size enhances this kind of self-purification.

A third, though less important, aspect of physical self-purification is *flocculation*, i.e., destabilization of colloids and formation of flocs which may subsequently settle (see, for example, Singley, 1971). Flocculation is enhanced by high turbulence, and by neutralizing with ions the repulsive charges on the colloidal particles.

There are also purely chemical phenomena which may be said to contribute to self-purification of rivers, the main one being *precipitation*. The insoluble compounds may be formed through both polar and covalent bonds. Precipitation, like ion exchange, is particularly important for the removal of heavy metals. Under the usual chemical conditions in rivers, hydroxides, carbonates, and sulfides of heavy metals have very low solubility, and the formation of these compounds in natural rivers is fairly likely (Förstner and Müller, 1974). The numerous other chemical processes which may possibly influence river water quality, for example, the escape of ammonia (NH_3) in the case of increase in the pH value (see page 41), are not believed to be important in rivers.

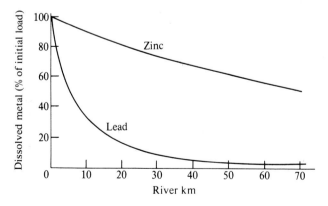

Figure 2-4-1 Removal of dissolved lead and zinc along a section of the Ruhr river (after Koppe, 1973).

No general statements can be made about the relative importance of the phenomena described in this section so far, because they are heavily dependent on the physical and chemical characteristics of the river. The main factors are turbulence, temperature, and pH value. And if one or more of these characteristics change, pollutants which have already been fixed in the sediments may be redissolved or resuspended. Heavy metals, for example, which have been precipitated may easily be dissolved when the pH value is lowered.

An example of how the phenomena described affect the concentration of dissolved heavy metals (zinc and lead) along a river is shown in Fig. 2-4-1 (Koppe, 1973). The measurements were taken on the Ruhr river in West Germany. They show that heavy metals may be fixed to suspended or settled solids almost completely over relatively short distances (lead), but in other cases removal may be much slower (zinc). If a heavy metal is removed quickly the enrichment of that metal in the sediments may yield concentrations which are comparable to concentrations in mineable ores.

Finally, it should be mentioned that all the phenomena described in this section are utilized in both water and wastewater treatment plants (see Sec. 7-1).

2-5 INTERRELATIONSHIPS BETWEEN THE PHENOMENA

All the phenomena presented in the preceding sections will be described in the next chapter by means of differential equations. The resulting set of relationships (the river quality model) may be partitioned into three subsets (namely, the hydrologic, the thermal, and the biochemical submodels), coupled together as

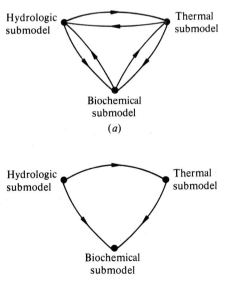

Figure 2-5-1 The links among the three submodels:
(*a*) in the general case
(*b*) when minor effects are neglected.

shown in Fig. 2-5-1a. The state variables of the hydrologic submodels influence all other variables, since all phenomena take place in the fluid medium. The water temperature strongly influences the biochemical processes. It also has an impact on hydrology because of evaporation and buoyancy. The biochemical processes (which are understood to also comprise the processes described in Sec. 2-4) may influence the hydrology through sedimentation or growth of water weed, for example. They may also affect the temperature through heat production or changes of the heat transfer characteristics. In most practical cases, however, the influences of the biochemical processes on hydrology and temperature, and of the temperature on hydrology may be neglected. Then the relations between the submodels are the ones indicated in Fig. 2-5-1b. One can see that the three groups of equations can be solved in cascade, starting from the hydrologic submodel and ending with the biochemical one.

REFERENCES

Section 2-1

Gray, D. M. (1970). *Handbook on the Principles of Hydrology*. Water Information Center, Inc., Port Washington, New York.

Section 2-2

Heidt, F. D. (1975). Heat Exchange Processes at the Surface of Stagnant and Flowing Waters (in German). In *Wärmeeinleitungen in Strömungen* (C. Zimmermann, H. Kobus, and P. Geldner, eds.). Technischer Verlag Resch, München, W. Germany.

Jobson, H. E. and Yotsukura, N. (1972). Mechanics of Heat Transfer in Nonstratified Open-Channel Flows. In *Environmental Impact on Rivers* (*River Mechanics III*) (H. W. Shen, ed.). H. W. Shen, Fort Collins, Colorado.

Krenkel, P. A. and Parker, F. L. (1969). Engineering Aspects, Sources, and Magnitude of Thermal Pollution. In *Biological Aspects of Thermal Pollution* (P. A. Krenkel and F. L. Parker, eds.). Proceedings of the National Symposium on Thermal Pollution, Portland, Oregon, 3–5 June 1968. Vanderbilt University Press, Nashville, Tennessee.

Section 2-3

ARW (Arbeitsgemeinschaft der Rheinwasserwerke) (1969). Annual Report No. 24. Karlsruhe, W. Germany.

Bhatla, M. N. and Gaudy, A. F. (1965). Role of Protozoa in the Diphasic Exertion of BOD. *J. San. Eng. Div., Proc. ASCE*, **91**, 68–87.

Bick, H. (1964). *Succession of Organisms during Self-Purification of Organically Polluted Water under Various Conditions* (in German). Ministerium für Ernährung, Landwirtschaft und Forsten des Landes Nordrhein-Westfalen, Düsseldorf, W. Germany.

Burkhead, C. E. and McKinney, R. E. (1969). Energy Concepts of Aerobic Microbial Metabolism. *J. San. Eng. Div., Proc. ASCE*, **96**, 253–268.

Eichelberger, J. W. and Lichtenberg, J. J. (1971). Persistence of Pesticides in River Water. *Environ. Sci. Technol.*, **5**, 541–544.

Frobisher, M. (1974). *Fundamentals of Microbiology*. W. B. Saunders, Philadelphia.

Gaudy Jr., A. F. (1962). Studies on Induction and Repression in Activated Sludge Systems. *Applied Microbiology*, **10**, 264–271.
Gaudy Jr., A. F. (1972). Biochemical Oxygen Demand. In *Water Pollution Microbiology* (R. Mitchell, ed.). John Wiley, New York.
Goering, J. J. (1972). The Role of Nitrogen in Eutrophic Processes. In *Water Pollution Microbiology* (R. Mitchell, ed.). John Wiley, New York.
Hopton, J. W. (1970). A Survey of some Physiological Characteristics of Stream-Borne Bacteria. *Water Research*, **4**, 493–499.
Javornický, P. and Prokešová, V. (1963). The Influence of Protozoa and Bacteria upon the Oxidation of Organic Substances in Water. *Int. Revue ges. Hydrobiol.*, **48**, 335–350.
Kolkwitz, R. and Marsson, M. (1902). Principles of Biological Evaluation of Water according to its Flora and Fauna (in German). *Mitt. Prüf. Anst. Wasserversorg. Abwasserbeseit.* Berlin, **1**, 33–72.
Laidler, K. J. (1958). *The Chemical Kinetics of Enzyme Action.* Oxford University Press, Oxford, U.K.
Liebmann, H. (1962). *Handbook of Fresh and Waste Water Biology*, vol. I and II (in German). R. Oldenbourg, München, W. Germany.
McCarty, P. L. (1972). Energetics of Organic Matter Degradation. In *Water Pollution Microbiology* (R. Mitchell, ed.). John Wiley, New York.
Münch, F. (1970). The Influence of Temperature on the Decomposition of Peptone and the Concomitant Succession of Organisms with Special Regard to the Dynamics of the Population of the Ciliates (in German). *Int. Revue ges. Hydrobiol.*, **55**, 559–594.
Pollock, M. R. (1962). Exoenzymes. In *The Bacteria: A Treatise on Structure and Function*, vol. IV (J. C. Gunsalus and R. Y. Stanier, eds.). Academic Press, New York.
Round, F. E. (1965). *The Biology of the Algae.* Edward Arnold, London.
Schulze-Rettmer, R. and Böhnke, B. (1973). Contributions to the Oxygen Budget of the Rhine River (in German). *Vom Wasser*, **41**, 187–208.
Servizi, J. A. and Bogan, R. H. (1964). Thermodynamic Aspects of Biological Oxidation and Synthesis. *J. WPCF*, **36**, 607–618.
Stehfest, H. (1972). Definition and Realization of Quality Standards for Rivers (in German). In *Application of the Polluter-Pays-Principle in the Field of Water Management, Proc. of an International Symposium, Kernforschungszentrum Karlsruhe, 20–21 Nov. 1972.* KFK 1804 UF, 146–160, Kernforschungszentrum Karlsruhe, Karlsruhe, W. Germany.
Straskrabová-Prokešová, V. and Legner, M. (1966). Interrelations Between Bacteria and Protozoa during Glucose Oxidation in Water. *Int. Revue ges. Hydrobiol.*, **51**, 279–293.
Stumm-Zollinger, E. (1966). Effects of Inhibition and Repression on the Utilization of Substrates by Heterogeneous Bacterial Communities. *Applied Microbiology*, **14**, 654–664.
Wezernak, C. T. and Gannon, J. J. (1970). Evaluation of Nitrification in Streams. *J. San. Eng. Div., Proc. ASCE*, **94**, 883–895.
Wilderer, P. (1969). *Enzyme Kinetics as Basis of the BOD Reaction* (in German). Thesis, University of Karlsruhe, Karlsruhe, W. Germany.
Wolf, P. (1971). Incorporation of Latest Findings into Oxygen Budget Calculations for Rivers (in German). *GWF—Wasser/Abwasser*, **112**, 200–203 and 250–254.
Wuhrmann, K. (1972). Stream Purification. In *Water Pollution Microbiology* (R. Mitchell, ed.). John Wiley, New York.

Section 2-4

Benoit, R. J. (1971). Self-Purification in Natural Waters. In *Water and Water Pollution Handbook*, vol. I (L. L. Ciacco, ed.). Marcel Dekker, New York.
Förstner, U. and Müller, G. (1974). *Heavy Metals in Rivers and Lakes* (in German). Springer-Verlag, Berlin.
Koppe, P. (1973). Investigation of the Behavior of Waste Water Constituents from Metal-Working Industries in the Water Cycle and their Influence on Drinking Water Supply (in German). *GWF-Wasser/Abwasser*, **114**, 170–175.

Singley, J. E. (1971). Chemical and Physical Purification of Water and Waste Water. In *Water and Water Pollution Handbook*, vol. I (L. L. Ciacco, ed.). Marcel Dekker, New York.

Velz, C. J. (1958). *Significance of Organic Sludge Deposits*. U.S. Public Health Service, Robert A. Taft Sanitary Engineering Center, Technical Report W 58-2, Cincinnati, Ohio.

Section 2-5

No references.

CHAPTER
THREE
STRUCTURE OF THE MODELS

3-1 BALANCE EQUATIONS

After having described the relevant hydrologic, thermal, and biochemical phenomena and identified the corresponding variables, the dynamic laws governing the evolution of the variables can now be derived. These laws are all based on the conservation principles for mass, momentum, and energy, regardless of which of the phenomena described in Chapter 2 is dealt with. In other words, the water quality model results from bookkeeping processes for all the relevant variables. Accordingly, the development of model equations follows the same scheme for all variables. This scheme is discussed now, while Secs. 3-3–3-5 are devoted to the more specific problems of deriving the model equations.

Densities and Velocities in a Fluid

There are several points of view from which one may look at a fluid. One possibility for instance, is the *molecular approach*, which considers the fluid to be a huge number of molecules moving around and colliding with each other in the vacuum. The approach best suited to river quality considerations, however, is the *continuum approach*. Following this approach each point in space is associated with the average value of the property considered (for example, energy, momentum, bacterial mass) over a small reference volume. The advantage of using the continuum approach is that a heterogeneous multicomponent fluid can be described as being composed of different continua, interacting with each other and occupying the same position in space at the same instant of time, so that properties of any of these continua may be assigned to every point of the space (*interpenetrating continua*).

In order to clarify this idea, consider a multicomponent fluid, made up of N different components. Let dM denote the mass contained in a small volume dV, and dm_i the mass of the i-th component in the same volume (the subscript will be omitted later on, if there is no possibility of confusion). The reference volume dV is assumed to be large compared with the mean distance between the particles which are to be described as continuum (e.g., molecules, bacteria); but it has to be small compared with the spatial scale of the phenomena being studied. Then the *mass density* (or *concentration*) ρ_i of the i-th component is defined as

$$\rho_i = \frac{dm_i}{dV} \qquad i = 1, 2, \ldots, N$$

and

$$\sum_{i=1}^{N} \rho_i = \sum_{i=1}^{N} \frac{dm_i}{dV} = \frac{dM}{dV} = \rho$$

is the *bulk mass density*.

These notions may be generalized to other characteristics of the fluid which are proportional to the volume, like momentum or energy (*extensive quantities* as opposed to *intensive quantities* like temperature): for any extensive quantity the *concentration* π_i, which is an intensive quantity, may be defined as

$$\pi_i = \frac{d\Pi_i}{dV}$$

so that the amount Π_i of the extensive property contained in any volume V is given by

$$\Pi_i = \int_V \pi_i \, dV \tag{3-1-1}$$

The average velocity ω_i of the particles of the i-th component in a reference volume dV may, in general, be different from those of the other components. Then, according to the continuum viewpoint, it is necessary to consider, for any point in space, a set of velocities $\{\omega_i\}_{i=1}^{N}$. The *bulk velocity* Ω may then be defined as

$$\Omega = \frac{\sum_{i=1}^{N} \rho_i \omega_i}{\sum_{i=1}^{N} \rho_i} = \sum_{i=1}^{N} \frac{\rho_i}{\rho} \omega_i$$

which, under normal circumstances, is practically equal to the water velocity. The physical meaning of Ω is momentum per unit of mass, so that $\rho \Omega$ is the *bulk momentum density* of the fluid.

Three-Dimensional Balance Equation

In a multicomponent fluid consider an arbitrary volume V, invariable in time, and the total amount Π of an extensive property of a given component (e.g., mass of a given pollutant) contained at time t within volume V. As is well known, the rate of change of quantity Π is given by

$$\frac{\partial \Pi}{\partial t} = -\int_S \pi\omega\mathbf{n}\,dS + \int_V \zeta\,dV \tag{3-1-2}$$

where S is the surface surrounding V, $\mathbf{n}$ is the unit normal vector directed outward from the surface S, and ζ is the rate of production of the extensive property per unit volume. The first term on the right hand side of Eq. (3-1-2) is the flow of the property out of the volume V, while the second term is the total amount of property generated within the volume. Recalling Eq. (3-1-1) and using Gauss' theorem, Eq. (3-1-2) can be written in the form

$$\int_V \left[\frac{\partial \pi}{\partial t} + \operatorname{div}(\pi\omega) - \zeta\right]dV = 0$$

from which it follows

$$\frac{\partial \pi}{\partial t} + \operatorname{div}(\pi\omega) = \zeta \tag{3-1-3}$$

since the volume V is arbitrary. Equation (3-1-3) is the formal expression of the *general conservation principle*.

Equation (3-1-3) describes the characteristics of any component of the fluid at a microscopic time scale. However, one is usually not interested in the very rapid fluctuations of densities and velocities which are due to the Brownian movement of the particles and to the turbulent nature of the river flow. Thus, a simplification may be introduced by averaging all the variables over a representative time interval Δt, which is sufficiently large to filter out the high frequency variations, but sufficiently small not to damp the low frequency variations one is interested in.

More precisely, let the average of a variable x over Δt be denoted by

$$\langle x \rangle = \frac{1}{\Delta t} \int_{\Delta t} x\,dt$$

Then, by definition, for any density π

$$\pi = \langle \pi \rangle + \pi^0 \tag{3-1-4a}$$

and for the velocity of any component

$$\omega = \langle \omega \rangle + \omega^0 \tag{3-1-4b}$$

where π^0 and ω^0 are the random fluctuations around the mean. Inserting Eq. (3-1-4) into Eq. (3-1-3), taking the averages, and using

$$\langle \pi^0 \rangle = \left\langle \frac{\partial \pi^0}{\partial t} \right\rangle = 0$$

and

$$\langle \omega^0 \rangle = \langle \langle \omega \rangle \pi^0 \rangle = \langle \omega^0 \langle \pi \rangle \rangle = 0$$

we obtain

$$\frac{\partial \langle \pi \rangle}{\partial t} + \text{div}\,(\langle \pi \rangle \langle \omega \rangle) + \text{div}\,(\langle \pi^0 \omega^0 \rangle) = \langle \zeta \rangle \qquad (3\text{-}1\text{-}5)$$

The velocity ω can be considered the sum of two terms, the bulk velocity Ω and the *diffusive (or migration) velocity* $(\omega - \Omega)$,

$$\omega = \Omega + (\omega - \Omega)$$

i.e., any movement is seen as a migration with respect to the fluid superimposed on the bulk fluid advection. For simplicity of notation, new symbols are used for the following average values

$$p = \langle \pi \rangle$$
$$\mathbf{v} = \langle \Omega \rangle$$
$$\mathbf{u} = \langle \omega - \Omega \rangle$$
$$I = \langle \zeta \rangle$$

Then Eq. (3-1-5) can finally be written as

$$\frac{\partial p}{\partial t} + \text{div}\,(\mathbf{v}p) + \text{div}\,(\mathbf{u}p) - \mathcal{D} = I \qquad (3\text{-}1\text{-}6a)$$

where

$$\mathcal{D} = -\left[\text{div}\,(\langle \Omega^0 \pi^0 \rangle) + \text{div}\,(\langle (\omega - \Omega)^0 \pi^0 \rangle)\right] \qquad (3\text{-}1\text{-}6b)$$

Equation (3-1-6) is the *general three-dimensional balance equation*.

The role of the different terms in Eq. (3-1-6) is easily identified as follows:

- div $(\mathbf{v}p)$ represents the advection by the bulk fluid motion. The *flow* $\mathbf{v}p$ introduces a coupling with the hydrologic submodel if p is a variable of the thermal or biochemical submodel.
- div $(\mathbf{u}p)$ represents the so-called *migration term*. The *flux* $\mathbf{u}p$ is remarkably different from zero only for nondissolved matter. This term may express the sedimentation movements of particulate matter, the rising of bubble gases towards the surface, or the movements due to "the free will" of animals. Usually it is expressed as a function of the particle characteristics by means of empirical formulae.
- $\mathcal{D}$ represents the total *dispersion and diffusion term* which is due partly to the random variations of the bulk velocity field (*turbulent dispersion* term div $(\langle \Omega^0 \pi^0 \rangle)$) and partly to the random movements at molecular level (*molecular diffusion* term div $(\langle (\omega - \Omega)^0 \pi^0 \rangle)$).

Mixing through molecular diffusion is usually small compared to turbulent mixing, so that the second term of $\mathcal{D}$ can be neglected. The turbulent dispersion term is usually expressed by (see, for example, Sayre, 1972; Holly, 1975; and Yotsukura and Sayre, 1976)

$$\mathscr{D} = \text{div } \mathbf{D} \text{ grad } p \qquad (3\text{-}1\text{-}7)$$

with

$$\mathbf{D} = \begin{bmatrix} D_x & 0 & 0 \\ 0 & D_y & 0 \\ 0 & 0 & D_z \end{bmatrix} \qquad (3\text{-}1\text{-}8)$$

I represents the chemical, biochemical, and ecological interactions. This term, in general, introduces a coupling between the density p and other interacting densities.

When Eq. (3-1-7) is used to express dispersion and a dissolved component is considered ($\mathbf{u} = 0$), Eq. (3-1-6) reduces to

$$\frac{\partial p}{\partial t} + \text{div } (\mathbf{v}p) - \text{div } \mathbf{D} \text{ grad } p = I \qquad (3\text{-}1\text{-}9)$$

Initial and boundary conditions must be given in order to solve Eqs. (3-1-6) or (3-1-9). The boundary conditions are expressed as functions of the production of the property under consideration outside the fluid (e.g., exchange of oxygen with atmosphere at the surface, oxygen release by macrophytes on the bottom, BOD discharge on river banks, or BOD sedimentation on the bottom). Not only the boundary conditions for p and $\mathbf{v}$ must be given but also the existence of physical frontiers for the fluid must be expressed using boundary conditions. Of course, these frontiers are idealized, i.e., most of the intricacy of the real frontiers is smoothed out. For instance, given a fixed coordinate system x, y, z, with $z = \xi$ and $z = h$ being the equations of the air–water interface and of the bottom, respectively, it is assumed that these *boundary surfaces* satisfy the equations (see Nihoul, 1975)

$$\frac{\partial \xi}{\partial t} + v_x \frac{\partial \xi}{\partial x} + v_y \frac{\partial \xi}{\partial y} = v_z \qquad \text{at } z = \xi \qquad (3\text{-}1\text{-}10\text{a})$$

$$\frac{\partial h}{\partial t} + v_x \frac{\partial h}{\partial x} + v_y \frac{\partial h}{\partial y} = v_z \qquad \text{at } z = h \qquad (3\text{-}1\text{-}10\text{b})$$

Equation (3-1-10) expresses the "model" boundaries as surfaces of an idealized fluid moving with the average bulk velocity $\mathbf{v}$. An equation analogous to Eq. (3-1-10) must also be used to express the existence of the *lateral boundaries* (the river banks).

Two- and One-Dimensional Balance Equations

In rivers, estuaries, and lakes the variations of concentrations in all directions may be important. However, the direct solution of a system of coupled three-dimensional balance equations is beyond the capacity of present-day computers, so that a reduction to the two- or one-dimensional case is necessary. Fortunately,

52 STRUCTURE OF THE MODELS

in many situations, variations in one direction are less important than in the others. In rivers, for example, the variations along the vertical line can usually be neglected. This is related to the fact that the width of rivers is much greater than the depth. In these cases one can reduce the dimensionality of the model by averaging the equations over that direction. Even in the most recently developed three-dimensional models of estuaries and lakes (see for instance, Leendertsee et al., 1973, or Chen and Orlob, 1975), the water body is cut horizontally into slices and the equations averaged over the slices are solved slice by slice. Thus, the reduction of three-dimensional balance equations to two or one dimension on the basis of certain simplifying assumptions will now be discussed. An important remark must be made beforehand. A reduction in the dimensionality of the model also curtails the frontiers of the system. If, for instance, one averages over depth, the resulting system has only lateral boundaries, while the surface boundaries have been incorporated in the new balance equation.

Neglecting molecular diffusion and making assumptions (3-1-7, 3-1-8) $\mathscr{D}$ may be written in the form

$$\mathscr{D} = \mathbf{V}_h(\mathbf{D}_h \mathbf{V}_h p) + \frac{\partial}{\partial z} D_z \left(\frac{\partial p}{\partial z} \right)$$

where

$$\mathbf{V}_h = \frac{\partial}{\partial x} \mathbf{i} + \frac{\partial}{\partial y} \mathbf{j}$$

is the two-dimensional gradient operator (**i** and **j** are the unit vectors in the x and y directions) and

$$\mathbf{D}_h = \begin{bmatrix} D_x & 0 \\ 0 & D_y \end{bmatrix}$$

The z-axis of the coordinate system is assumed to be vertical, while the x-axis is pointing in the longitudinal direction. Let $\mathbf{v}_h$ denote the horizontal component of the velocity $\mathbf{v}$

$$\mathbf{v}_h = v_x \mathbf{i} + v_y \mathbf{j}$$

and assume the horizontal migration term equal to zero (this is always the case except for organisms which can move independently of the stream). Then Eq. (3-1-6) can be written as

$$\frac{\partial p}{\partial t} + \mathbf{V}_h(p\mathbf{v}_h) + \frac{\partial}{\partial z}(p v_z) = I - \frac{\partial}{\partial z}(p u_z) + \mathbf{V}_h(\mathbf{D}_h \mathbf{V}_h p) + \frac{\partial}{\partial z}\left(D_z \frac{\partial p}{\partial z} \right) \quad (3\text{-}1\text{-}11)$$

Moreover, let $\bar{p}$ and $\bar{\mathbf{v}}$ denote the depth-averaged concentration and velocity respectively, i.e.,

$$\bar{p} = \frac{1}{H} \int_h^\xi p \, dz \qquad \bar{\mathbf{v}} = \frac{1}{H} \int_h^\xi \mathbf{v} \, dz$$

where $H = \xi - h$ is the river depth. Then concentration and velocity components in each point can be expressed as the mean value plus a fluctuation:

$$p = \bar{p} + \hat{p}$$

$$\mathbf{v}_h = \bar{\mathbf{v}}_h + \hat{\mathbf{v}}_h$$

These expressions may now be inserted into Eq. (3-1-11) which is then integrated over depth. Application of the rules for changing the sequence of differentiation and integration and use of the boundary conditions (3-1-10) yields the following *two-dimensional balance equation*

$$\frac{\partial(H\bar{p})}{\partial t} + \nabla_h(H\bar{p}\bar{\mathbf{v}}_h) = \mathscr{D}_h + HS_h \qquad (3\text{-}1\text{-}12a)$$

where

$$\mathscr{D}_h(x, y) = -\nabla_h \int_h^\xi \hat{p}\hat{\mathbf{v}}_h \, dz + \int_h^\xi \nabla_h(\mathbf{D}_h \nabla_h p) \, dz \qquad (3\text{-}1\text{-}12b)$$

is the global dispersion term, and

$$S_h(x, y) = H^{-1}\left[\int_h^\xi I \, dz + \left(D_z \frac{\partial p}{\partial z} - u_z p\right)_{z=\xi_-} - \left(D_z \frac{\partial p}{\partial z} - u_z p\right)_{z=h_+}\right] \qquad (3\text{-}1\text{-}12c)$$

is the average source term.

The last two terms on the righthand side of Eq. (3-1-12c) represent the fluxes at the air–water interface and at the bottom and must be expressed using the surface boundary conditions for p. Thus, the term HS_h in Eq. (3-1-12a) constitutes the total input in a water column of unit base; if I is a linear function of the concentrations of the different components this term can be expressed as a function of $\bar{p}$ and of the mean concentrations of the interacting constituents. As far as the first term on the righthand side of Eq. (3-1-12b) is concerned, it can be seen that this term describes a dispersion, which is due to the velocity fluctuations in the vertical direction. It would be zero if the velocity were uniform over the depth, but it is well known that this is never the case (the velocity is maximum near the surface and vanishes at the bottom). Assume, for example, the density p to be greater than zero only in a small vertical column, located at the point (x_0, y_0), at a certain instant of time t_0. The distribution of the depth-averaged concentration $\bar{p}$, which is just an impulse in (x_0, y_0) at t_0, would become broader and broader as time goes on, even if there were no turbulent mixing, because the horizontal velocities are different at different depths. Accordingly this term has been incorporated into the dispersion term, and, in analogy with Eq. (3-1-7), this transport can be described by a Fickian type of flow (for example, Sayre, 1972), i.e.,

$$\frac{1}{H}\int_h^\xi \hat{p}\hat{\mathbf{v}}_h \, dz = -\varepsilon \nabla_h \bar{p} \qquad (3\text{-}1\text{-}13)$$

54 STRUCTURE OF THE MODELS

with

$$\varepsilon = \begin{bmatrix} \varepsilon_x & 0 \\ 0 & \varepsilon_y \end{bmatrix}$$

Since the variations of v_x are particularly large, *longitudinal dispersion* is dominated by this mechanism, rather than by longitudinal eddy diffusion. For practical purposes, some further simplifying assumptions are usually made for Eq. (3-1-12), which may lead to (see, for example, Holley et al., 1972)

$$\frac{\partial(Hp)}{\partial t} + \mathbf{V}_h(Hp\mathbf{v}_h) = \mathbf{V}_h(H(\varepsilon + \mathbf{D}_h)\mathbf{V}_h p) + HS_h \qquad (3\text{-}1\text{-}14)$$

where the over-bars have been left out for the sake of simplicity.

Integrating Eq. (3-1-12) over the river width one can derive, in a very similar manner, the *one-dimensional balance equation*

$$\frac{\partial(Ap)}{\partial t} + \frac{\partial(Avp)}{\partial l} = \mathscr{D} + AS \qquad (3\text{-}1\text{-}15a)$$

where p and v now denote concentration and velocity averaged over the river cross section A, and the symbol l is used now for the distance along the axis of the river. The terms $\mathscr{D}$ and S are expressed by formulae analogous to Eqs. (3-1-12b) and (3-1-12c) (see also Sec. 3-5). Similarly, as in Eq. (3-1-14), one can write approximately

$$\mathscr{D} \doteq \frac{\partial}{\partial l}\left(AD\frac{\partial p}{\partial l}\right) \qquad (3\text{-}1\text{-}15b)$$

where D is now a scalar, called *longitudinal dispersion coefficient*. As already indicated above, longitudinal dispersion is mainly due to the combined effects of vertical and transverse mixing and the variations of the longitudinal velocity over the cross section. In other words, the transfer coefficient D in Eq. (3-1-15b) is much greater than D_x in Eq. (3-1-8). Since, for practical river studies the one-dimensional balance equation is mainly used, much effort has been devoted to the prediction or experimental determination of the longitudinal dispersion coefficient (see, for example, Bansal, 1971; Sayre, 1972; and McQuivey and Keefer, 1974).

The lateral boundary conditions expressing the exchange of p with the river banks are now included in the source term AS. Thus, considering a river stretch of length L, the only boundary conditions for Eq. (3-1-15a) are now $p(0, t)$ and $p(L, t)$, i.e., the concentrations at the upstream and the downstream end respectively. The cross-sectional average source $S(\text{mg/h m}^3)$ is given by

$$S = S_v + \frac{S_s}{H} + \frac{S_l}{A} \qquad (3\text{-}1\text{-}15c)$$

where S_v is the average volume source (mg/h m^3), S_s the average surface source (mg/h m²) and S_l the lateral source (mg/h m). Finally, it is worth knowing that Eq. (3-1-15) can be rearranged in the form

$$\frac{\partial p}{\partial t} + v\frac{\partial p}{\partial l} = -\frac{S_Q}{A}p + \frac{\partial}{A\,\partial l}\left(AD\frac{\partial p}{\partial l}\right) + S \qquad (3\text{-}1\text{-}16a)$$

where

$$S_Q = \frac{\partial A}{\partial t} + \frac{\partial(vA)}{\partial l} \qquad (3\text{-}1\text{-}16b)$$

is the rate of water inflow (see Sec. 3-3).

Limitations of One-Dimensional Models

In river quality studies Eq. (3-1-15) is generally believed to be sufficiently detailed, while the two-dimensional equation (3-1-14) is generally used for estuaries (Chen and Orlob, 1975; Orlob, 1976). However, it must be kept in mind that the one-dimensional equation is an approximation which may lead to severe errors if its limitations are not carefully considered (Ward and Fischer, 1971). For example, only if a quasi-equilibrium between lateral and vertical dispersion on the one hand, and differential convection on the other, has been established, are the average concentrations good indices and does Eq. (3-1-15b) correctly describe the dispersion effects (Sayre, 1972). In general, wastewaters are injected in single points, so that Eq. (3-1-15) is not adequate in the immediate vicinity of the outlet. Hence, before using a one-dimensional equation it is necessary to estimate the distance required by a tracer, injected in a point, to become well mixed and only if this distance comes out to be short with respect to the spatial scale of the phenomena under analysis, can the use of a one-dimensional model be considered to be justified.

Often the application of Eq. (3-1-15) is unjustified since there are pockets of little or no flow on the river banks. A tracer pulse released in the river can be partially trapped in these pockets. Thus, a sensor located in the stream sees first a large bulk of material and then a long tail until the pockets along the side are completely emptied of the tracer. In this case, the river may be divided into a main stream and a *stagnant zone* and the dynamics of each variable be described by two coupled one-dimensional equations (see Hays, 1966; Thackston and Schnelle, 1970; Sayre, 1972):

$$\frac{\partial(A_m p_m)}{\partial t} + \frac{\partial(A_m v p_m)}{\partial l} = \frac{\partial}{\partial l}\left(A_m D \frac{\partial p_m}{\partial l}\right) - \alpha(p_m - p_s) + A_m S_m$$

$$\frac{\partial(A_s p_s)}{\partial t} = \alpha(p_m - p_s) + A_s S_s$$

where indices m and s refer to main stream and stagnant zone respectively.

Benthal Constituents

So far only constituents flowing with the stream (*planctonic constituents*) have been considered, but sometimes matter and organisms settling on the bottom (*benthal constituents*) are also important (e.g., water weed and sediments). Their main characteristic is the absence of transportation mechanism, so that the evolution of the cross-sectional mean concentration of a benthal constituent is described by an equation of the type (3-1-15) with v and D equal to zero (see also the above mentioned equation for stagnant zones). Sources and sinks for benthal constituents, in addition to internal processes, are sedimentation and scour.

The Method of Characteristics

Solutions of the various balance equations can usually be obtained only numerically, and much scientific effort has been devoted to the development of efficient and reliable solution techniques. With most of them a discretization of independent variables is introduced, and hence, the main problem these techniques have to cope with is numerical dispersion. Since the numerical integration schemes usually have no physical interpretation, they are not discussed here; the interested reader is referred to the literature (see, for example, Hirsch, 1975) for a general discussion of the particular properties of the different solution schemes.

A particular integration method, the *method of characteristics*, is described briefly because of its stimulating interpretation and because it is the only one which allows analytical results to be obtained sometimes (see, for instance, Li, 1962; Di Toro, 1969; and Chapter 4). The method of characteristics can be applied only if all model equations are of type (3-1-15) and if the dispersive components of the flow are negligible with respect to the advective component, i.e., if $\mathscr{D} = 0$. Under this hypothesis the one-dimensional balance equation (3-1-15) can be written as

$$\frac{\partial p}{\partial t} + v \frac{\partial p}{\partial l} = -\frac{S_Q}{A} p + S \qquad (3\text{-}1\text{-}17\text{a})$$

where

$$S_Q = \frac{\partial A}{\partial t} + \frac{\partial (vA)}{\partial l} \qquad (3\text{-}1\text{-}17\text{b})$$

Equation (3-1-17) can be solved if the boundary conditions at $l = 0$

$$p(0, t) = p_b(t) \qquad t \geq 0 \qquad (3\text{-}1\text{-}18\text{a})$$

and the initial conditions at $t = 0$

$$p(l, 0) = p_i(l) \qquad 0 \leq l \leq L \qquad (3\text{-}1\text{-}18\text{b})$$

are given. Since Eqs. (3-1-17) and (3-1-18) are a Cauchy problem a solution exists for a large class of functions v, A, S_Q, and S.

The boundary conditions $p_b(t)$ are known if the cross-sectional average concentration p at $l = 0$ for $t \geq 0$ can be measured, while the initial conditions $p_i(l)$, corresponding to a "picture" of the river at time $t = 0$, are rarely known in practice.

The central idea of the method of characteristics is to substitute Eqs. (3-1-17) and (3-1-18) by the following three ordinary differential equations

$$\frac{dt}{d\tau} = 1 \tag{3-1-19a}$$

$$\frac{dl}{d\tau} = v(l, t) \tag{3-1-19b}$$

$$\frac{dp}{d\tau} = -\frac{1}{A(l, t)} S_\varrho(l, t) p + S(p, l, t) \tag{3-1-19c}$$

and the initial conditions (see point A of Fig. 3-1-1)

$$t(0) = 0, \quad l(0) = l_0, \quad p(0) = p_i(l_0) \quad \text{for } 0 \leq l_0 \leq L \tag{3-1-20a}$$

and (see point B of Fig. 3-1-1)

$$t(0) = t_0, \quad l(0) = 0, \quad p(0) = p_b(t_0) \quad \text{for } t_0 > 0 \tag{3-1-20b}$$

The line $(t(\tau), l(\tau), p(\tau))$ in the (t, l, p) space, which is a solution of Eq. (3-1-19) with a given initial condition $t(0) = t_0$, $l(0) = l_0$, $p(0) = p_0$, is called the *characteristic line*. In the hydrologic literature the term is also used to denote the line $(t(\tau), l(\tau))$ in the (t, l) plane, a practice which will be followed in this book.

It is obvious that Eq. (3-1-19) subject to conditions (3-1-20) is equivalent to Eq. (3-1-17) subject to conditions (3-1-18) (see Fig. 3-1-1). The main point is that

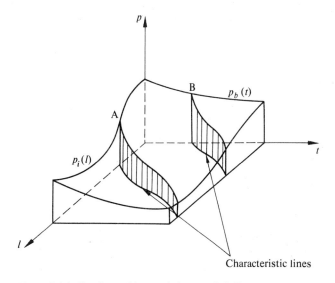

Figure 3-1-1 Cauchy problem and characteristic lines.

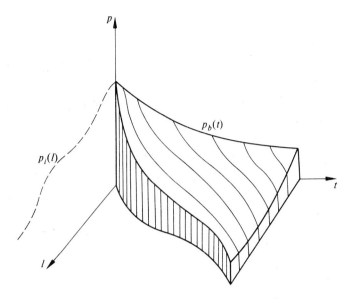

Figure 3-1-2 The solution of Eq. (3-1-17), when only boundary conditions are known.

Eq. (3-1-17), in contrast to the more general balance equations, has only one family of characteristic lines, all of which propagate in the downstream direction. Hence, the solution along each characteristic line can be obtained independently of the solution along a neighboring characteristic line. From a physical point of view this is quite understandable since dispersion was assumed to be insignificant, i.e., plugs of water of differential thickness in the l-direction retain their identity as they flow downstream. Hence, models (3-1-17, 3-1-19) are called *plug flow models*. Equation (3-1-19) describes the dynamics of the variable p as an observer, moving with this plug, sees them. The variable τ, which is called *flow time*, represents the time elapsed since the start of the plug at $l(0)$.

The consequence of unknown initial conditions $p_i(l)$ is now evident and is shown in Fig. 3-1-2: the solution $p(l, t)$ can be determined only for the pairs (l, t) lying on characteristic lines starting from a point on the t axis ($l = 0$).

Later on, when the method of characteristics is employed, Eqs. (3-1-19a) and (3-1-19b) are often omitted for the sake of simplicity. Only Eq. (3-1-19c) is used and the corresponding models, called *flow time models*, are written in the form

$$\dot{p} = -\frac{S_Q}{A} p + S \qquad (3\text{-}1\text{-}21)$$

3-2 ATTRIBUTES OF THE MODELS

The river quality models analyzed in this book are composed, as already pointed out, of several equations of the kind (3-1-15) coupled together. As shown in Sec.

2-5, this set of equations can be partitioned into three subsets constituting the hydrologic, the thermal, and biochemical submodels. The hydrologic submodel is coupled with all other models since the velocity v and the cross-sectional area A enter all other equations. The temperature interacts with the biochemical variables through the source terms S. Finally, biochemical variables are mutually coupled through the source terms S.

According to the assumptions made for specifying the model equations various attributes of the model or submodels result. Since these attributes are used quite often below, it seems worthwhile listing the definitions of all the corresponding types of models although some of those have already been introduced.

linear model: all equations are linear in the state variables
nonlinear model: at least one equation is nonlinear
plug flow model: the diffusion coefficient D is equal to zero in all equations
dispersion model: the diffusion coefficient D is different from zero
steady state model: partial derivatives with respect to time are missing (equal to zero)
unsteady state model: at least one equation contains a time derivative
time-invariant model: all the parameters are constant in time
time-varying model: at least one parameter is time varying
lumped parameter model: the equations are ordinary differential equations (the independent variable may be time or space)
distributed parameter model: at least one equation is a partial differential equation
flow time model: a lumped parameter model with flow time as independent variable
ecological model: at least one state variable is the biomass of some living compartment of the food web
chemical model: no state variable of the biochemical submodel is the biomass of a living compartment of the food web

It is important to notice that some of these attributes are not independent. For instance, time-varying models are unsteady state models and steady state models are lumped parameter models. Moreover, these attributes can be different when looking at the model as a whole or at the single submodels. For example, a river quality model is always nonlinear because of the terms Ap and Avp appearing in Eq. (3-1-15) (recall that v and A are functions of the state variables of the hydrologic submodel), while the biochemical submodel, in which v and A have to be considered as parameters, may be a linear model (see, for instance, Sec. 4-1).

3-3 THE HYDROLOGIC SUBMODEL

The hydrologic submodel is, in the most general case, constituted by two coupled partial differential equations, namely the water *mass* and *momentum balance equations*. These equations can be written in three, two, and one dimensions as already shown in Section 3-1. Nevertheless, the following discussion is restricted to one-dimensional models since only these models have been used extensively in river pollution modeling, while higher dimensional models have been employed for studies of estuaries and lakes (see, for instance, Chen and Orlob, 1975, and Orlob, 1976).

In the general one-dimensional balance equation (3-1-15)

$$\frac{\partial(Ap)}{\partial t} + \frac{\partial(Apv)}{\partial l} - \frac{\partial}{\partial l}\left(DA\frac{\partial p}{\partial l}\right) = AS \tag{3-3-1}$$

p is identified with the water density ρ (constant in space and time). Defining flow rate $Q = Av$

$$\frac{\partial A}{\partial t} + \frac{\partial Q}{\partial l} = AS$$

is obtained, which is the so-called *continuity equation*. The term

$$S_Q = \frac{\partial A}{\partial t} + \frac{\partial Q}{\partial l}$$

appears in all one-dimensional balance equations (see Eq. (3-1-16)), where it can be replaced by the rate AS of water inflow. If there is no lateral inflow and/or outflow along the river stretch ($S = 0$), the continuity equation becomes

$$\frac{\partial A}{\partial t} + \frac{\partial Q}{\partial l} = 0 \tag{3-3-2}$$

Equation (3-3-2) can be integrated only if it is coupled with the momentum balance equation or if the river discharge Q at any point l is assumed to be a known function of the depth H of the river at that point, i.e.,

$$Q = Q^*(H, l) \tag{3-3-3}$$

This function, called *stage-discharge relation*, is in general a convex function as shown in Fig. 3-3-1 and usually corresponds to the real behavior of the stream when the river flow is varying relatively slowly in time. From Eq. (3-3-3) and from

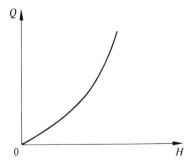

Figure 3-3-1 Stage-discharge relationship at a given point.

the knowledge of the geometry of the river bed a function

$$Q(l, t) = \tilde{Q}(A(l, t), l) \tag{3-3-4}$$

can be derived which is also convex. If the continuity equation (3-3-2) is multiplied by the velocity

$$w = \frac{\partial \tilde{Q}}{\partial A}$$

which is well defined because of Eq. (3-3-4), the new equation

$$\frac{\partial Q}{\partial t} + w \frac{\partial Q}{\partial l} = 0 \tag{3-3-5}$$

is obtained. This expression corresponds to the so-called *kinematic approximation* and implies that the river discharge is constant along the characteristic line $dl/dt = w$, i.e., the wave is propagating downstream at velocity w and there is no spreading effect. In other words, the stretch can be considered, from a functional point of view, as a pure time delay: an impulse variation of flow rate at one point produces a delayed impulse variation downstream, while a bell-shaped hydrograph as shown in Fig. 2-1-4 should be expected. Nevertheless, this model reproduces the observed fact that the velocity of propagation increases with depth. Indeed, since $\tilde{Q}(A, l)$ is convex with respect to A and A is increasing with H, it follows that w is an increasing function of the depth.

If the kinematic approximation described by Eq. (3-3-5) seems to be unreasonable, the natural alternative is to make use of the momentum balance equation which can be derived from Eq. (3-3-1) by neglecting the dispersion term ($D = 0$). Thus, Eq. (3-3-1) with $p = \rho v$ becomes

$$\frac{\partial (Av)}{\partial t} + \frac{\partial (Av^2)}{\partial l} = \frac{AS}{\rho}$$

or, by eliminating v,

$$\frac{\partial Q}{\partial t} + \frac{\partial}{\partial l}\left(\frac{Q^2}{A}\right) = \frac{AS}{\rho} \tag{3-3-6}$$

The term S at the righthand side of Eq. (3-3-6) takes into account the distributed sources and losses of momentum which are, respectively, the component of the gravity forces in the direction of the river and the force induced by hydrodynamic friction. The first of these two terms, say S_1, can be given the form

$$S_1 = -\rho g \left[\frac{dh}{dl} + \frac{1}{A} \frac{\partial}{\partial l}(AH_c) \right] \tag{3-3-7}$$

where g is the acceleration of gravity, h is the elevation of the bottom from a reference level, dh/dl is the bottom slope and H_c is the depth from the free surface to the centroid of the cross section. The second term S_2 is usually written as

$$S_2 = -\rho g i_f \tag{3-3-8}$$

where i_f is the so-called *friction slope* and is, in general, a function of the river flow Q and of the river depth H.

If we now deal, for simplicity, with rectangular cross sections (rectangular channel) of constant width, then

$$A = YH \qquad H_c = \tfrac{1}{2}H$$

where Y is river width, so that Eq. (3-3-7) becomes

$$S_1 = -\rho g \left(\frac{dh}{dl} + \frac{\partial H}{\partial l} \right)$$

which, together with Eqs. (3-3-6) and (3-3-8), gives rise to the momentum equation

$$\frac{\partial Q}{\partial t} + \frac{\partial}{\partial l}\left(\frac{Q^2}{A}\right) = -gA\left(\frac{dh}{dl} + \frac{\partial H}{\partial l} + i_f\right) \tag{3-3-9}$$

It can easily be shown that if the friction slope i_f is constant the momentum equation together with the continuity equation (3-3-2) degenerate into Eq. (3-3-5) (kinematic approximation). This can be easily understood, since if the friction phenomena are independent of Q and H, then a one to one relationship must exist between Q and H at any point (*stage-discharge relation*). Thus, if the model is to describe the attenuation and spreading out of the wave during its propagation, the friction slope i_f cannot be assumed to be constant in Eq. (3-3-9).

Both theoretical and experimental investigations indicate that the terms appearing on the right hand side of the momentum equation are, in general, quite large with respect to the kinetic terms appearing on the left hand side of the equation. Thus, if the kinetic terms are neglected a new approximation of the momentum equation is obtained in the form

$$\frac{\partial H}{\partial l} = -i_f - \frac{dh}{dl}$$

which, when derived with respect to time, gives

$$\frac{\partial}{\partial t}\left(\frac{\partial H}{\partial l}\right) = -\frac{\partial i_f}{\partial t} \tag{3-3-10}$$

This approximation (called *parabolic approximation*) is not so crude as the kinematic one since the friction slope i_f is allowed to vary in time. From the continuity equation and Eq. (3-3-10) a new equation is obtained in the form

$$\frac{\partial Q}{\partial t} + w\frac{\partial Q}{\partial l} = D^* \frac{\partial^2 Q}{\partial l^2} \tag{3-3-11}$$

where the velocity w and the diffusion D^* are suitable positive functions of Q and A. This equation shows that when moving downstream at velocity w the time derivative of Q has the sign of $\partial^2 Q/\partial l^2$. Therefore, a bell-shaped impulse propagating downstream at velocity w will be attenuated (at the peak $\partial^2 Q/\partial l^2 < 0$) and consequently spread out.

In summary, it can be stated that relatively simple versions of the continuity and momentum equations describe quite satisfactorily at least the two main characteristics of the phenomenon, namely propagation and attenuation of the peaks. For this reason, such equations have been employed for years (the first derivation was made by Barré de Saint Venant in 1871) and are still used in simulation studies despite three main disadvantages.

The first disadvantage is that these equations can be integrated (after discretization) only if the function $A(H, l)$ is known in a suitably high number of points l, and obviously this information is not always easily obtainable. Secondly, the solution of the continuity and momentum equations requires a relatively high computational effort. In fact, the method of characteristics (in a version slightly modified with respect to that described in Sec. 3-1) or some *explicit* or *implicit scheme* for integrating partial differential equations must be used. The integration along characteristic lines after discretization in time and space (the river segment is divided into reaches) implies the need to work on a nonregular grid in the (l, t) plane. The space step Δl (length of the reach) must be relatively small in order to give a sufficiently accurate description of the geometry of the river, while the time step Δt is related to Δl through the propagation velocity w ($\Delta t = \Delta l/w$). In real cases this relation implies a very small Δt with

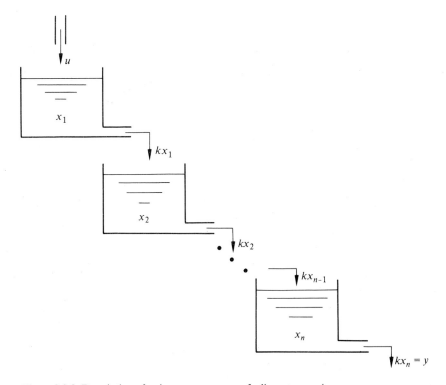

Figure 3-3-2 Description of a river as a sequence of n linear reservoirs.

respect to the duration of the phenomenon so that the method of characteristics turns out to be time consuming. On the other hand, explicit and implicit schemes use regular grids in the (l, t) plane and are therefore more attractive. Explicit schemes are very simple to program but unfortunately they require very small time steps Δt ($\Delta t < \Delta l/w$), since otherwise they become numerically unstable (see Greco and Panattoni, 1977). Implicit schemes are stable for any choice of the space and time intervals, and for this reason they are usually preferred to explicit schemes, even though the latter lead to algebraic equations which are easier to solve. Finally, the third disadvantage which is worth mentioning is that the friction slope i_f is, in general, unknown and must therefore be identified by applying a suitable parameter estimation scheme. Thus, in order to solve this problem it is necessary for some hydrographs to be available.

For all these reasons, simpler models have been proposed and used. Some of these models are derived from the general mass and momentum balance equations following a suitable criterion of approximation, such as space discretization, time discretization, linearization, frequency domain approximation or others. For example, the model obtained with the parabolic approximation (Eq. (3-3-11)) can be further simplified by assuming that w and D are two constant parameters to be estimated by fitting some available hydrograph. Many other models, actually the majority of the models, have been proposed in a very naive way by postulating the mathematical structure of the different blocks constituting a heuristic conceptual model. For example, one of the most classical models of this kind proposed by Nash (1957) corresponds to the model shown in Fig. 3-3-2, where the river is thought of as a sequence of n equal reservoirs.

If the outlet of each reservoir is assumed to be proportional to the volume of the water stored in the reservoir, then the system is described by the following linear equations (mass balance equations):

$$\dot{x}_1 = u - kx_1$$
$$\dot{x}_2 = kx_1 - kx_2$$
$$\vdots$$
$$\dot{x}_n = kx_{n-1} - kx_n$$
$$y = kx_n$$

where x_i is the volume of water stored in the i-th reservoir, u is the input flow rate of the first reservoir and y is the output flow rate of the last reservoir. In a more compact form the model can be considered to be a finite dimensional linear system of the kind

$$\dot{\mathbf{x}}(t) = \mathbf{F}\mathbf{x}(t) + \mathbf{g}u(t) \tag{3-3-12a}$$

$$y(t) = \mathbf{h}^T\mathbf{x}(t) \tag{3-3-12b}$$

where

$$\mathbf{x} = \begin{bmatrix} x_1 & x_2 & x_3 & \cdots & x_{n-1} & x_n \end{bmatrix}^T$$

$$\mathbf{F} = \begin{bmatrix} -k & 0 & 0 & \cdots & 0 & 0 \\ k & -k & 0 & \cdots & 0 & 0 \\ 0 & k & -k & \cdots & 0 & 0 \\ \vdots & \vdots & \vdots & & \vdots & \vdots \\ 0 & 0 & 0 & \cdots & k & -k \end{bmatrix} \quad \mathbf{g} = \begin{bmatrix} 1 \\ 0 \\ 0 \\ \vdots \\ 0 \end{bmatrix}$$

$$\mathbf{h}^T = \begin{bmatrix} 0 & 0 & 0 & \cdots & 0 & k \end{bmatrix}$$

The impulse response (*unit hydrograph*)

$$\tilde{y}(t) = \mathbf{h}^T e^{\mathbf{F}t} \mathbf{g}$$

of this model can be explicitly determined since the matrix $\mathbf{F}$ has a very simple form (Jordan canonical form), and turns out to be

$$\tilde{y}(t) = \frac{1}{T(n-1)!} \left(\frac{t}{T}\right)^{n-1} e^{-t/T} \qquad (3\text{-}3\text{-}13)$$

where $T = 1/k$ is the so-called *time constant* of the reservoirs. The impulse response $\tilde{y}(t)$ is plotted in Fig. 3-3-3 for different values of n. Obviously the curves in Fig. 3-3-3 can also be interpreted as the flows at the end of the intermediate reaches (reservoirs), so that by comparison with Fig. 2-1-4 it can be concluded that this model can take into account both propagation and attenuation of the peaks. The time of occurrence of the peak can immediately be derived from Eq. (3-3-13) and is given by

$$t^* = (n-1)T$$

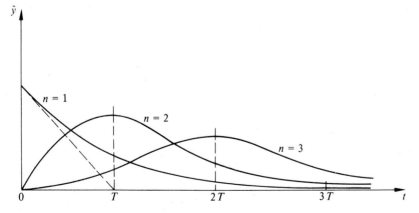

Figure 3-3-3 The impulse response of the Nash model for different values of n.

while the value y^* of the peak is

$$y^* = \frac{1}{T} \frac{(n-1)^{n-1}}{(n-1)!} e^{-(n-1)}$$

If the impulse response of the real system were available, these two expressions could be used to estimate the two parameters, n and T, such that the time of arrival t^* and the height y^* of the peak would be exactly the same in reality and in the model. In practice, however, the two parameters are selected such that a good fit with some available hydrograph is obtained.

Although there is no particular advantage in doing so, the *Nash model* and similar linear models are often presented in terms of *transfer functions*. The transfer function $M(s)$ associated with a linear system of the kind (3-3-12) is the *Laplace transform* of its impulse response and is given by

$$M(s) = \frac{Y(s)}{U(s)} = \mathbf{h}^T(s\mathbf{I} - \mathbf{F})^{-1}\mathbf{g} \qquad (3\text{-}3\text{-}14)$$

where s is the complex variable, $U(s)$ and $Y(s)$ are the Laplace transforms of $u(t)$ and $y(t)$, and $\mathbf{I}$ is the identity matrix. Substituting the expressions for $\mathbf{F}$, $\mathbf{g}$, and $\mathbf{h}^T$ corresponding to the Nash model into Eq. (3-3-14)

$$M(s) = \frac{1}{(1+sT)^n}$$

is obtained, which states that the transfer function of the system is simply the product of n transfer functions $m_i(s)$, which, in this particular case, are all equal to $1/(1+sT)$. On the other hand, the transfer function of a single reservoir described by

$$\dot{x} = -kx + u$$
$$y = kx$$

is given by (see Eq. (3-3-14))

$$m(s) = k \cdot (s+k)^{-1} \cdot 1 = \frac{1}{1+sT}$$

which proves that the transfer function of the Nash model is the product of the transfer functions of the n reservoirs, as should be expected, recalling that the transfer function of two systems connected in series is the product of the two transfer functions (see Sec. 1-2). This means that if the reservoirs were allowed to be different a transfer function would be obtained of the kind

$$M(s) = \prod_{i=1}^{n} m_i(s) = \prod_{i=1}^{n} \frac{1}{1+sT_i}$$

which is characterized by n parameters to be estimated.

Moreover, if the conceptual model is modified and the river considered a sequence of channels and reservoirs, and each channel characterized by a transport time τ_i, then a transfer function would be obtained of the following form (recall

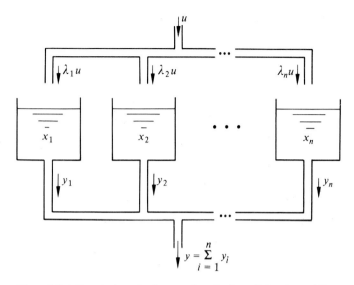

Figure 3-3-4 Description of a river as a bunch of parallel connected linear reservoirs.

that a system characterized by a pure time delay τ_i has a transfer function given by $\exp(-\tau_i s))$

$$M(s) = \frac{\prod_{i=1}^{n} e^{-\tau_i s}}{\prod_{i=1}^{n}(1+sT_i)} = \frac{e^{-\tau s}}{\prod_{i=1}^{n}(1+sT_i)}$$

where τ is the sum of all time delays introduced by the channels.

Another possible conceptual model consists of n reservoirs connected in parallel as shown in Fig. 3-3-4, where the coefficients λ_i satisfy the conditions

$$0 \le \lambda_i \le 1 \qquad \sum_{i=1}^{n} \lambda_i = 1$$

in order to conserve the total mass. Since each reservoir is characterized by a transfer function

$$m_i(s) = \frac{\lambda_i}{1+sT_i}$$

and the transfer function of n systems connected in parallel is the sum of the single transfer functions we have

$$M(s) = \sum_{i=1}^{n} m_i(s) = \frac{\prod_{j=1}^{n-1}(1+st_j)}{\prod_{i=1}^{n}(1+sT_i)} \qquad (3\text{-}3\text{-}15)$$

where the time constants t_j are suitable functions of the λ_i's and T_i's. The transfer function given by Eq. (3-3-15) is the most general one which has the property of conserving the mass between input and output of finite dimensional systems. In fact, given a function $f(t)$ its Laplace transform is given by

$$F(s) = \int_0^\infty e^{-st} f(t)\, dt$$

so that

$$F(0) = \int_0^\infty f(t)\, dt$$

Thus, the conservation of mass between input and output of a given system can be expressed as $U(0) = Y(0)$, which implies (recall that $Y(s) = M(s)U(s)$) that $M(0) = 1$, and this property is indeed satisfied by Eq. (3-3-15). This shows that the model structure given in Fig. 3-3-4 does not really impose any particular constraint on the input-output relationship of the system, since the corresponding transfer function is the most general one satisfying the principle of mass conservation. Hence, even if a model structure has been postulated which seems to be highly specific, it may turn out that the corresponding mathematical model is the most general one.

A discussion of other hydrological models such as those in which overland flow and/or underground flow are involved, which would somehow make the survey complete, is not undertaken since it would represent too great a departure from the main goal of the book. However, the problems, the difficulties, and the different approaches already discussed with reference to wave propagation, are quite representative of any other field in hydrology and the interested reader should therefore be able to enlarge his knowledge by referring to the most recent books and treatises on this fascinating field (Gray, 1970). Of course, after this short and crude introduction to the art of modeling hydrological phenomena the reader cannot be expected to be familiar with this topic. For this reason, detailed descriptions of the dynamics of river flow will not be used in the remainder of the book. This is actually quite normal, since almost all studies on river pollution make reference to steady state hydrological conditions.

3-4 THE THERMAL SUBMODEL

Heat can be considered to be one of the dispersing substances for which the equations of Sec. 3-1 were derived. This means that the distribution of heat in a river is governed by an equation of the type (3-1-6) with

$$p = \theta \rho T$$

where θ, ρ, and T are the specific heat capacity, the density, and the temperature of the water respectively. This equation establishes the thermal submodel. As discussed in Chapter 2, the hydrological variables in it may be considered to be externally imposed, since the influence of the temperature distribution on hydrology is small. Moreover, as in Sec. 3-3, our considerations will be confined to the one-dimensional balance equation which is usually sufficient for temperature investigations in ordinary rivers. Only for estuaries and river impoundments is the solution of the full two- or three-dimensional model often necessary (see Orlob and Selna, 1970; Leendertsee et al., 1973; Orlob, 1976). (The expressions given below for the components of the air–water heat exchange could also be used in the higher dimensional case if T is understood to be the local surface temperature rather than the cross-sectional mean temperature.) Hence, if it is assumed, for the sake of simplicity, that there is no water added to or subtracted from the river section considered, the *one-dimensional heat balance equation* reads (see Eq. (3-1-16))

$$\frac{\partial T}{\partial t} + v \frac{\partial T}{\partial l} - \frac{1}{A} \frac{\partial}{\partial l}\left(AD \frac{\partial T}{\partial l}\right) = \frac{1}{\theta \rho}\left(\frac{\lambda}{A} - \frac{Y}{A} F(T, M)\right) \qquad (3\text{-}4\text{-}1)$$

where λ = amount of heat added per length and time unit by man
Y = river width
F = flux of heat from the river to the atmosphere per unit surface area
M = set of meteorological conditions which determine the heat exchange between water and atmosphere

The less important components of the heat budget, like heat exchange with the river bed and frictional heat (see Sec. 2-2), are omitted, and θ and ρ are assumed to be constant.

The simplest way to specify the function $F(T, M)$ in Eq. (3-4-1) is to assume that the net heat flux through the air–water interface is proportional to the deviation of the water temperature from the equilibrium temperature T_E (see Sec. 2-2). If, in addition, the longitudinal dispersion is disregarded and the characteristic method applied by introducing the flow time τ (see Sec. 3-1) the equation

$$\frac{dT}{d\tau} = \frac{1}{\theta \rho}\left(\frac{\lambda}{A} - \frac{Y}{A} k(T - T_E)\right) \qquad (3\text{-}4\text{-}2)$$

is obtained, which has been used quite often (see, for example, Edinger et al., 1968). However, the simplicity of Eq. (3-4-2) is deceptive, because T_E is a complicated function of the meteorological conditions, and because k cannot be considered constant if the model is to be realistic.

Therefore, the most common approach to specify $F(T, M)$ is to develop separate expressions for each component of the heat flux through the air–water interface and to replace $F(T, M)$ by the sum of those expressions, i.e.,

$$F(T, M) = -F_s(M) + F_{sr}(M) - F_a(M) + F_{ar}(M) + F_w(T) + F_l(T, M) + F_c(T, M)$$

$$(3\text{-}4\text{-}3)$$

where F_s = incoming solar radiation
F_{sr} = reflected solar radiation
F_a = long wave atmospheric radiation
F_{ar} = reflected long wave atmospheric radiation
F_w = long wave radiation of the water
F_l = latent heat flux
F_c = sensible heat flux

Expressions for the incoming *solar radiation* F_s are usually based on empirical formulae which describe the diurnal and seasonal variations of this heat budget component in the case of clear sky. The data necessary to set up such a formula can easily be measured. The values which the formula yields are then corrected for cloudiness, for example, according to the following equation

$$F_s = (1 - 0.65b^2)F_{so} \qquad (3\text{-}4\text{-}4)$$

where F_{so} is the solar radiation from the clear sky and b is the proportion of the sky covered with clouds (*cloudiness ratio*). The *reflected proportion of the solar radiation* depends on the altitude of the sun and on cloudiness and is also calculated through empirical formulae which give the ratio F_{sr}/F_s as a function of cloudiness and altitude of the sun (see, for example, Anderson, 1954). Quite often the reflectivity is simply assumed to be constant, 6 percent being a typical value.

The *long-wave radiation of the atmosphere* F_a is mainly governed by the *Stefan-Boltzmann law*, which says that the radiation intensity is proportional to the fourth power of the temperature, i.e.,

$$F_a = \varepsilon_a \sigma T_a^4 \qquad (3\text{-}4\text{-}5)$$

where ε_a is the *emissivity* of air, which depends strongly on humidity e_a, σ is the Stefan-Boltzmann constant, and T_a is the air temperature. As emphasized in Sec. 2-2, temperature and humidity distribution over a relatively thick layer are relevant for the long-wave atmospheric radiation, and these distributions ought to appear in the expression for F_a. But values of T_a and e_a measured at a height of a few meters (usually 2 m) have proved to be sufficiently representative for the calculation of F_a. The dependence of ε_a on humidity e_a (measured as water vapor pressure in the air) is again expressed with reference to clear sky, and then corrected for cloudiness. A typical example of an expression of this kind is (see Anderson, 1954)

$$\varepsilon_a = (0.74 + 0.0065 e_a)(1 + 0.17b^2) \qquad (3\text{-}4\text{-}6)$$

where e_a is measured in mmHg and b is the cloudiness ratio. The proportion of the long wave atmospheric radiation which is reflected is usually assumed to be constant; a typical value is 3 percent.

The *long-wave radiation of the water* F_w is also proportional to the fourth power of the absolute temperature. The emissivity ε_w is constant and slightly smaller than 1 because the river water is not a black body. Hence a reasonable expression for F_w is

$$F_w = 0.95\sigma T^4 \qquad (3\text{-}4\text{-}7)$$

This component of the heat budget is the least problematic one.

The heat exchange through evaporation is governed by the turbulent mixing processes in the air layer above the water surface (see Sec. 2-2). Therefore, it should be described in principle by equations like the balance equations of Sec. 3-1. In practice, however, simple, semiempirical formulae are used, which give the heat flux as a function of a very few meteorological parameters. The most common formula for the flux of *latent heat* is

$$F_l = \beta(w)\bigl(e_s(T) - e_a\bigr) \qquad (3\text{-}4\text{-}8)$$

where $\beta(w)$ is a function describing the influence of the wind velocity w on evaporation, and $e_s(T)$ is the saturated water vapor pressure at the water temperature T. The proportionality between F_l and the water vapor difference can be justified in the following way: the flux of water vapor is proportional to the gradient of the water vapor pressure (see Sec. 3-1), and this gradient is approximately proportional to the difference in Eq. (3-4-8). (See also Sec. 3-5, where a formula for physical reaeration which is analogous to Eq. (3-4-8) will be explained.) Many different expressions for the wind function $\beta(w)$ have been reported in the literature (LAWA, 1971); a linear one is mostly used:

$$\beta = a + bw \qquad (3\text{-}4\text{-}9)$$

where a and b are suitable constants and w is usually measured at the same height as e_a and T_a. The ratio between a and b is about 1 if w is measured in m/s. The saturated water vapor pressure $e_s(T)$ is very well known and may be expressed in the model by a convenient empirical formula. An example, given by Jobson and Yotsukura (1972), is

$$e_s(T) = 0.75 \exp(54.721 - 6788.6 T^{-1} - 5.0016 \ln T)$$

where e_s is in mmHg and T in Kelvin. In analogy to Eq. (3-4-8) the flux of *sensible heat* is written as

$$F_c = \alpha(w)(T - T_a) \qquad (3\text{-}4\text{-}10)$$

where α is the wind function for convective heat transport. Since the transport mechanisms for both latent and sensible heat are the same, a close correlation between α and β can be expected. In fact, Bowen (1926) has shown that approximately

$$\frac{F_c}{F_l} = 0.524 \frac{T - T_a}{e_s - e_a} \qquad (3\text{-}4\text{-}11)$$

(*Bowen ratio*), where T is in °C and e_a in mmHg. This relationship has been confirmed by many experiments.

Considering the expressions (3-4-4–3-4-10) it can be seen that Eq. (3-4-1) is nonlinear because of expressions (3-4-7) and (3-4-8), and that the meteorological parameters involved are air temperature, humidity, and wind speed at a certain height, cloudiness, and altitude of the sun. The local conditions around the river are reflected by the parameter values in the various expressions.

Sometimes only the difference $\Delta T = T' - T$ between the river temperature T' which results from anthropogenic waste heat discharges and the natural river temperature T is of interest. This is the case for example, if a river temperature standard is formulated as a maximum permissible deviation from the natural river temperature (which is common practice). For ΔT a model can be derived which is considerably simpler than Eq. (3-4-1), because all terms which do not depend on T are eliminated. If diffusion is disregarded and the balance equation written in flow time

$$\frac{d\Delta T}{d\tau} = \frac{v}{\theta \rho Q}(\lambda - Y\{0.95\sigma[(T+\Delta T)^4 - T^4] + \beta(w)[e_s(T+\Delta T) - e_s(T) + 0.524\,\Delta T]\}) \quad (3\text{-}4\text{-}12)$$

is obtained, by subtracting the balance equation for T from the one for T'. If ΔT is small (which is usually the case) $(T + \Delta T)^4$ and $e_s(T + \Delta T)$ can be linearized around T and Eq. (3-4-12) becomes

$$\frac{d\Delta T}{d\tau} = \frac{v}{\theta \rho Q}\left\{\lambda - \Delta T Y\left[3.8\sigma T^3 + \beta(w)\left(\frac{de_s}{dT} + 0.524\right)\right]\right\} \quad (3\text{-}4\text{-}13)$$

Equation (3-4-13) is a linear ordinary differential equation which can easily be solved. It can be used to calculate the effect of the heat discharge λ on river temperature T if measurements of T are available (Faude et al., 1974). Equation (3-4-13) even remains a good approximation if instead of the "natural" temperature T a reasonable reference temperature is used, because the first derivatives of the functions which were linearized in Eq. (3-4-12) do not vary greatly over the range over which T usually varies. Even in the full equation (3-4-1) the terms dependent on T are sometimes linearized around a reasonable reference temperature (see, for example, Jobson and Yotsukura, 1972), although in this case the variations of the linearized functions themselves (rather than the derivatives) are important.

Finally, it should be stressed that the empirical and semiempirical formulae (3-4-4–3-4-10) are only typical examples from a great variety of similar expressions (see, for instance, Eckel and Reuter, 1950; Anderson, 1954; Parker and Krenkel, 1970; LAWA, 1971; Jobson and Yotsukura, 1972; Harleman et al., 1973). There is no set of formulae which could clearly be considered superior to all others. In view of their simplicity the degree of accuracy with which they can predict the actual river temperature is surprising. Figure 3-4-1 compares, for example, the predictions of a temperature model of type (3-4-1) for the Rhine river with measured data at Andernach, which is about 500 km downstream from the point where the boundary condition is imposed. The results shown in Fig. 3-4-1 are taken from Motor Columbus (1971), while the model is described in Bøgh and

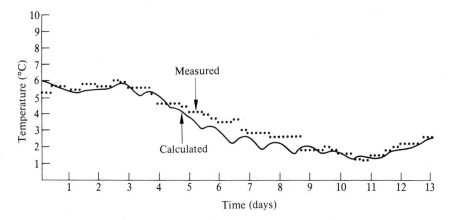

Figure 3-4-1 Comparison of measured Rhine river temperatures at Andernach with computer simulation results (after Motor Columbus, 1971).

Zünd (1970). The figure shows that the calculated values follow the actual measurements almost perfectly. The way in which diffusion is taken into account in this model is worth mentioning (see also Jaske and Spurgeon, 1968): the river is appropriately split up into two fictitious channels which flow parallel with different velocities. The flow within the channels is without longitudinal dispersion, but the two channels are completely mixed with each other transversely from time to time. This computational scheme corresponds directly to the statement in Sec. 3-1 that longitudinal dispersion is mainly due to the combined effect of longitudinal stream velocity differences across the river and transverse turbulent mixing.

3-5 THE BIOCHEMICAL SUBMODEL

As pointed out in Sec. 1-3, one of the first steps in building a mathematical model is the selection of the variables relevant to the problem. For the self-purification processes this is not as easy as for the thermal submodel, where temperature is the natural choice. The only variable which occurs as naturally in self-purification models is the oxygen concentration c, which was explained in Sec. 2-3 to be an important criterion for water quality. As in the previous section, the discussion in this section is confined to the one-dimensional case without addition and extraction of water, and since the source term of Eq. (3-1-16) is the essential point the dispersion term is even omitted. Then the *balance equation for the oxygen concentration* c can be written as

$$\frac{\partial c}{\partial t} + v \frac{\partial c}{\partial l} = r(c) - R + P \qquad (3\text{-}5\text{-}1)$$

where r is the *reaeration rate* and the terms R and P stand, respectively, for consumption and production of oxygen by all kinds of biochemical processes in the river, which may under certain circumstances also be a function of c. The variable $c(l, t)$ denotes the cross-sectional mean of the oxygen concentration. When, in the following, the oxygen concentration averaged over other space coordinates is referred to the same symbol c is used and the coordinates which have not been averaged out are listed. Anthropogenic source terms have not been included in Eq. (3-5-1) and will not be included in the model equations throughout this section.

Physical Aeration

The term $r(c)$ in Eq. (3-5-1) reads

$$r(c) = \frac{1}{H}\left[D_z \frac{\partial c(l, z)}{\partial z}\right]_{z=\xi_-} \tag{3-5-2}$$

where H = water depth
D_z = vertical diffusion coefficient
ξ = mean z value of the water surface at (l, t)

As shown in Sec. 3-1, the source term (3-5-2) appears after averaging the three-dimensional balance equation over y and z. Equation (3-5-2) corresponds to the second term on the right hand side of Eq. (3-1-12c) (the migration velocity u_z is zero since the oxygen is truly dissolved). The term $r(c)$ has to be expressed now in terms of the actual boundary condition

$$c(l, \xi) = c_s \tag{3-5-3}$$

where c_s is the *oxygen saturation concentration*.

The usual way to account for condition (3-5-3) through expression (3-5-2) is

$$\frac{1}{H}\left[D_z \frac{\partial c(l, z)}{\partial z}\right]_{z=\xi_-} = k_2(c_s - c) \equiv k_2 d \tag{3-5-4}$$

where k_2 is called the *reaeration coefficient* and d denotes the *oxygen deficit*. A possible justification of (3-5-4) is based on the assumption that there is a thin *water surface layer* in which the diffusion of oxygen is much slower than in the remaining body of water. In order to keep the discussion as simple as possible the aeration process in an open tank, which is assumed to be completely homogeneous in any horizontal cross section, is considered. If the volume of the layer is small compared to the total water volume we can assume that the oxygen transport through the layer is governed by a steady state *diffusion equation*

$$\frac{\partial^2 c(z)}{\partial z^2} = 0 \tag{3-5-5}$$

with boundary conditions

$$c(\xi) = c_s \qquad c(\xi - \Delta) = c_\Delta \tag{3-5-6}$$

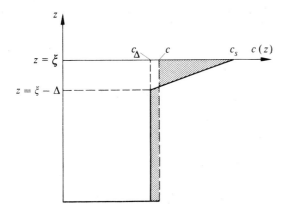

Figure 3-5-1 Idealized vertical oxygen profile in a river.

where Δ is the thickness of the layer, and c_Δ is the oxygen concentration at $z = \xi - \Delta$. The assumption of a steady state within the layer is equivalent to assuming that during the time in which the steady state in the layer is reached the c-value at $\xi - \Delta$ does not change significantly, which is reasonable under the assumptions made above. The solution of the diffusion equation (3-5-5) satisfying the boundary conditions (3-5-6) is a linear function, therefore the vertical oxygen profile is as shown in Fig. 3-5-1 (remember the assumption of practically perfect mixing underneath the surface layer) and

$$\frac{\partial c}{\partial z} = \frac{c_s - c_\Delta}{\Delta}$$

in the layer. Since the volume of the layer was assumed to be very small

$$c_\Delta \simeq c$$

and Eq. (3-5-4) results.

There is some experimental evidence for the existence of that surface layer in which the turbulent mixing is much less effective than in greater depths, but until now no satisfactory quantitative theory for it has been developed. This means that the value of k_2 in Eq. (3-5-4) cannot be calculated accurately from more fundamental river parameters like velocity, depth, etc. There are instead a lot of approximate formulae, derived through different theoretical approaches or directly through regression analysis of measured values of k_2. In the latter case k_2 is usually expressed as

$$k_2 = \alpha_1 v^{\beta_1} H^{-\gamma_1} \quad (3\text{-}5\text{-}7a)$$

or as

$$k_2 = \alpha_2 D^{\beta_2} H^{-\gamma_2} \quad (3\text{-}5\text{-}7b)$$

where D is the longitudinal dispersion coefficient (see Sec. 3-1), and quantities symbolized by Greek letters are positive parameters. If such a formula is given with specific values for the parameters it is usually said to be valid only for certain classes of rivers.

Since the theoretical basis for the evaluation of expression (3-5-4) is so weak, its dependence on river flow rate Q and temperature T is practically also only empirically known. The saturation concentration c_s is independent of Q. The *temperature dependence of* c_s is very well known from experiments and may be described by many suitable functions. A second order approximation, used by Beck and Young (1976), for example, is

$$c_s = 14.54 - 0.39T + 0.01T^2 \tag{3-5-8}$$

with T in °C and c_s in mg/l. For the *temperature dependence of* k_2 an expression of the kind

$$k_2(T) = k_2(20)\alpha^{(T-20)} \tag{3-5-9}$$

is quite often used (see, for instance, Sec. 5-2 and Isaacs and Gaudy (1968)), where α is a parameter slightly greater than 1, and T is the temperature in °C. The

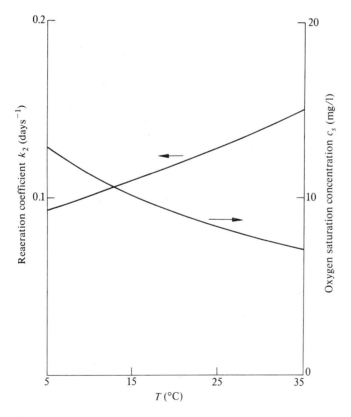

Figure 3-5-2 Temperature dependence of the oxygen saturation concentration and of the reaeration coefficient for a particular river (after Krenkel and Parker, 1969).

temperature dependence of c_s and of k_2 for a particular river (Krenkel and Parker (1969)) is given in Fig. 3-5-2. The dependence of k_2 on Q for a particular river can easily be derived from formulae of type (3-5-7a), if the functions $v(Q)$ and $H(Q)$ are known, which is usually the case (see Secs. 3-3 and 5-3). In general, for a given river this dependence is quite weak. The physical reasons for this is that there are two effects which partly compensate each other: on the one hand the depth H increases with increasing Q, which would result in a smaller k_2 value (see Eq. (3-5-4)); on the other hand, turbulence also increases, which enhances oxygen transport from the surface into greater depths (i.e., the surface layer mentioned above becomes thinner).

Reviews of the different approaches to the calculation of k_2 are given by Thackston and Krenkel (1969), Negulescu and Rojanski (1969), Wilson and MacLeod (1973), and Müller (1975). By and large one can state that the expression (3-5-4) has been proved to describe the physical reaeration term in Eq. (3-5-1) well, but that there is no satisfactory way to calculate k_2 from easily measurable river characteristics like depth and velocity, not to mention nonhydrologic influences like wind or pollution. Hence k_2 should be considered a parameter which has to be estimated from observations of river quality variables (see Chapter 5).

Chemical Models

The processes summarized by R in Eq. (3-5-1) are so complex that it is clear from the beginning that one cannot introduce a state variable for each pollutant and each species, so that the problem emerges how to select appropriately aggregated variables. On the other hand, there may be other quality criteria (beside the oxygen concentration) which are of final interest; for instance the concentration of certain compounds, like phenols or nitrite.

The most direct approach to the problem is to postulate the existence of a chemical reaction between oxygen and the oxidizable matter in the river, without worrying about the organisms involved in the degradation (*Chemical Models*). The great variety of compounds usually present in rivers is reduced to one or a few classes of oxidizable substances, which are treated as fictitious reactants. Because of the actual differences among the constituents of these fictitious reactants, the amount present is best characterized by the amount of oxygen needed for their complete biochemical oxidation. This measure of pollution is the *biochemical oxygen demand* (BOD) (see, for example, Gaudy, 1972) already mentioned in Sec. 2-3, and the corresponding state variable is denoted by b (with an index if there are several components) in the following.

However, this quantity is hard to measure directly, even if all oxidizable matter is looked upon as one reactant, since it takes a very long time until all degradation processes have died away in a river or wastewater sample. Figure 3-5-3 illustrates this fact: the curve, which was given in Wilderer et al. (1970), and which is based on measurements by Gotaas (1948), shows the accumulated oxygen consumption of a wastewater sample over time. After more than a month the

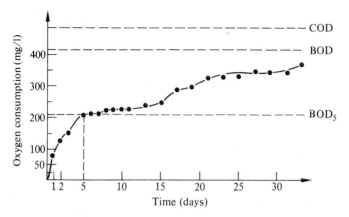

Figure 3-5-3 Oxygen consumption of a wastewater sample over time and relationship to BOD, COD, and BOD_5 (curve from Wilderer et al., 1970).

ultimate oxygen demand cannot be ascertained, even though with the temperature as high as 20 °C the reaction rates were fairly high (see page 88). The biodegradable matter originally present may well have disappeared completely after a much shorter time, but part of their chemical energy has been converted into chemical energy of living biomass, which is then released through oxidation very slowly (see Sec. 2-3).

One way to avoid this long measurement time is to observe the oxygen consumption over a reasonable time and then to estimate the BOD by extrapolation. However, this requires a model for the oxygen consumption in the measurement experiment, which is subject to the same kind of discussion as the oxygen consumption models for the river itself (see page 79). If the parameters of this model are assumed to be known it suffices to measure the total oxygen consumption during the measurement period in order to estimate the BOD. Therefore, quite often the oxygen consumption of samples during 2 or 5 days, abbreviated BOD_2 or BOD_5, is taken as a measure of b. But the assumption that all parameters of the BOD kinetics are known a priori is hardly fulfilled, since, for instance, the oxygen consumption over 2 or 5 days depends strongly on the biological community caught randomly with the sample. And indeed BOD_2 or BOD_5 measurements are notorious for their bad reproducibility.

Another way of easily obtaining an approximate BOD value is to oxidize chemically all matter and measure the oxygen consumption. This measure is called *chemical oxygen demand* (COD). Since in the ideal case all components are completely oxidized the COD can be higher than the BOD because there may be oxidizable compounds present which are not biodegradable (see Fig. 3-5-3). (The boundary between very slowly degradable and nondegradable cannot be defined precisely, however.) On the other hand, there may be compounds which are not oxidized by the oxidant used. As far as organic matter is concerned this is quite unlikely with *potassium dichromate* ($K_2Cr_2O_7$), which is used nowadays for standard COD measurements. With potassium permanganate however, which was formerly used, often even less than 50 percent of the theoretical COD of the organic matter was measured. The inorganic matter relevant to the oxygen budget, which is essentially ammonium and nitrite, is usually determined separately. Its oxygen demand can be determined stoichiometrically and subtracted from the measured COD value if necessary. (However, ammonium, for example, is not oxidized by either of the oxidants mentioned.)

The simplest and most widely used chemical model reflecting the major characteristics of BOD reaction, such as the one in Fig. 3-5-3, is the first order reaction model

$$\frac{\partial b}{\partial t} + v\frac{\partial b}{\partial l} = -k_1 b \qquad (3\text{-}5\text{-}10)$$

which is part of the well-known *Streeter-Phelps model* (see Sec. 4-1). The parameter k_1 is called *degradation* or *deoxygenation coefficient*, and the biodegradable matter is assumed to be dissolved or suspended. Depending on the circumstances, nonlinear models have also been found useful, for instance,

$$\frac{\partial b}{\partial t} + v\frac{\partial b}{\partial l} = -k_1 bc \qquad (3\text{-}5\text{-}11)$$

$$\frac{\partial b}{\partial t} + v\frac{\partial b}{\partial l} = -k_1 b^2 \qquad (3\text{-}5\text{-}12)$$

(see Sec. 4-2). Because of the definition of BOD, the righthand sides of Eqs. (3-5-10)–(3-5-12), which represent volume source terms (see Sec. 3-1), must directly enter the model equation for c. Hence, for example, Eq. (3-5-1) becomes

$$\frac{\partial c}{\partial t} + v\frac{\partial c}{\partial l} = k_2(c_s - c) - k_1 b \qquad (3\text{-}5\text{-}13)$$

if Eq. (3-5-10) is used. If the total BOD is divided into several components the degradation dynamics are described by as many equations of the types given above with different values of the parameters (see Sec. 4-2). Usually no measurements of the single components are available, so that the number of output variables is smaller than the number of state variables. Nevertheless, *parameter and state estimation* of those models may still be possible, as will be shown in Chapter 5. Sometimes, however, measurements for single components of BOD may be available. An example already mentioned is the oxygen demand of ammonium, which is oxidized to nitrate. Measurements of the concentration of ammonium can be used for estimating a model containing a chemical reaction equation for a fictitious reaction between ammonium and oxygen.

The chemical models given so far were for dissolved or suspended matter only. Their application to benthal BOD, for instance BOD of sediments, is straightforward: if, for the moment, the change of the benthal BOD by sedimentation or resuspension is disregarded, the same kinds of equations described above may be used but without derivatives with respect to l (see Sec. 3-1). The first-order reaction as in Eq. (3-5-10), however, is hardly realistic in this case, since the BOD degradation of sediments is taking place essentially at the water–sediment interface. Conditions in deeper sediment layers are usually anaerobic, which means that degradation there proceeds very slowly. Expressions more appropriate are

$$\frac{\partial b}{\partial t} = -\alpha c^\beta \tag{3-5-14}$$

(Edwards and Rolley, 1965) or even the zero-th-order reaction

$$\frac{\partial b}{\partial t} = -\alpha \tag{3-5-15}$$

Thus, using Eq. (3-5-10) for the dynamics of the planctonic BOD (b_1) and Eq. (3-5-14) for the benthal BOD (b_2), the model would read

$$\frac{\partial b_1}{\partial t} + v\frac{\partial b_1}{\partial l} = -k_{11}b_1 \tag{3-5-16a}$$

$$\frac{\partial b_2}{\partial t} = -k_{12}c^{k_{13}} \tag{3-5-16b}$$

$$\frac{\partial c}{\partial t} + v\frac{\partial c}{\partial l} = k_2(c_s - c) - k_{11}b_1 - k_{12}c^{k_{13}} \tag{3-5-16c}$$

where dispersion and external sources have again been left out. Sedimentation and resuspension can be dealt with by adding suitable terms to the righthand sides of Eqs. (3-5-16a) and (3-5-16b), as discussed on page 90 (a review of different approaches to the problem of benthal BOD is given by Müller, 1975).

Little more can be said on the *temperature dependence of the degradation* rate constants used in chemical models than that they increase as T increases. There are numerous empirical formulae to express this dependence, in most cases a formula of type (3-5-9) is used. The dependence on Q is usually neglected.

Ecological Models

It is already obvious from Fig. 3-5-3 that the simple chemical reaction models described so far only very roughly describe the actual oxygen dynamics in a river: the many plateaus which are turnovers of chemical energy from one link of the food chain to another certainly cannot be adequately explained as reactions between BOD components and oxygen. Figure 3-5-4, taken from Gates et al., (1966), illustrates that even the qualitative behavior of the oxygen concentration may be unreproducible by a simple chemical reaction model like the frequently used Streeter-Phelps model, which is given by Eqs. (3-5-10) and (3-5-13) (see Sec. 4-1). In the figure, observed oxygen values from a laboratory self-purification experiment are compared with the model solution. In the course of the experiment organic pollution was added at three time instants, which can easily be identified from Fig. 3-5-4. The parameter k_2 in Eq. (3-5-13) was known, while k_1 in Eq. (3-5-10) was estimated separately for each of the three stages such that the fit to the

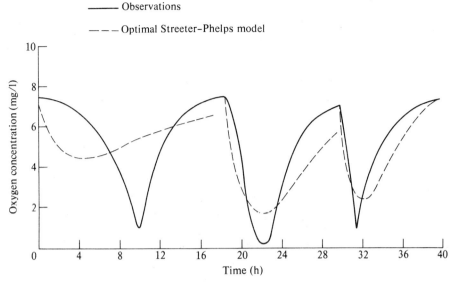

Figure 3-5-4 Reproduction of measured oxygen sag curves by means of the Streeter-Phelps model (after Gates et al., 1966).

measured data was optimal in the least square sense (see Chapter 5). The figure shows that in particular the slow rise of the self-purification in the first stage (which is probably due to the low initial bacterial mass) cannot be simulated by the model.

Also, for more sophisticated models, which explicitly take into account living organisms, suitably aggregated variables have first to be selected. For single, dissolved pollutants consumed by bacteria the mass concentration will be the natural variable (or any equivalent to it). But for particulate pollutants the surface is more important, since the bacteria can attack the pollutant only at the surface. Hence in this case the available surface is a more appropriate variable (see, for instance, Boling et al., 1975). Usually the number of pollutants in a river is tremendous; for the Rhine river, for instance, the number of organic compounds is estimated to be higher than 10^6 (Kölle et al., 1972). Therefore, one has to aggregate these compounds and use measures which are more meaningful for bacterial growth than just mass or surface. Such a measure could again be the BOD (this time of the bacterial nutrients only), which is correlated with the *free energy* of the reactions the bacteria catalyze (McCarty, 1965, and 1972). Another measure sometimes used is the *total organic carbon* (TOC).

Also, the organisms cannot be dealt with by species, since this would entail a huge number of state variables and severe measurement problems. For the bacteria, for instance, it seems reasonable to lump together all heterotrophic bacteria, as already explained in Sec. 2-3. Having many species in one group the question arises whether one may use the number of individuals as the lumped variable (which is sometimes easier to measure) or whether the biomass has to be used. An example of how different these measures may be is given in Fig. 3-5-5

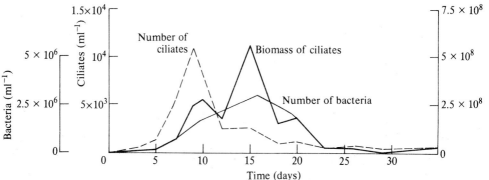

Figure 3-5-5 Development of ciliates feeding on bacteria (after Bick, 1964).

(see Bick, 1964), where the development of the ciliates number and biomass is compared with the abundance of their prey, namely the bacteria. Looking only at the protozoan number it would be hard to understand why the maximum of the curve occurs before the bacterial maximum, while the biomass curve is in agreement with the concept developed in Sec. 2-3 for the ciliata–bacteria interaction. Generalizing these findings, it is important to characterize groups of organisms in the model by their biomass, rather than by their number, because it describes better the activities of the organisms both as predator and prey.

As explained in Sec. 2-3 the most important step in the self-purification process is the interaction between the pollutants and the bacteria. In the case of a single dissolved energy donor, which is assumed to be the growth limiting nutrient, and whose concentration is denoted by s, the degradation can be modeled in the following way

$$\dot{s} = -\frac{\alpha_1 s}{\alpha_2 + s} B \qquad (3\text{-}5\text{-}17)$$

where B is the bacterial mass concentration and α_1 and α_2 are parameters. Eq. (3-5-17) is written in flow time (see Eq. (3-1-21)) since this simplifies the notation considerably. An equation of type (3-5-17) can be derived for a single enzymatic reaction from the *law of mass action*, provided that the enzyme–substrate complex disintegrates slowly into the reaction products and the enzyme (see Laidler, 1958). Thus, the concentration of the enzyme appears first, instead of the bacterial mass density B. For a sequence of enzymatic reactions, the same expression for the rate of degradation of the original substrate can be used given certain assumptions; the reaction parameters and the enzyme concentration in it are those of the slowest reaction in the sequence (see Wilderer, 1969; Boes, 1970). Equation (3-5-17) results if we further assume that the substrate is degraded along a single metabolic pathway, and that the bacterial concentration is proportional to the enzyme concentration. Equation (3-5-17), which is named after *Michaelis-Menten* (1913), is often used in cases where the assumptions which led to it are not fulfilled with certainty. Then, it has to be interpreted as a two-parameter approximation of the real expression of the degradation rate which is assumed to be proportional to the product Bs for low values of s (probability of enzyme–substrate molecular collision) and independent of s and proportional to B for $s \to \infty$ (existence of an upper limit for the feeding activity of a bacterium).

Assuming the absence of predators, the dynamic equation for the bacterial mass may be written as

$$\dot{B} = -\beta_1 \dot{s} - \beta_2 B \qquad (3\text{-}5\text{-}18)$$

where β_1 and β_2 are parameters. The first term on the righthand side of Eq. (3-5-18) is equivalent to the assumption that the ratio between the amount of newly formed biomass and the amount of nutrient degraded (*yield factor*) is constant (see Gunsalus and Shuster, 1961; McCarty, 1965 and 1972). The derivative of s on the righthand side of Eq. (3-5-18) has to be understood only as an abbreviation of the righthand side of Eq. (3-5-17), which facilitates surveying. The term $\beta_2 B$ in Eq. (3-5-18), which may be called *maintenance rate*, accounts for bacterial mass decrease through endogenous respiration, death, and possibly predation (Dawes and Ribbons, 1964; McCarty, 1972). It is obviously only an approximation, since, for example, the death rate ought to depend on the nutrient supply in the past.

The oxygen balance equation in flow time which results from the processes described by Eqs. (3-5-17) and (3-5-18) is

$$\dot{c} = k_2(c_s - c) + \gamma_1 \dot{s} - \gamma_2 \beta_2 B \qquad (3\text{-}5\text{-}19)$$

where γ_1 and γ_2 are the specific oxygen consumptions for removal of s and B, respectively. If s is given as BOD and $\beta_2 B$ in Eq. (3-5-18) includes only endogenous respiration (i.e., self-oxidation of bacterial mass) the following relationship must hold:

$$\gamma_1 + \beta_1 \gamma_2 = 1 \qquad (3\text{-}5\text{-}20)$$

A typical solution of model (3-5-17–3-5-19) is shown in Fig. 3-5-6, together with measurements taken by Gates et al. (1969) from a laboratory experiment in which the energy donor was glucose. The parameters were optimally selected using the *quasilinearization technique* described in Sec. 5-3. The variance of some of the parameters was very high, however, because no measurements of B were available (Stehfest, 1973).

Equations (3-5-17)–(3-5-19) for the pollutant–bacteria interaction may also be used if many pollutants are lumped together in one variable s, and indeed for part of the model described in detail in Sec. 4-4 this has been done. If several pollutants or groups of pollutants are to be taken into account separately one has to distinguish between the various possible interactions mentioned in Sec. 2-3. The pollutants may be degraded independently according to Eq. (3-5-17). This has frequently been observed (Wilderer, 1969), especially if the degradation processes are quite dissimilar, as, for example, with a nutrient combination of carbohydrates and proteins. To obtain the corresponding model one has to replace Eq. (3-5-17)

84 STRUCTURE OF THE MODELS

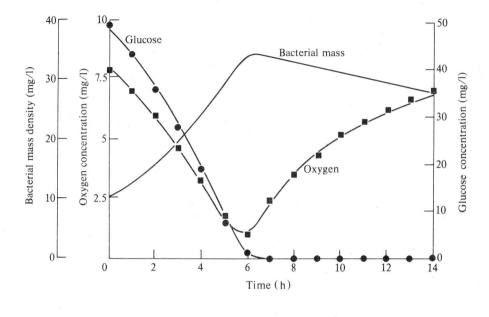

■ ● Measurements from Gates et al.(1969)

—— Optimal model solution

Figure 3-5-6 Simple self-purification model confronted with measurements.

by as many equations of the same kind as there are different pollutants and to add the corresponding terms in Eqs. (3-5-18) and (3-5-19).

In the case of inhibition it is necessary to distinguish between *competitive and allosteric inhibition*. Following the same kind of arguments used to derive the Michaelis-Menten expression one can easily derive expressions to describe the kinetics of single enzymatic reactions which are inhibited by another compound (Laidler, 1958). Similarly, it can be argued that these expressions may be applied to the inhibition of bacterial degradation. The resulting degradation rates are

$$\dot{s} = -\frac{\alpha_1 s}{\alpha_2 + s + \alpha_3 I} B \qquad (3\text{-}5\text{-}21)$$

for competitive inhibition, and

$$\dot{s} = -\frac{\alpha_1 s}{(\alpha_2 + s)(1 + \alpha_3 I)} B \qquad (3\text{-}5\text{-}22)$$

for allosteric inhibition, where I is the concentration of the inhibitor, which may be either a persistent compound (e.g., toxin) or another substrate whose degradation is governed by the Michaelis-Menten equation (Hartmann and Laubenberger, 1968; Stehfest, 1973). The degradation dynamics for large substrate concentrations

s in Eqs. (3-5-21) and (3-5-22) can clearly be seen to be as described in Sec. 2-3: the effect of a competitive inhibitor disappears for $s \to \infty$, while for allosteric inhibition the maximum degradation rate depends on I. Equations (3-5-18) and (3-5-19) are, of course, also valid with inhibition in the degradation equation. The dynamics of a self-purification system where degradation of one pollutant is inhibited by another is shown in Fig. 3-5-7 (Stehfest, 1973). The measurements are essentially from an experiment carried out by Gaudy et al. (1963), using sorbitol and glucose as energy donors and a heterogeneous bacterial population acclimatized to glucose. Inspection of the metabolic pathways and of the observations led to the assumption of allosteric inhibition (Stehfest, 1973); hence the model

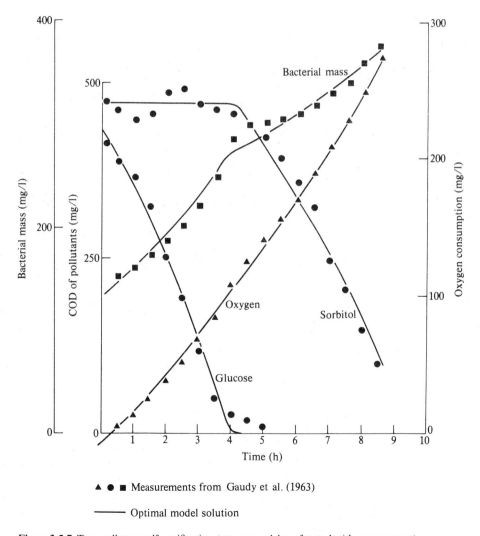

▲ ● ■ Measurements from Gaudy et al. (1963)

——— Optimal model solution

Figure 3-5-7 Two-pollutant self-purification system: model confronted with measurements.

used consists of the following equations

$$\dot{s}_1 = -\frac{\alpha_1 s_1}{\alpha_2 + s_1} B$$

$$\dot{s}_2 = -\frac{\beta_1 s_2}{(\beta_2 + s_2)(1 + \beta_3 s_1)} B$$

$$\dot{B} = -\gamma_1 \dot{s}_1 - \gamma_2 \dot{s}_2 - \gamma_3 B$$

$$\dot{O} = -\delta_1 \dot{s}_1 - \delta_2 \dot{s}_2 + \delta_3 \gamma_3 B$$

where O is oxygen consumption and α_i, β_i, γ_i, and δ_i are parameters which are all positive. Again the parameters were selected optimally through the quasi-linearization technique described in Sec. 5-3. A model including Eq. (3-5-21) for competitive inhibition is described in Sec. 4-4.

In the case of *repression* the most natural modeling approach is to introduce an additional equation for the concentration of the enzyme whose formation is repressed. For an endoenzyme one possibility could be

$$\dot{E} = \frac{\alpha_1 s}{(\alpha_2 + s)(1 + \alpha_3 R)} (\alpha_4 B - E) B - \alpha_5 E - \Delta \frac{E}{B}$$

where E and R are the concentrations of enzyme and repressor, respectively, and Δ is an expression giving the rate of decrease of the bacterial mass due to decay and predation, which may be taken from the model equation for B. The α_i are parameters, α_4 being the maximum of the enzyme–biomass ratio.

For particulate pollutants very sophisticated models could easily be developed using, for instance, variables for different classes of particle size with the particles passing into classes of smaller size as they are degraded. But the data needed for estimating such models are usually lacking, so that a less refined approximation is only possible. Equation (3-5-17) for instance, could also be used since the justification given above applies to particles of small size.

Similar statements can be made about modeling the degradation activity of *bacteria attached to the river bottom*. Here also colonizable surface plays an important role. For a dissolved nutrient which is taken up by attached bacteria as well as by suspended bacteria a possible model (without oxygen balance) is the following

$$\frac{\partial s}{\partial t} + v \frac{\partial s}{\partial l} = -\frac{\alpha_1 s}{\alpha_2 + s} B_1 - \frac{\alpha_3 s}{\alpha_4 + s} \cdot \frac{B_2}{\alpha_5 + B_2} \quad (3\text{-}5\text{-}23a)$$

$$\frac{\partial B_1}{\partial t} + v \frac{\partial B_1}{\partial l} = \beta_1 \frac{\alpha_1 s}{\alpha_2 + s} B_1 - \beta_2 B_1 + \gamma_2 B_2^2 \quad (3\text{-}5\text{-}23b)$$

$$\frac{\partial B_2}{\partial t} = \gamma_1 \frac{\alpha_3 s}{\alpha_4 + s} \cdot \frac{B_2}{\alpha_5 + B_2} - \beta_2 B_2 - \gamma_2 B_2^2 \quad (3\text{-}5\text{-}23c)$$

where s is the energy donor concentration, B_1 and B_2 are the concentrations of planctonic and benthic bacterial biomass respectively, and all Greek letters denote parameters. The second term on the righthand side of Eq. (3-5-23a) accounts for pollutant removal by attached bacteria and may be explained as follows: the number of bacteria getting in touch with the pollutants in the water is for small values of B_2 equal to the number of bacteria present. As B_2 increases the bottom area becomes a limiting factor, bacteria grow in the third dimension, and the number of bacteria being in touch with nutrients and oxygen, i.e., being at the surface of the bacterial layer, becomes constant. Hence the biomass of the bacteria eliminating pollutants may be assumed to be proportional to $\alpha_1 B_2/(\alpha_5 + B_2)$, which is the second factor of the term under discussion. The first factor has an analogous form for the reasons mentioned in connection with Eq. (3-5-17). The same expression times a yield factor appears in Eq. (3-5-23c) for the growth of the attached bacteria. The second term on the righthand side of this equation describes endogenous respiration. The third term is to account for flaking off of bacteria, which only occurs if the bacterial layer becomes too thick. Since the bacteria flaked off continue to reproduce they contribute to B_1 and therefore the term $\gamma_2 B_2^2$ occurs also in Eq. (3-5-23b).

For bacterial degradation of bottom deposits, which also only takes place at the surface (because of anaerobic conditions in greater depths) a model may be derived in a similar way. But in view of the lack of data it may be necessary to use one of the simple chemical models described above. This is sufficient particularly if the processes in the bottom deposits are much less important than the processes going on in the water, which is often the case.

For the interaction between bacteria and protozoa feeding on the bacteria Eq. (3-5-17) may again be used (with B instead of s and protozoa mass density instead of B), and it may even be applied to prey–predator relationships of higher order. If one group of organisms is feeding on n others its *food preferences* for the different preys may be expressed by equations like

$$\dot{s}_i = -\frac{\alpha_i s_i}{1 + \sum_{j=1}^{n} \beta_{ij} s_j} p + (\mu_i - \sigma_i) s_i \qquad (3\text{-}5\text{-}24)$$

where s_i and p are prey and predator density, μ_i and σ_i are growth and decay coefficients of the prey, and α_i and β_{ij} are parameters. The similarity of the predation term on the righthand side of Eq. (3-5-24) with Eq. (3-5-21) for competitive inhibition is obvious.

In principle, all parameters determining the ecological processes described above are functions of both temperature and river flow rate. But in most cases the dependencies are small, so that in view of the model structure uncertainties they may be neglected. A temperature dependence which must be taken into account, however, concerns the *maximum specific growth rates*. In the case of model (3-5-17–3-5-19) the maximum specific growth rate of the bacteria is, for instance, $\alpha_1 \beta_1$, in the case of model (3-5-23) it is $\gamma_1(\alpha_3/\alpha_5)$ for the attached bacteria. (Note that endogenous respiration is excluded from the definition of maximum specific growth

88 STRUCTURE OF THE MODELS

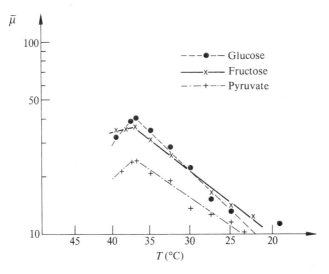

Figure 3-5-8 Maximum specific growth rate of *Escherichia coli* bacteria on various substrates vs. temperature (after Peters, 1973).

rate.) The rate coefficient α of a chemical reaction depends on T according to *Arrhenius' law*

$$\alpha(T) = \beta_1 \, e^{-\beta_2/T} \tag{3-5-25}$$

where β_1, β_2 are parameters and T is measured in K. Since the growth processes are chains of chemical reactions, their velocity should also depend on T according to Eq. (3-5-25), as long as the temperature is not so high that proteins become denatured. This has indeed been confirmed for many species relevant for self-purification. As an example, Fig. 3-5-8 shows the temperature dependence of the maximum specific growth rate $\bar{\mu}$ of *Escherichia coli* bacteria (Peters, 1973). The units are chosen such that a straight line results if Eq. (3-5-25) applies. It can be seen that Arrhenius' law applies, and similar curves have also been obtained for other organisms (see, for instance, Sudo and Aiba, 1972).

The Arrhenius law should also apply to the *temperature dependence of the endogenous respiration coefficient*, and this has indeed been observed, for example, by Benedek et al. (1972). Since the Arrhenius law describes a strong dependency on T in the temperature region we are interested in, the temperature dependence of endogenous respiration has also to be taken into account. Other temperature or flow rate dependencies are not considered here, although the parameter γ_2 in Eq. (3-5-23) in particular is highly sensitive to changes of Q.

Photosynthesis

Modeling the *biogenic aeration* represented by the term P in Eq. (3-5-1) is in itself a very large field. The underlying processes have been investigated, both experimentally and theoretically, mainly for lakes or impoundments. In polluted rivers, however, processes related to the degradation of anthropogenic wastes dominate. Therefore a few remarks on the subject will suffice.

The simplest way to take the activity of phototrophs into account is to look upon P in Eq. (3-5-1) as an external input. It could be a periodic function of time, in order to account for variations of P due to diurnal variations of light intensity (see Sec. 4-2), or it could even be a constant (see Sec. 5-3). If we want to explicitly model the population dynamics of photosynthesizing organisms, no completely new concepts (compared to the ones discussed above) have to be introduced. The main difference between growth models for bacteria and phototrophs is that for the latter several growth limiting factors have often to be considered, while for the bacteria only the concentration of energy donors is usually important (because in the pollutant degradation processes the other nutrients required by the bacteria occur as by-products anyway). The dependence of the specific growth rate on the limiting factors may again be described by the Michaelis–Menten expression, so that, for example, the growth of planctonic algae (neglecting dispersion) may be modeled through (Chen, 1970)

$$\frac{\partial A}{\partial t} + v \frac{\partial A}{\partial l} = \alpha_1 \cdot \frac{J}{\alpha_2 + J} \cdot \frac{[PO_4^{3-}]}{\alpha_3 + [PO_4^{3-}]} \cdot \frac{[NO_3^-]}{\alpha_4 + [NO_3^-]} \cdot \frac{[CO_2]}{\alpha_5 + [CO_2]} \cdot A - \frac{\beta_1 A}{\beta_2 + A} G - \gamma A$$

where A = algal density
  J = light intensity
  G = grazer density
  $\alpha_i, \beta_i, \gamma$ = parameters
  $[C]$ = concentration of inorganic compound C

The light intensity is also a function of A because of shading. The formulation of the corresponding equations for the nutrients and the grazers and of the terms in the oxygen balance equation is straightforward and need not be discussed here.

The specific growth rate of phototrophs may often be considered independent of temperature, as can be expected for photochemical reactions. On the other hand, endogenous respiration depends on temperature as for chemotrophic organisms, so that the ratio between assimilation and respiration increases as temperature decreases (see Sec. 2-3 and Round, 1965).

Other Self-Purification Phenomena

As in the case of biochemical degradation and synthesis, there is also a great variety of models for the processes of *sedimentation* and *resuspension*. They differ widely in complexity. At one end of the range there are sophisticated statistical models like the one by Shen and Cheong (1974), which try to describe how heavier particles are dragged along the river bottom. At the other end, there are simple first order models: for example, Wolf (1971) gives the following equation for sedimentation of suspended solids (in flow time)

$$\dot{s} = -\alpha_1 \left(s - s_0 \frac{v^4}{\alpha_2 + v^4} \right)$$

where s = concentration of suspended solids
s_0 = initial concentration of suspended solids
α_i = parameters

An analogous model may be formulated for resuspension. Relatively little work has been done on how to include in water quality models the other phenomena mentioned in Sec. 2-4, like *adsorption, precipitation, flocculation*, etc. Models for these processes have been developed, however, in the context of the theory of pollution control facilities (Keinath and Wanielista, 1975), and they may be adapted for inclusion into river quality models.

The discussion of this section may lead to the conclusion that it is easy to develop a fairly detailed water quality model for a given river. The reason for this is that the major problem of how to determine the numerous model parameters uniquely from observations has not yet been discussed. Because of this problem, to which Chapter 5 is devoted, it is usually necessary to use models which are less precise than those developed from a qualitative insight into the problem.

REFERENCES

Section 3-1

Bansal, M. K. (1971). Dispersion in Natural Streams. *J. Hydraulics Div., Proc. ASCE*, **97**, HY11, 1867–1886.

Chen, C. W. and Orlob, G. T. (1975). Ecologic Simulation for Aquatic Environments. In *Systems Analysis and Simulation in Ecology*, Vol. 3 (B. D. Patten, ed.). Academic Press, New York.

Di Toro, D. (1969). Stream Equations and Method of Characteristics. *J. San. Eng. Div., Proc. ASCE*, **95**, 699–703.

Hays, J. R. (1966). *Mass Transport Mechanisms in Open Channel Flow*. Ph.D. Thesis, Vanderbilt University, Nashville, Tennessee.

Hirsch, C. (1975). Methods of Analysis in Water Quality Simulation Models. In *Modelling and Simulation of Water Resources Systems, Proc. IFIP Working Conference on Computer Simulation of Water Resources Systems*. North-Holland, Amsterdam.

Holley, E. R., Siemons, J., and Abraham, G. (1972). Some Aspects of Analyzing Transverse Diffusion in Rivers. *J. Hydraulics Res.*, **10**, 27–57.

Holly, F. M. (1975). *Two-Dimensional Mass Dispersion in Rivers*. Hydrology Paper 78, Colorado State University, Fort Collins, Colo.

Leendertsee, J. J., Alexander, R. C., and Lin, S. K. (1973). *A Three Dimensional Model for Estuaries and Coastal Seas*. Vol. I: *Principles of Computation*. Report 4-1417-OW.RR, Rand Corporation, Santa Monica, California.

Li, W. H. (1962). Unsteady Dissolved Oxygen Sag in a Stream. *J. San. Eng. Div., Proc. ASCE*, **88**, 75–85.

McQuivey, R. S. and Keefer, T. N. (1974). Simple Method for Predicting Dispersion in Streams. *J. Env. Eng. Div., Proc. ASCE*, **100**, EE4, 997–1011.
Nihoul, J. (1975). Marine Systems Analysis. In *Modelling of Marine Systems* (J. Nihoul, ed.). Elsevier, Amsterdam.
Orlob, G. T. (1976). Estuarial Models. In *Systems Approach to Water Management* (A. K. Biswas, ed.). McGraw-Hill, New York.
Sayre, W. W. (1972). Natural Mixing in Rivers. In *Environmental Impact on Rivers (River Mechanics III)* (H. W. Shen, ed.). H. W. Shen, Fort Collins, Colorado.
Thackston, E. L. and Schnelle, K. B. (1970). Predicting Effects of Dead Zones on Stream Mixing. *J. San. Eng. Div., Proc. ASCE*, **96**, 319–331.
Ward, P. R. B. and Fischer, H. B. (1971). Some Limitations on Use of the One-Dimensional Dispersion Equation. *Water Resour. Res.*, **7**, 215–220.
Yotsukura, N. and Sayre, W. W. (1976). Transverse Mixing in Natural Channels. *Water Resour. Res.*, **12**, 695–704.

Section 3-2

No references

Section 3-3

Chen, C. W. and Orlob, G. T. (1975). Ecologic Simulation for Aquatic Environments. In *Systems Analysis and Simulation in Ecology*, Vol. 3 (B. C. Patten, ed.). Academic Press, New York.
Gray, D. M. (1970). *Handbook on the Principles of Hydrology*. Water Information Center, Port Washington, New York.
Greco, F. and Panattoni, L. (1977). Numerical Solution Methods of the Saint Venant Equations. In *Mathematical Models for Surface Water Hydrology* (T. A. Ciriani, U. Maione, and J. R. Wallis, eds.). John Wiley, New York.
Nash, J. E. (1957). The Form of the Instantaneous Unit Hydrograph. *I.A.S.H.*, vol. 3, n. 45.
Orlob, G. T. (1976). Estuarial Models. In *Systems Approach to Water Management* (A. K. Biswas, ed.). McGraw-Hill, New York.

Section 3-4

Anderson, E. R. (1954). Energy Budget Studies. In *Water Loss Investigations: Lake Hefner Studies*. Prof. Paper 269, 71-119, U.S. Geol. Survey, Washington, D.C.
Bøgh, P. and Zünd, H. (1970). THEDY: A Program for Digital Simulation of the Unsteady Heat Budget of River Systems (in German). *Neue Technik*, **1**, 27–31.
Bowen, I. S. (1926). The Ratio of Heat Losses by Conduction and by Evaporation from any Water Surface. *Phys. Rev.*, **27**, 785.
Eckel, O. and Reuter, H. (1950). On the Evaluation of Heat Turnover in Rivers during Summer (in German). *Geografiska Annaler Stockholm*, **32**, 188–209.
Edinger, J. E., Duttweiler, D. W., and Geyer, J. C. (1968). The Response of Water Temperatures to Meteorological Conditions. *Water Resour. Res.*, **4**, 1137–1143.
Faude, D., Bayer, A., Halbritter, G., Spannagel, G., Stehfest, H., and Wintzer, D. (1974). *Energy and Environment in Baden-Württemberg* (in German). Report No. KFK 1966 UF, Kernforschungszentrum Karlsruhe, Karlsruhe, W. Germany.
Harleman, D. R. F., Brocard, D. N., and Najarian, T. O. (1973). *A Predictive Model for Transient Temperature Distributions in Unsteady Flows*. Report No. 175, Ralph M. Parsons Laboratory for Water Resources and Hydrodynamics, MIT, Cambridge, Mass.
Jaske, R. T. and Spurgeon, J. L. (1968). A Special Case, Thermal Digital Simulation of Waste Heat Discharges. *Water Research*, **2**, 777–802.

Jobson, H. E. and Yotsukura, N. (1972). Mechanics of Heat Transfer in Nonstratified Open-Channel Flows. In *Environmental Impact on Rivers (River Mechanics III)* (H. W. Shen, ed.). H. W. Shen, Fort Collins, Colo.

LAWA (Länderarbeitsgemeinschaft Wasser) (1971). *Foundations for Evaluating Heat Discharges into Rivers* (in German). Koehler & Hennemann, Wiesbaden, W. Germany.

Leendertsee, J. J., Alexander, R. C., and Lin, S. K. (1973). *A Three-Dimensional Model for Estuaries and Coastal Seas.* Vol. I: *Principles of Computation.* Report 4-1417-OW.RR, Rand Corporation, Santa Monica, California.

Motor Columbus, Ingenieurunternehmung AG (1971). *Study on the Heat-Carrying Capacity of the Rhine River* (in German). Motor Columbus, Baden, Switzerland.

Orlob, G. T. (1976). Estuarial Models. In *Systems Approach to Water Management* (A. K. Biswas, ed.). McGraw-Hill, New York.

Orlob, G. T. and Selna, L. G. (1970). Temperature Variations in Deep Reservoirs. *J. Hydraulics Div., Proc. ASCE*, **96**, HY2, Proc. Paper 7063.

Parker, F. L. and Krenkel, P. A. (1970). *Physical and Engineering Aspects of Thermal Pollution.* Butterworths, London.

Section 3-5

Beck, M. B. and Young, P. C. (1976). Systematic Identification of DO–BOD model structure. *J. Env. Eng. Div., Proc. ASCE*, **102**, 909–927.

Benedek, P., Farkas, P., and Litheraty, P. (1972). Kinetics of Aerobic Sludge Stabilization. *Water Research*, **6**, 91–97.

Bick, H. (1964). *Succession of Organisms during Self-Purification of Organically Polluted Water under Various Conditions* (in German). Ministerium für Ernährung, Landwirtschaft und Forsten des Landes Nordrhein-Westfalen, Düsseldorf, W. Germany.

Boes, M. (1970). *Mathematical Analysis of BOD Kinetics* (in German). Thesis, University of Karlsruhe, Karlsruhe, W. Germany.

Boling, Jr., R. H., Peterson, R. C., and Cummins, K. W. (1975). Modelling of Small Woodland Streams. In *Systems Analysis and Simulation in Ecology*, Vol. III (B. C. Patten, ed.). Academic Press, New York.

Chen, C. W. (1970). Concepts and Utilities of Ecological Models. *J. Sanit. Eng. Div., Proc. ASCE*, **96**, 1085–1097.

Dawes, E. A. and Ribbons, D. W. (1964). Some Aspects of the Endogenous Metabolism of Bacteria. *Bact. Rev.*, **28**, 126–149.

Edwards, R. W. and Rolley, H. C. J. (1965). Oxygen Consumption of River Muds. *J. Ecol.*, **53**, 1–19.

Gates, W. E., Marlar, J. T., and Westfield, J. D. (1969). The Application of Bacterial Process Kinetics in Stream Simulation and Stream Analysis. *Water Research*, **3**, 663–686.

Gates, W. E., Pohland, F. G., Mancy, D. G., and Schafie, F. R. (1966). A Simplified Physical Model for Studying Assimilative Capacity. *Proc. 21st Ind. Waste Conf.*, 665–687. Purdue University Extension Service.

Gaudy, Jr., A. F. (1972). Biochemical Oxygen Demand. In *Water Pollution Microbiology* (R. Mitchell, ed.). John Wiley, New York.

Gaudy, Jr., A. F., Komolrit, K., and Bhatla, M. N. (1963). Sequential Substrate Removal in Heterogeneous Populations. *J. WPCF*, **35**, 903–922.

Gotaas, J. B. (1948). Effect of Temperature on Biochemical Oxidation of Sewage. *Sewage and Industrial Wastes*, **20**, 441–477.

Gunsalus, I. C. and Shuster, C. W. (1961). Energy Yielding Metabolism in Bacteria. In *The Bacteria: A Treatise on Structure and Function*, Vol. II (I. C. Gunsalus and R. Y. Stanier, eds.). Academic Press, New York.

Hartmann, L. and Laubenberger, G. (1968). Toxicity Measurements in Activated Sludge. *J. San. Eng. Div., Proc. ASCE*, **94**, 247–256. Corrections in *J. San. Eng. Div., Proc. ASCE*, **96**, 607–609.

Isaacs, W. P. and Gaudy, A. E. (1968). Atmospheric Oxygenation in a Simulated Stream. *J. San. Eng. Div., Proc. ASCE*, **94**, 319–344.
Keinath, T. M. and Wanielista, M. P. (1975). *Mathematical Modeling for Water Pollution Control Processes*. Ann Arbor Science, Ann Arbor, Mich.
Kölle, W., Ruf, H., and Stieglitz, L. (1972). The Burden of the Rhine River by Organic Pollutants (in German). *Die Naturwissenschaften*, **59**, 299–305.
Krenkel, P. A. and Parker, F. L. (1969). Engineering Aspects, Sources, and Magnitude of Thermal Pollution. In *Biological Aspects of Thermal Pollution* (P. A. Krenkel and F. L. Parker, eds.). *Proc. Nat. Symp. on Thermal Pollution, Portland, Oreg., June 3–5, 1968.* Vanderbilt University Press, Nashville, Tenn.
Laidler, K. J. (1958). *The Chemical Kinetics of Enzyme Action*. Oxford University Press, Oxford, U.K.
McCarty, P. L. (1965). Thermodynamics of Biological Synthesis and Growth. *Proc. 2nd International Conference on Water Pollution Research*, 169–199. Pergamon Press, New York.
McCarty, P. L. (1972). Energetics of Organic Matter Degradation. In *Water Pollution Microbiology* (R. Mitchell, ed.). John Wiley, New York.
Michaelis, L. and Menten, M. (1913). The Kinetics of Invertase Action (in German). *Biochem. Z.*, **49**, 333–369.
Müller, S. (1975). *Oxygen Budget in Rivers* (in German). Bayer. Landesamt für Umweltschutz, München, W. Germany.
Negulescu, M. and Rojanski, V. (1969). Recent Research to Determine Reaeration Coefficients. *Water Research*, **3**, 189–202.
Peters, H. (1973). Thermodynamical Analysis of Oxydative Degradation of Pure Substrates through E. coli (in German). Jahresbericht des Instituts für Ingenieurbiologie und Biotechnologie des Abwassers, Universität Karlsruhe, Karlsruhe, W. Germany.
Round, F. E. (1965). *The Biology of the Algae*. Edward Arnold, London.
Shen, H. W. and Cheong, H.-F. (1972). Dispersion of Contaminants Attached to Sediment Bed Load. In *Environmental Impact on Rivers* (*River Mechanics III*) (H. W. Shen, ed.). H. W. Shen, Fort Collins, Colo.
Stehfest, H. (1973). *Mathematical Modelling of Self Purification of Rivers* (in German; English translation available as Report IIASA PP-77-11 of the International Institute for Applied Systems Analysis, Laxenburg, Austria). Report KFK 1654 UF, Kernforschungszentrum Karlsruhe, Karlsruhe, W. Germany.
Sudo, R. and Aiba, S. (1972). Growth Rate of *Aspidiscidae* Isolated from Activated Sludge. *Water Research*, **6**, 137–144.
Thackston, E. L. and Krenkel, P. A. (1969). Reaeration Prediction in Natural Streams. *J. San. Eng. Div., Proc. ASCE*, **95**, 65–94.
Wilderer, P. (1969). *Enzyme Kinetics as a Basis of the BOD Reaction* (in German). Thesis, University of Karlsruhe, Karlsruhe, W. Germany.
Wilderer, P., Hartmann, L., and Janečková, J. (1970). Objections Against the Use of Long-Term BOD for Characterizing Raw Waste Water (in German). *Wasser- und Abwasser-Forschung*, **1**, 7–12.
Wilson, G. T. and MacLeod, N. (1973). A Critical Appraisal of Empirical Equations and Models for the Prediction of the Coefficient of Reaeration of Deoxygenated Water. *Water Research*, **8**, 341–366.
Wolf, P. (1971). Incorporation of Latest Findings into Oxygen Budget Calculations for Rivers (in German). *GWF–Wasser/Abwasser*, **112**, 200–203 and 250–254.

CHAPTER
FOUR

SOME PARTICULAR SELF-PURIFICATION MODELS

4-1 STREETER-PHELPS MODEL

The so-called *Streeter-Phelps model* is not only the oldest (1925) among the biochemical submodels, but also the one most widely used in river quality analysis. Therefore, the properties of this model are studied in detail in this section. As already pointed out in Sec. 3-5, the main assumption underlying the Streeter-Phelps model is that two variables, namely concentration of BOD (biochemical oxygen demand) and DO (dissolved oxygen) are sufficient to describe the biochemical processes. Moreover, Streeter and Phelps (1925) assumed that:

(a) the BOD decay rate is proportional to BOD concentration
(b) the deoxygenation and BOD decay rate are equal
(c) the reoxygenation rate is proportional to the oxygen deficit

The model equations resulting from these assumptions are

$$\frac{\partial b}{\partial t} + v \frac{\partial b}{\partial l} = -k_1 b \tag{4-1-1a}$$

$$\frac{\partial c}{\partial t} + v \frac{\partial c}{\partial l} = -k_1 b + k_2(c_s - c) \tag{4-1-1b}$$

with boundary conditions

$$b(0, t) = b_0(t) \qquad c(0, t) = c_0(t)$$

and initial conditions

$$b(l,0) = b_i(l) \qquad c(l,0) = c_i(l)$$

Longitudinal dispersion was neglected for the sake of simplicity. The parameter k_1 is called *deoxygenation coefficient* and k_2 *reaeration coefficient*.

Equation (4-1-1a) describes the dynamics of the BOD degradation. If finite time BOD_θ (e.g., BOD_5) is measured (see Sec. 3-5) a fixed relationship between BOD_θ and BOD is usually assumed; for example

$$BOD_\theta = (1 - e^{-K_L \theta})\, BOD$$

where $K_L (\text{day}^{-1})$ is the coefficient of BOD decay as evaluated by the standard laboratory BOD test. If the dynamics of BOD_θ (b_θ) are to be described, model (4-1-1) has to be modified in the following way

$$\frac{\partial b_\theta}{\partial t} + v \frac{\partial b_\theta}{\partial l} = -k_1 b_\theta$$

$$\frac{\partial c}{\partial t} + v \frac{\partial c}{\partial l} = -\frac{k_1}{(1 - e^{-K_L \theta})} b_\theta + k_2 (c_s - c)$$

Sometimes the difference between BOD and BOD_θ is assumed to be small and Eq. (4-1-1) is directly used with b defined as BOD_θ.

As pointed out in Sec. 3-1, model (4-1-1) can be converted into a set of ordinary differential equations

$$\frac{db}{d\tau} = -k_1 b \qquad (4\text{-}1\text{-}2a)$$

$$\frac{dc}{d\tau} = -k_1 b + k_2(c_s - c) \qquad (4\text{-}1\text{-}2b)$$

$$\frac{dl}{d\tau} = v \qquad (4\text{-}1\text{-}2c)$$

$$\frac{dt}{d\tau} = 1 \qquad (4\text{-}1\text{-}2d)$$

with initial conditions

$$b(0) = b_i(l_0) \qquad c(0) = c_i(l_0) \qquad l(0) = l_0 \qquad t(0) = 0 \quad \text{for} \quad 0 \le l_0 \le L$$

and

$$b(0) = b_0(t_0) \qquad c(0) = c_0(t_0) \qquad l(0) = 0 \qquad t(0) = t_0 \quad \text{for} \quad t_0 > 0$$

Equation (4-1-2) describes the self-purification process in *flow time* τ. The time t_0 is called *release time* (Di Toro and O'Connor, 1968).

If $v(l, t)$, $b_0(t)$, $c_0(t)$ are independent of time the steady state case applies, for which the equation could be derived directly from Eq. (4-1-1) by setting all the derivatives with respect to t equal to zero. If in those simplified equations l is replaced by τ according to Eq. (4-1-2c), Eqs. (4-1-2a) and (4-1-2b) result.

Equations (4-1-2a) and (4-1-2b) are linear differential equations which can be written in the form

$$\dot{\mathbf{x}}(\tau) = \mathbf{F}\mathbf{x}(\tau) + \mathbf{g}u$$

$$\mathbf{y}(\tau) = \mathbf{H}\mathbf{x}(\tau)$$

where the two-dimensional state vector $\mathbf{x}$ is given by

$$\mathbf{x}(\tau) = [b(\tau) \quad c(\tau)]^T$$

and

$$\mathbf{F} = \begin{bmatrix} -k_1 & 0 \\ -k_1 & -k_2 \end{bmatrix} \quad \mathbf{g}u = \begin{bmatrix} 0 \\ k_2 c_s \end{bmatrix}$$

The output transformation matrix is given by

$$\mathbf{H} = [0 \quad 1]$$

if DO is considered the only output variable and by

$$\mathbf{H} = \begin{bmatrix} 1 & 0 \\ 0 & 1 \end{bmatrix} \quad \text{or} \quad \mathbf{H} = \begin{bmatrix} 1 - e^{-5K_L} & 0 \\ 0 & 1 \end{bmatrix}$$

if BOD or BOD_5, respectively, are also considered output variables.

The solution of Eqs. (4-1-2a) and (4-1-2b) (see Sec. 1-2) is given by

$$b(\tau) = b(0) \, e^{-k_1 \tau} \tag{4-1-3a}$$

$$c(\tau) = c_s - (c_s - c(0)) \, e^{-k_2 \tau} + \frac{k_1 b(0)}{k_1 - k_2} (e^{-k_1 \tau} - e^{-k_2 \tau}) \tag{4-1-3b}$$

The trajectories shown in Fig. 4-1-1 are the solutions $l = l(t - t_0)$ of Eqs. (4-1-2c) and (4-1-2d) and they can be used to define a function $t_0 = t_0(l, t)$ giving the starting (release) time of a particle of water which is at point l at time t. Then, since $\tau = t - t_0$, the solution of Eq. (4-1-1) can be derived by means of Eq. (4-1-3)

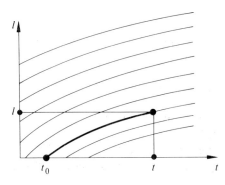

Figure 4-1-1 Trajectories of various water plugs and the function $t_0 = t_0(l, t)$.

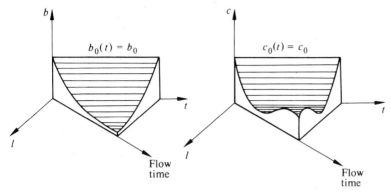

Figure 4-1-2 The surfaces $b(l, t)$ and $c(l, t)$ when the upstream boundary conditions $b_0(t)$, $c_0(t)$ and the stream velocity $v(l, t)$ are constant.

giving

$$b(l,t) = b_0(t_0(l,t))\, e^{-k_1(t-t_0(l,t))} \qquad (4\text{-}1\text{-}4a)$$

$$c(l,t) = c_s - \left[c_s - c_0(t_0(l,t))\right] e^{-k_2(t-t_0(l,t))}$$
$$+ \frac{k_1 b_0(t_0(l,t))}{k_1 - k_2}\left(e^{-k_1(t-t_0(l,t))} - e^{-k_2(t-t_0(l,t))}\right) \qquad (4\text{-}1\text{-}4b)$$

The surfaces $b(l,t)$ and $c(l,t)$ are shown in Fig. 4-1-2 for the steady state case in which v is constant in space and time, while in Fig. 4-1-3 the surface $b(l,t)$ is presented for the case in which a constant BOD load is discharged into a river whose flow rate Q and velocity v vary sinusoidally in time, i.e.,

$$Q = \bar{Q} + \Delta Q \sin(\bar{\omega} t) \qquad (\Delta Q < \bar{Q})$$
$$v = \bar{v} + \Delta v \sin(\bar{\omega} t) \qquad (\Delta v < \bar{v})$$

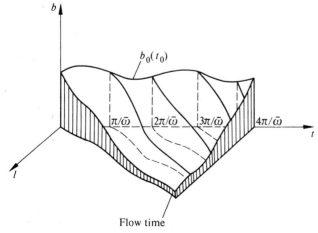

Figure 4-1-3 The surface $b(l, t)$ when a constant BOD load is discharged in a sinusoidally varying flow.

98 SOME PARTICULAR SELF-PURIFICATION MODELS

In the steady state case (Fig. 4-1-2) the form of the oxygen profile is that of a *sag curve*, both in flow time and in space, while BOD and DO are constant in time at any given point. For this reason the representation of Fig. 4-1-4 is usually preferred to that of Fig. 4-1-2, since it more clearly shows how biodegradable matter disappears and how oxygen concentration first decreases and then increases as water flows downstream (self-purification of the river). The latter property also applies to the unsteady state case since Eq. (4-1-4) implies that

$$\lim_{l \to \infty} b(l, t) = \lim_{l \to \infty} [c_s - c(l, t)] = 0$$

for all bounded boundary conditions $b_0(t)$ and $c_0(t)$. The self-purification property corresponds to model (4-1-2) to have a unique and asymptotically stable equilibrium state $\bar{\mathbf{x}} = [0 \ c_s]^T$ (the matrix $\mathbf{F}$ has negative eigenvalues).

Figure 4-1-4 shows the existence of a *critical point* where the DO deficit reaches its maximum value (*critical deficit*). The position l_c of the critical point and the critical deficit d_c are functions of the boundary conditions since

$$l_c = \frac{v}{k_1(f-1)} \ln \left\{ f \left[1 - (f-1) \frac{d_0}{b_0} \right] \right\} \qquad (4\text{-}1\text{-}5)$$

and

$$d_c = \frac{b_0}{f} \left\{ f \left[1 - (f-1) \frac{d_0}{b_0} \right] \right\}^{1/(1-f)} \qquad (4\text{-}1\text{-}6)$$

where $f = k_2/k_1$ is called the *self-purification rate* (Fair and Geyer, 1965) and $d_0 = c_s - c_0$ is the oxygen deficit at $l = 0$. Equations (4-1-5) and (4-1-6) can easily be found by annihilating the first derivative of $c(t)$ given by Eq. (4-1-3b). A number of interesting properties of these functions are given by Liebman (1965), Liebman and Loucks (1966) and Arbabi et al. (1974). Some of these properties are used in the following (see Sec. 8-2) and are now briefly mentioned. From Eq. (4-1-5), it can be proved that the critical point exists, i.e., that l_c is positive, if and only if

$$\frac{d_0}{b_0} \leq 1/f \qquad (4\text{-}1\text{-}7)$$

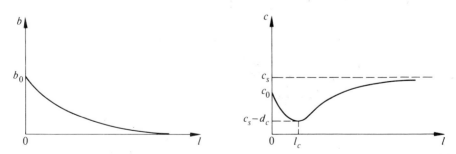

Figure 4-1-4 BOD and DO profiles along the river in steady state conditions.

Moreover, since f, b_0 and d_0 are non-negative, the critical point can never be located downstream the point

$$L_c = \frac{v}{k_1(f-1)} \ln f \qquad (4\text{-}1\text{-}8)$$

In particular l_c tends to L_c for increasing initial loads b_0, while $l_c = L_c$ if the river is perfectly oxygenated at the boundary ($d_0 = 0$). The critical deficit d_c is a non-linear function of the boundary conditions and this generates some difficulty as will be seen later on (pages 252–254). Equation (4-1-6) may also be written in the form

$$d_c = \frac{b_0}{f} e^{-\frac{k_1}{v} l_c}$$

which shows (recall that l_c is bounded by L_c) that the critical deficit d_c increases without limit with the initial load b_0. In other words, there are initial concentrations b_0, such that $d_c > c_s$, i.e., the oxygen concentration predicted by the model can be negative. This physical nonsense is obviously implied by assumption (a), which says that the deoxygenation rate is independent of the DO concentration. While this assumption is reasonable at high levels of DO, it is at the least doubtful at low values of DO, when anaerobic conditions can occur (see Secs. 2-3 and 3-5). A modified nonlinear model which does not have this disadvantage will be shown in the next section.

The influence on the model of variations in some of the main parameters can now be considered (*sensitivity analysis*). The approach differs depending upon whether large or small variations of the parameters are expected. (There are also other considerations which may be relevant for the selection of the suitable method of sensitivity analysis; see Stehfest, 1975.) In the case of large variations one has, in general, to simulate the model (or look at its analytic solution) for a few representative values of the parameters and then make the comparison (see Lin et al., 1973). On the other hand, if only small variations around some nominal conditions are of interest, the analysis can be carried out by the first order approximation suggested by the so-called *sensitivity theory* (see Rinaldi and Soncini-Sessa, 1976 and 1978). Since this approach is quite efficient and elegant it is now briefly presented and applied to the analysis of the Streeter-Phelps model.

Assume that a continuous, lumped parameter system is described by the differential equation

$$\dot{\mathbf{x}}(t) = \mathbf{f}(\mathbf{x}(t), \theta, t) \qquad (4\text{-}1\text{-}9)$$

where θ is a constant parameter with nominal value $\bar{\theta}$ and let the initial state $\mathbf{x}_0$ of the system depend upon the parameter, i.e.,

$$\mathbf{x}_0 = \mathbf{x}_0(\theta) \qquad (\bar{\mathbf{x}}_0 = \mathbf{x}_0(\bar{\theta})) \qquad (4\text{-}1\text{-}10)$$

The solution of Eq. (4-1-9) with the initial condition (4-1-10) is a function

$$\mathbf{x} = \mathbf{x}(t, \theta)$$

100 SOME PARTICULAR SELF-PURIFICATION MODELS

which, under very general conditions, can be expanded in series in the neighborhood of the nominal value of the parameter, i.e.,

$$\mathbf{x}(t, \theta) = \bar{\mathbf{x}}(t) + \left[\frac{\partial \mathbf{x}(t, \theta)}{\partial \theta}\right]_{\bar{\theta}} (\theta - \bar{\theta}) + \cdots$$

where $\bar{\mathbf{x}}(t) = \mathbf{x}(t, \bar{\theta})$ is the nominal solution. The vector $[\partial \mathbf{x}/\partial \theta]_{\bar{\theta}}$, namely the derivative of the state vector with respect to the parameter, is called *sensitivity vector* and will be denoted by **s** from now on, i.e.,

$$\mathbf{s}(t) = \left[\frac{\partial \mathbf{x}}{\partial \theta}\right]_{\bar{\theta}}$$

Thus, the perturbed solution of Eq. (4-1-9) can easily be obtained as

$$\mathbf{x}(t, \theta) \cong \bar{\mathbf{x}}(t) + \mathbf{s}(t)(\theta - \bar{\theta})$$

once the sensitivity vector is known.

When there are many parameters $\theta_1, \theta_2, \ldots, \theta_q$, the knowledge of the sensitivity vectors $\mathbf{s}_1, \mathbf{s}_2, \ldots, \mathbf{s}_q$ allows the association of certain characteristics of the system behavior with certain particular parameters. If, for example, the nominal solution $\bar{x}(t)$ of a first order system is the one shown in Fig. 4-1-5 where $s_1(t)$ and $s_2(t)$ are the sensitivity coefficients of x with respect to two parameters θ_1 and θ_2, then it is reasonable to say that the first parameter is responsible for the overshoot of $\bar{x}$, while the second is responsible for the asymptotic behavior of the system. This characterization of the parameters turns out to be very often of great importance in the phase of the validation of the structure of the system.

It is easy to prove that the sensitivity vector $\mathbf{s}(t)$ satisfies the following vector differential equation

$$\dot{\mathbf{s}} = \left[\frac{\partial \mathbf{f}(\mathbf{x}, \bar{\theta}, t)}{\partial \mathbf{x}}\right]_{\bar{\mathbf{x}}} \mathbf{s} + \left[\frac{\partial \mathbf{f}(\bar{\mathbf{x}}, \theta, t)}{\partial \theta}\right]_{\bar{\theta}} \qquad (4\text{-}1\text{-}11)$$

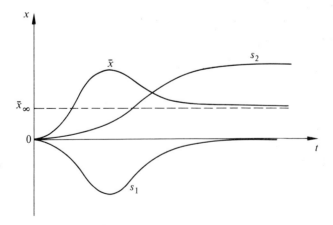

Figure 4-1-5 Nominal solution $\bar{x}$ and sensitivity coefficients s_1 and s_2 of a system.

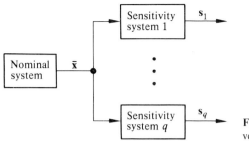

Figure 4-1-6 Computation of the sensitivity vectors.

with initial conditions

$$\mathbf{s}(0) = \left[\frac{\partial \mathbf{x}_0}{\partial \theta}\right]_{\bar{\theta}} \quad (4\text{-}1\text{-}12)$$

Thus, the sensitivity vector is the state vector of system (4-1-11), called *sensitivity system*, which is always a linear system, even if system (4-1-9) is nonlinear. Because of this property the sensitivity vectors can often be determined analytically. In any case, they can always be computed by means of simulation following the scheme shown in Fig. 4-1-6.

This methodology is now applied to some very particular but interesting sensitivity problems of river pollution. The model used is the steady state Streeter-Phelps model (4-1-2) or some suitable modification of it. The reader interested in the details of this analysis should refer to Rinaldi and Soncini-Sessa (1976, 1978).

BOD Load Variation

Let us first analyze the effects of a variation of the BOD load discharged at a particular point into the river. By measuring flow time from this point the system is described by

$$\dot{b} = -k_1 b$$
$$\dot{c} = -k_1 b + k_2(c_s - c)$$

and the initial conditions are ($\bar{\theta} = 0$)

$$b_0 = \bar{b}_0 + \theta \qquad c_0 = \bar{c}_0$$

Thus, the sensitivity system is given by

$$\dot{s}_b = -k_1 s_b \quad (4\text{-}1\text{-}13a)$$
$$\dot{s}_c = -k_1 s_b - k_2 s_c \quad (4\text{-}1\text{-}13b)$$

and its initial conditions are

$$s_b(0) = \left[\frac{\partial b_0}{\partial \theta}\right]_{\bar{\theta}} = 1 \qquad s_c(0) = 0$$

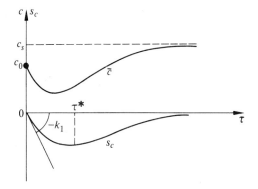

Figure 4-1-7 The sensitivity of dissolved oxygen concentration with respect to BOD load.

The solution of Eq. (4-1-13a) with $s_b(0) = 1$ is given by

$$s_b(\tau) = e^{-k_1 \tau}$$

which can be introduced in Eq. (4-1-13b) together with $s_c(0) = 0$, thus giving

$$s_c(\tau) = k_1 \frac{e^{-k_1 \tau} - e^{-k_2 \tau}}{k_1 - k_2} \tag{4-1-14}$$

The sensitivity coefficient s_c given by Eq. (4-1-14) is always negative. Its behavior is as shown in Fig. 4-1-7 and it has a minimum (see Eq. (4-1-8)) for

$$\tau^* = \frac{\ln (k_2/k_1)}{k_2 - k_1} = \frac{L_c}{v} \tag{4-1-15}$$

This means that a positive perturbation of the BOD load in a point implies that all the river downstream from that point becomes worse as far as its oxygen content is concerned. Of course, this is the conclusion that can be derived from the Streeter-Phelps model which does not represent a priori the behavior of a real river. Indeed, it will be shown in Sec. 5-3 that because of the mechanisms of the food web, it can be expected that the conditions of the river are sometimes improved by an increase of the BOD load (see also Sec. 10-5 for an interesting example of the implications of this fact).

Equilibrium Temperature Variations

The influence of the temperature on the dissolved oxygen of a river will now be discussed. In order to simplify the discussion a constant point source of BOD on a perfectly clean and oxygenated river is assumed. Moreover, suppose that the water temperature T is constant along the river (e.g., equal to the equilibrium temperature, see Sec. 3-4). Thus, the upstream boundary conditions of the stretch are given and depend upon the temperature T of the water since the oxygen saturation level c_s is a (decreasing) function of T. Under these assumptions, the system is described by

$$\dot{b} = -k_1(T)b \qquad (4\text{-}1\text{-}16a)$$
$$\dot{c} = -k_1(T)b + k_2(T)(c_s(T) - c) \qquad (4\text{-}1\text{-}16b)$$

with initial conditions

$$b_0 = \bar{b}_0 \qquad c_0 = c_s(T)$$

The corresponding sensitivity system (4-1-11) is given by

$$\dot{s}_b = -k_1 s_b - k_1' \bar{b}$$
$$\dot{s}_c = -k_1 s_b - k_2 s_c - k_1' \bar{b} + (k_2 c_s)' - k_2' \bar{c}$$

with initial conditions

$$s_b(0) = 0 \qquad s_c(0) = c_s'$$

where ′ denotes derivative with respect to T, and $\bar{b}$ and $\bar{c}$ are the nominal solutions of Eq. (4-1-16). The solution of the sensitivity system can be derived by taking Eq. (4-1-3) into account and is given by

$$s_b(\tau) = -k_1' \bar{b}_0 \tau \, e^{-k_1 \tau}$$

$$s_c(\tau) = c_s' + \frac{k_1 k_2' - k_1' k_2}{(k_2 - k_1)^2} \bar{b}_0 (e^{-k_1 \tau} - e^{-k_2 \tau})$$

$$+ \frac{k_1}{k_2 - k_1} \bar{b}_0 \tau (k_1' \, e^{-k_1 \tau} - k_2' \, e^{-k_2 \tau})$$

From these expressions it follows that

$$\dot{s}_c(0) = -k_1' \bar{b}_0 \qquad s_c(\infty) = s_c(0) = c_s'$$

Hence, since $k_1' > 0$ and $c_s' < 0$ (see Sec. 3-5), the DO sensitivity coefficient s_c is always characterized by the following three properties

$$s_c(0) < 0 \qquad \dot{s}_c(0) < 0 \qquad s_c(\infty) < 0$$

Two possible sensitivity curves s_c are shown in Fig. 4-1-8, the first one (a) being negative everywhere and the second one (b) showing that along a segment of the river (segment AB) the conditions are improved by an increment of the temperature. A curve of type (b) would result if the reaeration coefficient should increase drastically with temperature. Nevertheless, even under this hypothetical condition, the dominant effect is a decrease of the dissolved oxygen concentration with increasing temperature, and hence for reasons of safety, high temperature conditions are often selected as the reference conditions in the design of wastewater treatment plants or other river pollution control facilities. Actually, in these design problems reference is usually made to low flow–high temperature conditions, since similar results can also be proved for flow rate. The only difference is that the sensitivity analysis should be done for the system describing self-purification in space, because flow velocity varies with flow rate (Rinaldi and Soncini-Sessa, 1978).

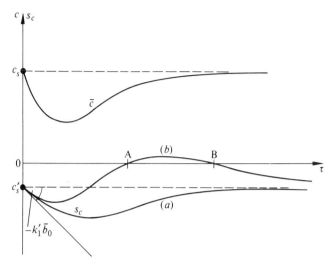

Figure 4-1-8 The sensitivity of dissolved oxygen concentration with respect to equilibrium temperature.

Heat Discharge

As a final example, the effects that heat pollution has on the self-purification processes, as described by the Streeter-Phelps model, is discussed in very simple terms.

The case studied is illustrated in Fig. 4-1-9a, where a river with flow rate Q_1 and temperature T_0^* receives a heat discharge with flow rate Q_2 and temperature $\left(T_0^* + (Q_1 + Q_2)/Q_2 \Delta T_0\right)$. Then, after mixing (at the point $l = 0$) a flow rate $Q = Q_1 + Q_2$ and a temperature $T_0 = T_0^* + \Delta T_0$ is obtained. The variation ΔT_0 induced in the river by the heat discharge is the parameter and its nominal value $\overline{\Delta T_0}$ is zero, thus meaning that the nominal conditions refer to the case in which there is no heat discharge. Moreover, the BOD concentration of the discharge is assumed to be the same as that of the river while both the river and the discharge are assumed to be saturated with oxygen (see Sec. 7-2 for justification), as shown in Fig. 4-1-9b, so that the initial conditions are

$$T_0 = T_0^* + \Delta T_0$$

$$b_0 = b_0^*$$

$$c_0 = \frac{Q_1}{Q_1 + Q_2} c_s(T_0^*) + \frac{Q_2}{Q_1 + Q_2} c_s\left(T_0^* + \frac{Q_1 + Q_2}{Q_2} \Delta T_0\right)$$

A thermal submodel must now be coupled with the Streeter-Phelps model considered so far. Thus, assuming that water temperature dynamics can be described by a differential equation (see Sec. 3-4), we obtain

$$\dot{T} = \phi(T) \tag{4-1-17a}$$

$$\dot{b} = -k_1(T)b \tag{4-1-17b}$$

$$\dot{c} = -k_1(T)b + k_2(T)(c_s(T) - c) \tag{4-1-17c}$$

with initial nominal conditions

$$\bar{T}_0 = T_0^* \qquad \bar{b}_0 = b_0^* \qquad \bar{c}_0 = c_s(T_0^*)$$

Thus, the sensitivity system is given by

$$\dot{s}_T = \phi' s_T \tag{4-1-18a}$$

$$\dot{s}_b = -k_1' \bar{b} s_T - k_1 s_b \tag{4-1-18b}$$

$$\dot{s}_c = (-k_1'\bar{b} + k_2'c_s + k_2 c_s' - k_2'\bar{c})s_T - k_1 s_b - k_2 s_c \tag{4-1-18c}$$

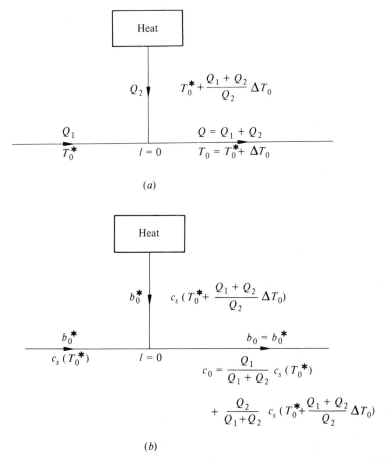

Figure 4-1-9 Balance equations at the discharge point: (a) flow rate and temperature (b) BOD and DO.

where $\phi' = [\partial \phi/\partial T]_{\bar{T}}$ and the initial conditions are

$$s_T(0) = 1 \qquad s_b(0) = 0 \qquad s_c(0) = c'_s$$

Equations (4-1-17) and (4-1-18) can easily be solved since they are of triangular structure. If $T = T_0^*$ is assumed to be a constant solution of Eq. (4-1-17a) this system of equations can be solved analytically and the solution gives the three sensitivity coefficients

$$s_T = e^{\phi' \tau}$$

$$s_b = \frac{k'_1}{\phi'} \bar{b}_0 e^{-k_1 \tau}(1 - e^{\phi' \tau})$$

$$s_c = A e^{-k_1 \tau} + B e^{-k_2 \tau} + C e^{\phi' \tau} + D e^{(\phi' - k_1)\tau} + E e^{(\phi' - k_2)\tau}$$

where the constants $A, B, \ldots, E$ are given by

$$A = -\frac{k_1 k'_1}{(k_2 - k_1)\phi'} \bar{b}_0$$

$$B = -\frac{k_2 c'_s}{k_2 + \phi'} + \frac{k_1 k'_2}{(k_2 - k_1 + \phi')\phi'} \bar{b}_0 + \frac{k'_1 k_2}{(k_2 - k_1)(k_2 - k_1 + \phi')} \bar{b}_0 + c'_s$$

$$C = \frac{k_2 c'_s}{k_2 + \phi'}$$

$$D = \left[\frac{k_1 k'_2}{k_2 - k_1} - \left(1 - \frac{k_1}{\phi'}\right) k'_1 \right] \frac{\bar{b}_0}{k_2 - k_1 + \phi'}$$

$$E = -\frac{k_1 k'_2}{(k_2 - k_1)\phi'} \bar{b}_0$$

The corresponding sensitivity curves are shown in Fig. 4-1-10 for realistic values of the parameters; the main conclusion is that the oxygen concentration is lowered everywhere and in particular around the minimum of the DO curve. However, the perturbation introduced by the heat discharge is absorbed along the river, this being the main distinction between this case of temperature perturbation and the preceding one.

4-2 OTHER CHEMICAL MODELS

Since the first work of Streeter and Phelps, the process of natural self-purification has been extensively studied and quite sophisticated mathematical models have been proposed for its description. Most of them are basically the classical Streeter-Phelps model with the source terms suitably modified in order to account for some neglected biochemical phenomena. In all these models the river is assumed to be a chemical reactor in which a reaction between BOD and DO takes place. Since this

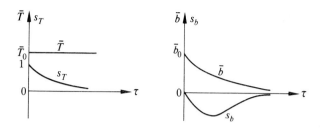
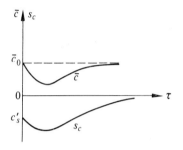

Figure 4-1-10 Sensitivity coefficients of temperature, BOD, and DO to heat discharge.

way of looking at the problem is the fundamental contribution of Streeter and Phelps, these models are often called *modified Streeter-Phelps models*.

As stated in the preceding section one of the hypotheses of Streeter and Phelps is that the deoxygenation rate and the BOD decay rate are equal. This is not always true. For example, the decay rate of the stream-borne BOD can be higher than the deoxygenation rate because of sedimentation, or lower because of resuspension. Consequently, Thomas (1948) proposed the following equation for the BOD decay rate

$$\frac{db}{d\tau} = -(k_1 + k_3)b$$

where the new coefficient $k_3 \gtrless 0$ takes into account factors such as sedimentation, flocculation, scour, and resuspension.

Moreover, BOD is entering the river not only from point sources (effluents), but also through distributed sources and local runoff. To account for these distributed sources a term $L(l,t)$ (*distributed BOD load*) has been added by Dobbins (1964) to the righthand side of the BOD equation, which then becomes (recall that A is the cross-sectional area)

$$\frac{\partial b}{\partial t} + v \frac{\partial b}{\partial l} = -(k_1 + k_3)b + \frac{L(l,t)}{A}$$

Again this equation can be written in flow time by defining the BOD load $L(\tau)$ along a characteristic line as (see Fig. 4-1-1 where $\tau = t - t_0$)

$$L(\tau) = L(l(\tau), t(\tau))$$

Thus, the BOD equation becomes

$$\frac{db}{d\tau} = -(k_1 + k_3)b + \frac{L(\tau)}{A} \qquad (4\text{-}2\text{-}1)$$

As far as the DO equation is concerned Camp (1963) and Dobbins (1964) introduced a constant term $(P - R)$ giving

$$\frac{dc}{d\tau} = -k_1 b + k_2(c_s - c) + (P - R) \qquad (4\text{-}2\text{-}2)$$

which represents the difference between P the addition of oxygen due to photosynthetic production (assumed constant in time) and R the DO removal from benthal oxygen demand. The new model can be considered as an ordinary Streeter-Phelps model with a changed oxygen saturation concentration.

O'Connor (1967) proposed the use of two independent terms $P(l,t)$ and $R(l,t)$ to represent oxygen production and oxygen removal, respectively, where $P(l,t)$ is assumed to be a periodic function of time. Moreover, O'Connor (1967) suggested a more relevant modification of the classical Streeter-Phelps model which could explain the existence of different phases appearing in the BOD decay (see Fig. 3-5-3). He assumed that the total BOD is the sum of two components, the carbonaceous BOD (b_c) and the nitrogeneous BOD (b_n), thus writing

$$b = b_c + b_n$$

In accordance with other authors (see, for instance, Gameson, 1959, and Courchaine, 1963) he conjectures that the first two phases of the BOD decay can be explained by assuming that the decomposition reactions for the two different types of BOD proceed with different rates (k_c and k_n respectively) and that a time lag, of increasing length for decreasing degree of treatment, separates these reactions. This time lag could be explained either by an inhibition of carbonaceous BOD over nitrogenous BOD, or by the low growth rate of nitrifiers (see Sec. 2-3). Thus, for the particular case in which the time lag between the two reactions is negligible, as it would be in a stream where all the BOD load comes from biologically treated effluents, O'Connor's model (in flow time) is the following

$$\frac{db_c}{d\tau} = -(k_c + k_3)b_c$$

$$\frac{db_n}{d\tau} = -k_n b_n$$

$$\frac{dc}{d\tau} = -k_c b_c - k_n b_n + k_2(c_s - c) + P(\tau) - R(\tau)$$

where $P(\tau)$ and $R(\tau)$ represent oxygen production and removal along a characteristic line. Obviously the nitrogenous BOD is assumed to be present in dissolved form, since no sedimentation effect is considered in the second equation.

In order to improve the fit to observed data, some authors proposed models with nonlinear source terms. First, on the basis of a suggestion given by Thomas (1953), Young and Clark (1965) proposed a second order reaction equation to model the BOD decay (see Eq. (3-5-12))

$$\frac{db}{d\tau} = -\theta_1 b^2$$

Later on Braun and Berthouex (1970) proposed a Michaelis-Menten expression (see Eq. (3-5-17)) for the BOD decay rate:

$$\frac{db}{d\tau} = -\frac{\theta_1 b}{\theta_2 + b}[B_0 + \theta_3(b_0 - b)]$$

where the bacterial biomass $B(\tau) = [B_0 + \theta_3(b_0 - b(\tau))]$ is assumed to be a linear function of the excerpted BOD. This model has been proved to be of particular interest when the BOD data show an initial lag phase (see Fig. 4-2-1).

Following the chemical engineering tradition, Shastry et al. (1975) suggested modeling the deoxygenation process as a second order reaction where the reaction rate is proportional to the concentration of the two reactants (see Eq. (3-5-11))

$$\frac{db}{d\tau} = -\theta_1 bc \tag{4-2-3a}$$

$$\frac{dc}{d\tau} = -\theta_1 bc + \theta_2(c_s - c) \tag{4-2-3b}$$

Rinaldi and Soncini-Sessa (1974) have shown why this equation can be of particular

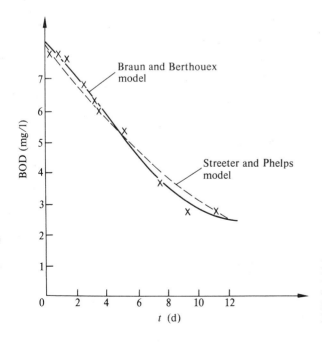

Figure 4-2-1 Comparison of BOD decay profiles of the Braun and Berthouex model and the Streeter-Phelps model on laboratory data.

interest. System (4-2-3) is a nonlinear system of the form

$$\dot{\mathbf{x}} = \mathbf{f}(\mathbf{x}, \mathbf{u})$$

and is still characterized by the unique equilibrium point

$$\bar{\mathbf{x}} = \begin{bmatrix} 0 & c_s \end{bmatrix}^T$$

since $\dot{\mathbf{x}} = 0$ implies $b = 0$ and $c = c_s$. Moreover, the linearized system associated with this equilibrium point is given by (see Sec. 1-2)

$$\delta\dot{\mathbf{x}} = \left[\frac{\partial \mathbf{f}}{\partial \mathbf{x}}\right]_{\mathbf{x}=\bar{\mathbf{x}}} \delta\mathbf{x} = \begin{bmatrix} -\theta_1 c_s & 0 \\ -\theta_1 c_s & \theta_2 \end{bmatrix} \delta\mathbf{x}$$

Letting $\theta_1 = k_1/c_s$, $\theta_2 = k_2$ the matrix $[\partial \mathbf{f}/\partial \mathbf{x}]_{\mathbf{x}=\bar{\mathbf{x}}}$ is equal to the **F** matrix of the Streeter-Phelps model. Then, it can be concluded that the two models have the same behavior in the vicinity of the equilibrium point. Nevertheless, model (4-2-3) never generates negative DO concentrations as the Streeter-Phelps model does (see Fig. 4-2-2).

In any one of these modified Streeter-Phelps models, the steady state DO concentration profile is a sag curve. In particular, for a Dobbins' model (see Eqs. (4-2-1)(4-2-2)) with $L(\tau) = \text{const.}$, Liebman and Loucks (1966) have proved that the critical deficit d_c is again a nonlinear function of the initial BOD and DO concentrations, while the limit position ($b_0 \to \infty$ or $c_0 \to c_S$) for the critical point l_c is

$$L_c = \frac{v \ln f}{(k_1 + k_3)(f - 1)}$$

with $(f = k_2/(k_1 + k_3))$ (see Sec. 4-1). However, the limit position L_c is no longer an upper bound of l_c, rather the critical point is located upstream from the point

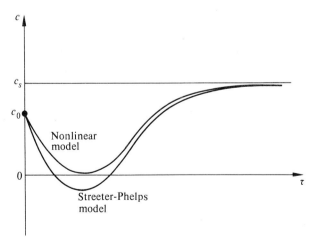

Figure 4-2-2 Comparison of DO sag curves of the Streeter-Phelps model and the nonlinear model (4-2-1).

L_c if the distributed BOD load L satisfies the following relationship

$$\frac{L}{A} < \frac{k_1 + k_3}{k_1}(P - R + k_2 d_0)$$

and downstream from the point L_c if the opposite inequality holds.

Numerous particular problems, such as photosynthetic oxygen production, effects of time-varying distributed BOD load, effects of constant BOD load discharged in an oscillating flow, DO deficit due to unsteady point sources of BOD have been extensively discussed in the literature (see, for instance, Di Toro and O'Connor, 1968; O'Connor and Di Toro, 1970; Li, 1972; Li and Kozlowski, 1974; and Rinaldi and Soncini-Sessa, 1974). The first two problems mentioned are discussed in this section in greater detail, not only because they are of a certain interest per se, but also to show how different mathematical techniques can usefully be applied to gain insights into model behavior. In both cases temperature, flow rate, and velocity are assumed to be constant in time and space in order to simplify the discussion.

Photosynthetic Oxygen Production

Downstream from the effluent of a biological treatment plant it is often observed that the growth of benthic algae is stimulated by the nutrients contained in the treated wastewater. The algae may act as both a source and a sink of oxygen, owing to photosynthetic oxygen production and respiration. For the sake of simplicity, the spatial distribution of the algal population is assumed to begin abruptly at $l = 0$ and to remain constant for $l > 0$. Then, the photosynthetic oxygen production P and the oxygen removal R due to algal respiration are constant in space for $l > 0$. Moreover, it can reasonably be assumed that the respiration R is time-invariant, while the photosynthetic oxygen production P varies in the same way as the incident solar radiation (Westlake, 1968). Thus, the oxygen source term $P(t) - R$ can be assumed to be periodic and to have the shape shown in Fig. 4-2-3 (O'Connor and Di Toro, 1970). In other words,

$$P(t) - R = \bar{P} - R + \delta P(t)$$

where $\bar{P}$ is the mean value of $P(t)$ and $\delta P(t)$ is a periodic function with zero mean and period $T = 2\pi/\omega_0 = 1$ day. If sedimentation effects are negligible, the modified Streeter-Phelps model describing the phenomenon is

$$\frac{\partial b}{\partial t} + v\frac{\partial b}{\partial l} = -k_1 b \tag{4-2-4a}$$

$$\frac{\partial c}{\partial t} + v\frac{\partial c}{\partial l} = -k_1 b + k_2(c_s - c) + \bar{P} - R + \delta P(t) \tag{4-2-4b}$$

In order to solve these equations easily assume time-invariant boundary conditions $(b(0, t) = b_0$ and $c(0, t) = c_0)$. Obviously the solution of Eq. (4-2-4a) is the same as

112 SOME PARTICULAR SELF-PURIFICATION MODELS

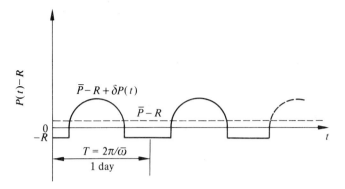

Figure 4-2-3 Idealized daily fluctuations of photosynthetic oxygen production.

that of Eq. (4-1-1a)

$$b(l, t) = b_0 \, e^{-k_1 l/v} \tag{4-2-5}$$

As far as the dissolved oxygen is concerned, let us assume that it can be written in the form

$$c(l, t) = \bar{c}(l) + \delta c(t) \tag{4-2-6}$$

with $\delta c(t)$ being a periodic function with zero mean. The validity of this assumption is now verified by actually determining the functions $\bar{c}(l)$ and $\delta c(t)$ in such a way that Eq. (4-2-6) is a solution of Eq. (4-2-4b). By substituting Eq. (4-2-6) into Eq. (4-2-4b)

$$\delta \dot{c}(t) + v \frac{d\bar{c}(l)}{dl} = -k_1 b + k_2 (c_s - \bar{c}(l)) - k_2 \delta c(t) + \bar{P} - R + \delta P(t) \tag{4-2-7}$$

is obtained. Thus, averaging this equation over the period of one day and remembering that $\delta P(t)$ and $\delta c(t)$ have zero mean yields

$$v \frac{d\bar{c}(l)}{dl} = -k_1 b + k_2 (c_s - \bar{c}(l)) + \bar{P} - R \tag{4-2-8}$$

which is of type (4-2-2). This equation can be explicitly integrated, using Eq. (4-2-5) and the boundary condition $\bar{c}(0) = c_0$ (see page 116). The asymptotic value of $\bar{c}(l)$ (note that system (4-2-8) is an asymptotically stable linear system since its eigenvalue is $-k_2$), is

$$c_S + \frac{\bar{P} - R}{k_2}$$

i.e., in points relatively far downstream from the discharge point oversaturated oxygen values are on the average obtained. Substituting Eq. (4-2-8) into Eq. (4-2-7)

$$\delta \dot{c}(t) = -k_2 \delta c + \delta P(t) \tag{4-2-9}$$

is obtained, which has a unique periodic solution for each periodic function

4-2 OTHER CHEMICAL MODELS

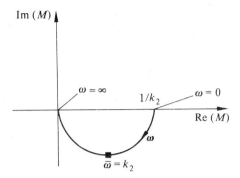

Figure 4-2-4 The DO frequency response to a uniformly distributed oxygen source.

$\delta P(t)$. This periodic solution can easily be obtained by developing the input $\delta P(t)$ in a *Fourier series* (a sum of sine waves) and by determining the corresponding series of δc via frequency response. The transfer function $M(s)$ of system (4-2-9) is given by (see Eq. (1-2-13))

$$M(s) = \frac{1}{s + k_2}$$

and the corresponding frequency response is shown in Fig. 4-2-4 (see Eq. (1-2-18)). Note that $|M(i\omega)|$ is a decreasing function of ω, i.e., the river is acting as a *low pass filter* since high frequency components of $\delta P(t)$ ($\omega > k_2$) are attenuated. Recalling Eq. (4-2-6) it can be concluded that, in the presence of algal population, the classical shape of the dissolved oxygen concentration may still be observed at any instant of time and that the whole sag curve has a diurnal fluctuation around the daily mean concentration $\bar{c}(l)$.

Distributed BOD Load

In this second particular case the effects induced by a time-varying distributed BOD load are analyzed. Again, for the sake of simplicity, it is assumed that downstream of the initial point $l = 0$ the distributed BOD load is constant in space ($L(l,t) = L(t)$) and that at the initial time $t = 0$ the river is perfectly clean and oxygenated ($b(l,0) = d(l,0) = 0$), while at the initial point $b(0,t) = 0$ for $t > 0$.

The following equations may be used to describe this situation:

$$\frac{\partial b}{\partial t} + v\frac{\partial b}{\partial l} = -k_1 b + \frac{L(t)}{A} \tag{4-2-10a}$$

$$\frac{\partial d}{\partial t} + v\frac{\partial d}{\partial l} = k_1 b - k_2 d \tag{4-2-10b}$$

Even if $L(t)$ is a periodic function, Eq. (4-2-10) can no longer be integrated as in the preceding case, since the form of $c(l,t)$ is now more complex. The notion of *transfer function* can now be of greater help. Applying the Laplace transformation

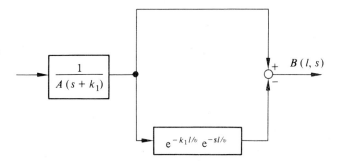

Figure 4-2-5 Block diagram of the transfer function between a uniformly distributed BOD load $L(t)$ and the BOD concentration $b(l, t)$ at a given point l.

to Eq. (4-2-10a) gives (recall that $b(l, 0) = 0$)

$$sB(l, s) + v\frac{dB(l, s)}{dl} = -k_1 B(l, s) + \frac{\hat{L}(s)}{A} \tag{4-2-11}$$

where $B(l, s)$ and $\hat{L}(s)$ are the Laplace transforms of $b(l, t)$ and $L(t)$, respectively. Equation (4-2-11) is an ordinary linear differential equation which gives

$$M_l(s) = \frac{B(l, s)}{\hat{L}(s)} = \frac{1}{A(s + k_1)} (1 - e^{-k_1 l/v} e^{-sl/v}) \tag{4-2-12}$$

where $M_l(s)$ is the transfer function between the distributed BOD load and the BOD concentration at a given point l. The block diagram corresponding to this transfer function is shown in Fig. 4-2-5, while the corresponding frequency response $M_l(i\omega)$ is shown in Fig. 4-2-6, which shows that the river is again acting as a low pass filter.

It is interesting to note that, if the variations of the load are purely sinusoidal, i.e., $L(t) = \bar{L} + \Delta L \sin(\omega_0 t)$, the amplitude and phase of the induced BOD fluctuations are functions of l (see the second block in Fig. 4-2-5). Therefore, this phenomenon is more complex than the photosynthetic oxygen production, where the DO fluctuations were independent of l. Moreover, in this case there are points at which the amplitude of the fluctuation reaches a maximum and points

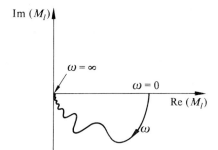

Figure 4-2-6 The BOD frequency response to a uniformly distributed load.

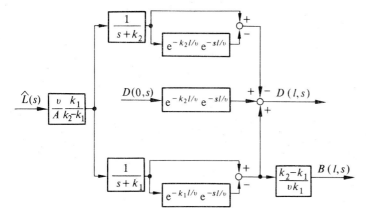

Figure 4-2-7 Block diagram of the transfer function between a uniformly distributed BOD load $L(t)$ and the DO deficit at a given point.

at which it has a minimum. The coordinates of these points can easily be computed by letting the derivative of $|M_l(i\omega)|$ with respect to l be zero

$$k_1 \cos\left(\omega_0 \frac{l}{v}\right) + \omega_0 \sin\left(\omega_0 \frac{l}{v}\right) - k_1 e^{-k_1 l/v} = 0$$

The function $k_1 e^{-k_1 l/v}$ tends to zero for increasing values of l, so that, for parts of the river sufficiently downstream, the points characterized by maximum and minimum fluctuations are equally spaced.

Applying the same procedure to Eq. (4-2-10b) as to Eq. (4-2-10a) the following is obtained

$$D(l,s) = e^{-k_2 l/v} e^{-sl/v} D(0,s) + \frac{vk_1}{A(k_2-k_1)} \left[\frac{1 - e^{-k_1 l/v} e^{-sl/v}}{s + k_1} - \frac{1 - e^{-k_2 l/v} e^{-sl/v}}{s + k_2}\right] \hat{L}(s)$$

where $D(l,s)$ and $D(0,s)$ are the Laplace transforms of the deficit $d(l,t)$ and of the boundary deficit $d(0,t)$. The block diagram corresponding to the preceding expression is shown in Fig. 4-2-7. Again, it is possible to show that particular points exist at which DO fluctuations reach maximum and minimum values.

Sensitivity Analysis

Finally, it will now be shown how important information on the model behavior can be obtained by employing the simple sensitivity analysis technique presented in the preceding section. Consider, for instance, computing the variations of DO concentration due to an increase in the benthic algal population considered on page 111. This information can easily be obtained by determining the sensitivity of Eq. (4-2-8) with respect to the parameter $(\bar{P} - R)$, since the main consequence of the population increase is a corresponding increase in this parameter.

The sensitivity system associated with Eq. (4-2-8) is

$$v\frac{ds_c(l)}{dl} = -k_2 s_c(l) + 1 \qquad (4\text{-}2\text{-}13)$$

If the initial DO concentration c_0 is not affected by the algal bloom $(s_c(0) = 0)$ the solution of Eq. (4-2-13) is

$$s_c(l) = \frac{1}{k_2}(1 - e^{-k_2 l/v})$$

which is represented in Fig. 4-2-8. The conclusion follows that the algal bloom induces a general improvement of the daily mean DO concentration, and that the greatest effects appear far downstream of the BOD source.

As a second example, consider now the case in which the effluent at $l = 0$ contains some non-biodegradable surfactants. These substances do affect the re-aeration process so that when they are present the reaeration coefficient k_2 has values which are lower than the normal ones. Thus, the effect of the removal of the surfactants from the wastewater discharge can be analyzed by determining the DO sensitivity with respect to the reaeration coefficient k_2. Using Eq. (4-2-8) once more, the sensitivity system turns out to be

$$v\frac{ds_c(l)}{dl} = -k_2 s_c(l) + (c_s - \bar{c}(l)) \qquad (4\text{-}2\text{-}14)$$

where

$$\bar{c}(l) = c_s + \frac{\bar{P} - R}{k_2} - \left(c_s + \frac{\bar{P} - R}{k_2} - \bar{c}_0\right)e^{-k_2 l/v}$$
$$+ \frac{k_1 b_0}{k_1 - k_2}(e^{-k_1 l/v} - e^{-k_2 l/v})$$

is the nominal average DO concentration. Integrating Eq. (4-2-14) with the initial condition $s_c(0) = 0$ (the initial DO level cannot depend upon the surfactant contents of the wastewater discharge), the following equation is obtained

$$s_c(l) = \left[\left(c_s + \frac{\bar{P} - R}{k_2} - \bar{c}_0\right) + \frac{k_1 b_0}{k_1 - k_2}\right]\frac{l}{v}e^{-k_2 l/v}$$
$$+ \frac{k_1 b_0}{(k_1 - k_2)^2}(e^{-k_1 l/v} - e^{-k_2 l/v}) - \frac{\bar{P} - R}{k_2^2}(1 - e^{-k_2 l/v})$$

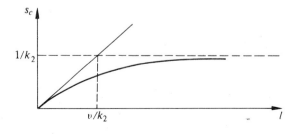

Figure 4-2-8 Sensitivity coefficient of dissolved oxygen with respect to average photosynthetic oxygen production.

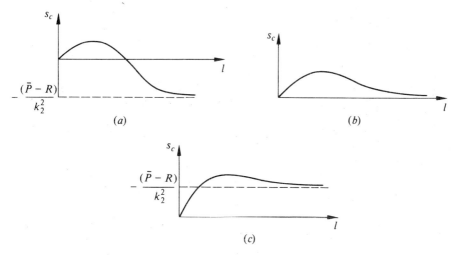

Figure 4-2-9 Sensitivity coefficient of dissolved oxygen with respect to the reaeration coefficient k_2: (a) when photosynthetic oxygen production is dominant ($\bar{P} > R$) (b) in absence of sources and sinks of oxygen ($\bar{P} = R$) (c) when benthal oxygen demand is dominant ($\bar{P} < R$).

Figure 4-2-9 shows three typical profiles of $s_c(l)$ for different values of the oxygen source $(P - R)$: positive value (predominance of photosynthetic oxygen production), zero value (absence of sources and sinks), negative value (predominance of benthal oxygen demand). In all cases, the removal of the surfactants generates an improvement in the DO concentration immediately downstream of the discharge. Nevertheless, if photosynthesis predominates, the DO concentration can decrease downstream. This singular phenomenon can be better understood by remembering that, for high values of l, algae create oversaturated DO concentrations. Then, if the oxygen interchange between air and water is improved, lower oversaturation values will be obtained.

4-3 APPROXIMATED STREETER-PHELPS DISPERSION MODELS

The aim of this section is to present BOD–DO models which to some extent take dispersion into account. More precisely, dispersion models are analyzed both in the steady state and in the unsteady state case and, finally, simple lumped parameter models are derived which can be interpreted as approximations of the general dispersion model.

The *Streeter-Phelps model with dispersion* can be written in the form (see Secs. 3-1 and 4-1)

$$\frac{\partial b}{\partial t} + v\frac{\partial b}{\partial l} - D\frac{\partial^2 b}{\partial l^2} = -k_1 b \qquad (4\text{-}3\text{-}1a)$$

$$\frac{\partial d}{\partial t} + v\frac{\partial d}{\partial l} - D\frac{\partial^2 d}{\partial l^2} = k_1 b - k_2 d \tag{4-3-1b}$$

if the cross-sectional area A and the dispersion coefficient D do not depend on l, and if there is no BOD load along the river (recall that d is the DO deficit). The analysis which follows makes reference to this model, but other chemical models like the ones described in Sec. 4-2 could be discussed along the same lines.

Equation (4-3-1) can be simplified by defining the auxiliary variable

$$a = d + \frac{k_1}{k_1 - k_2} b \tag{4-3-2}$$

since from Eq. (4-3-1)

$$\frac{\partial a}{\partial t} + v\frac{\partial a}{\partial l} - D\frac{\partial^2 a}{\partial l^2} = -k_2 a$$

is obtained. Therefore, once Eq. (4-3-1a) has been analytically solved, the solution $a = a(l, t)$ can immediately be obtained by replacing k_1 with k_2, and hence the solution of the DO deficit equation can be deduced from Eq. (4-3-2).

Steady State Analysis

For steady state conditions the following BOD dispersion model is obtained by splitting the second order differential equation (4-3-1a) with $\partial b/\partial t = 0$ into the two first order equations

$$\frac{\partial b}{\partial l} = \beta \tag{4-3-3a}$$

$$\frac{\partial \beta}{\partial l} = \frac{k_1}{D} b + \frac{v}{D} \beta \tag{4-3-3b}$$

while the corresponding *plug flow* model ($D = 0$) is simply

$$\frac{\partial b}{\partial l} = -\frac{k_1}{v} b \tag{4-3-4}$$

Model (4-3-3) is a simple linear model of the form

$$\frac{d\mathbf{x}(l)}{dl} = \mathbf{F}\mathbf{x}(l)$$

with

$$\mathbf{x}(l) = \begin{bmatrix} b(l) \\ \beta(l) \end{bmatrix} \qquad \mathbf{F} = \begin{bmatrix} 0 & 1 \\ \dfrac{k_1}{D} & \dfrac{v}{D} \end{bmatrix}$$

The eigenvalues of the matrix **F** are given by

$$\lambda_1 = \frac{v}{2D}\left(1 + \sqrt{1 + 4\frac{k_1 D}{v^2}}\right) \qquad \lambda_2 = \frac{v}{2D}\left(1 - \sqrt{1 + 4\frac{k_1 D}{v^2}}\right)$$

and the corresponding eigenvectors are

$$\mathbf{x}^{(1)} = \begin{bmatrix} 1 \\ \lambda_1 \end{bmatrix} \qquad \mathbf{x}^{(2)} = \begin{bmatrix} 1 \\ \lambda_2 \end{bmatrix}$$

Recalling that any *eigenvector* $\mathbf{x}^{(i)}$ is, by definition, a vector such that $\mathbf{F}\mathbf{x}^{(i)} = \lambda_i \mathbf{x}^{(i)}$, then if the state $\mathbf{x}(l)$ of the system in a point l is given by $\mathbf{x}^{(i)}$, $d\mathbf{x}(l)/dl = \lambda_i \mathbf{x}(l)$, i.e., the tangent to the trajectory describing the evolution of the state vector in the state space is proportional to the vector itself. This implies that in the state space there are two particular straight lines through the origin which correspond to trajectories of the system. The trajectory corresponding to $\mathbf{x}^{(1)}$ is directed away from the origin since $\lambda_1 > 0$, while the trajectory corresponding to $\mathbf{x}^{(2)}$ is directed toward the origin since $\lambda_2 < 0$. Thus, the evolution of BOD (b) and its gradient (β) along the river is that of a *saddle point* in state space as shown in Fig. 4-3-1.

Since there is no BOD load downstream of the initial point ($l = 0$)

$$\lim_{l \to \infty} b(l) = 0$$

must hold, and this can be obtained if and only if the initial state is proportional to the second eigenvector, i.e., if and only if

$$\beta(0) = \lambda_2 b(0)$$

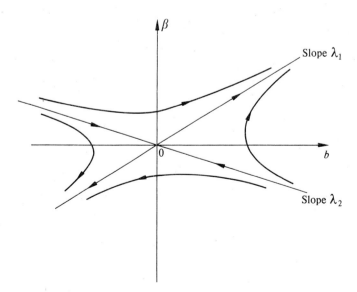

Figure 4-3-1 Trajectories in the state space (b, β).

which implies

$$\beta(l) = \lambda_2 b(l) \tag{4-3-5}$$

for all $l > 0$. Thus, from Eqs. (4-3-3a) and (4-3-5)

$$\frac{\partial b}{\partial l} = \lambda_2 b \tag{4-3-6}$$

is obtained, and by comparing Eq. (4-3-6) with Eq. (4-3-4) the following conclusions can be drawn: the plug flow model (4-3-4) can be used as an equivalent dispersion model under the condition that the stream velocity v is substituted by an *equivalent flow velocity* v_1 such that $\lambda_2 = -k_1/v_1$. This equivalent velocity v_1 turns out to be given by

$$v_1 = v \frac{2k_1 D/v^2}{\sqrt{1 + \frac{4k_1 D}{v^2}} - 1} \cong v\left(1 + 2\frac{k_1 D}{v^2}\right)$$

and justifies *Dobbins' criterion* (Dobbins, 1964) which says that if $2(k_1 D/v^2)$ is smaller than $\frac{1}{100}$, the effect of dispersion is negligible. Similarly, for the auxiliary variable a given by Eq. (4-3-2) an equivalent velocity can be defined

$$v_2 = v \frac{2k_2 D/v^2}{\sqrt{1 + \frac{4k_2 D}{v^2}} - 1} \cong v\left(1 + 2\frac{k_2 D}{v^2}\right)$$

In conclusion, as far as steady state conditions are concerned the dispersion model (4-3-1) has an equivalent plug flow model given by

$$\begin{aligned} v_1 \frac{\partial b}{\partial l} &= -k_1 b \\ v_2 \frac{\partial d}{\partial l} &= \frac{k_1}{k_1 - k_2}\left(\frac{v_2}{v_1}k_1 - k_2\right)b - k_2 d \end{aligned} \tag{4-3-7}$$

where the velocities v_1 and v_2 depend upon the dispersion coefficient D. The equivalent plug flow model given by Eq. (4-3-7) can be used to analyze steady state situations, since it is much simpler than the dispersion model (4-3-1) with $\partial b/\partial t = \partial d/\partial t = 0$.

Unsteady State Analysis

To obtain approximate dispersion models the dispersion model (4-3-1) is analyzed via transfer function techniques. Let

$$B(l, s) = \mathscr{L}[b(l, t)]$$

be the Laplace transform of $b(l, t)$ with respect to time. Then the Laplace transform of Eq. (4-3-1a) is given by

4-3 APPROXIMATED STREETER-PHELPS DISPERSION MODELS

$$sB - b(l,0) + v\frac{dB}{dl} - D\frac{d^2B}{dl^2} = -k_1 B \qquad (4\text{-}3\text{-}8)$$

with boundary conditions

$$B(0, s) = B_0(s) \;(= \mathcal{L}[b_0(t)])$$

$$\lim_{l \to \infty} B(l, s) = 0$$

In order to compute the transfer function of the system it is necessary to assume that the initial conditions are zero (see Sec. 1-2), i.e.,

$$b(l, 0) = 0 \qquad \text{for all } l$$

Then, the second order differential equation (4-3-8) can be split into the following two first order differential equations

$$\frac{dB}{dl} = \beta$$

$$\frac{d\beta}{dl} = \frac{k_1 + s}{D} B + \frac{v}{D} \beta \qquad (4\text{-}3\text{-}9)$$

which are formally similar to Eq. (4-3-3). From the preceding results and Eq. (4-3-9) the following is obtained

$$\frac{dB}{dl} = \lambda_2(s) B$$

where

$$\lambda_2(s) = \frac{v}{2D}\left(1 - \sqrt{1 + 4\frac{(k_1 + s)D}{v^2}}\right)$$

Therefore, the Laplace transform of $b(l, t)$ at point l can be written as

$$B(l, s) = B_0(s) e^{\lambda_2(s) l}$$

and the transfer function $M_l(s)$ specifying the input–output relationship between $b_0(t)$ and $b(l, t)$ is given by

$$M_l(s) = e^{\lambda_2(s)l} = e^{\frac{vl}{2D}\left(1 - \sqrt{1 + 4\frac{(k_1 + s)D}{v^2}}\right)}$$

Its inverse, which is the impulse response (see Sec. 1-2), is

$$m_l(t) = \frac{1}{2} \frac{lt^{-3/2}}{\sqrt{D\pi}} e^{-\frac{1}{4Dt}(l - vt)^2 - k_1 t} \qquad (4\text{-}3\text{-}10)$$

It should be noted, however, that in reality the input function $b_0(t)$ can never be an impulse function. Even if a BOD impulse is discharged at $l = 0$, dispersion will cause $b_0(t)$ to be a function which dies away relatively slowly. But for sufficiently large l expression (4-3-10) is a good approximation to the pollution distribution induced by a BOD impulse discharge.

If a stretch of the river composed of a sequence of n equal reaches of length L/n is considered its transfer function can be written as

$$M_L(s) = [m(s)]^n = [M_{L/n}(s)]^n$$

where

$$m(s) = e^{\frac{vL/n}{2D}\left(1 - \sqrt{1 + 4\frac{(k_1 + s)D}{v^2}}\right)}$$

is the transfer function of each reach. Figure 4-3-2 shows the stretch of the river, the n reaches and the corresponding block diagram representation.

The structure of this model is exactly the same as that of the Nash model described in Sec. 3-3. This similarity suggests to approximate the transfer function $m(s)$ of each reach by means of a simple transfer function of the form $\mu/(1 + Ts)$ in order to obtain n serially connected first order lumped parameter models for the whole river stretch. The static gain μ and the time constant T can be selected in many different ways, but if the transfer function $m(s)$ is written in the form

$$m(s) = \frac{m(0)}{1 + \dfrac{L/n}{\sqrt{v^2 + 4k_1 D}} s + \dfrac{1}{2}\dfrac{L/n}{\sqrt{v^2 + 4k_1 D}}\left(\dfrac{L/n}{\sqrt{v^2 + 4k_1 D}} - \dfrac{2D}{v^2 + 4k_1 D}\right)s^2 + \cdots}$$

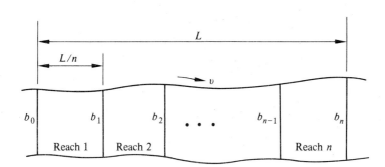

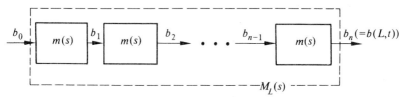

Figure 4-3-2 Subdivision of a stretch of river into n reaches and corresponding block diagram representation.

the most natural approximation, called *low frequency approximation* because it approximates the frequency response of the system (see Sec. 1-2) at low values of the frequency $s = i\omega$, is given by

$$m(s) \cong \frac{m(0)}{1 + \dfrac{L/n}{v\sqrt{1 + \dfrac{4k_1 D}{v^2}}} s}$$

Assuming that $2k_1 D/v^2 \ll 1$, this expression can be further modified to give

$$m(s) \cong \frac{m(0)}{1 + \dfrac{L/n}{v\left(1 + 2\dfrac{k_1 D}{v^2}\right)} s}$$

Similarly, the gain $m(0)$ can be approximated by

$$m(0) \cong \frac{v}{v + k_1 L/n}$$

The *approximate dispersion model* can then be written as

$$B_i(s) = \frac{\mu}{1 + Ts} B_{i-1}(s)$$

$$A_i(s) = \frac{\mu'}{1 + T's} A_{i-1}(s) \qquad (4\text{-}3\text{-}11)$$

$$D_i(s) = A_i(s) - \frac{k_1}{k_1 - k_2} B_i(s)$$

where

$$\mu = \frac{v}{v + k_1 L/n} \qquad T = \frac{L/n}{v\left(1 + 2\dfrac{k_1 D}{v^2}\right)}$$

$$\mu' = \frac{v}{v + k_2 L/n} \qquad T' = \frac{L/n}{v\left(1 + 2\dfrac{k_2 D}{v^2}\right)}$$

and $B_i(s)$, $A_i(s)$ and $D_i(s)$ are the Laplace transforms of the BOD, of the auxiliary variable $a(t)$, and of the DO deficit, respectively, at the downstream end of the i-th reach. Equation (4-3-11) completely specifies the BOD–DO deficit dynamics of the dispersion model in terms of transfer functions.

It is interesting to note that this approximate dispersion model has the same structure as the so-called *CSTR* (*Continuously Stirred Tank Reactor*) *model* heuristically proposed by Young and Beck (1974) who joined together two of the most well known simplifying assumptions from hydrology and from chemical engineering. The first consists of imagining the river to be constituted by a sequence of pools (the reaches), and the second corresponds to assuming that each pool is a perfectly mixed reactor. Thus, the mass balance for a single reach (*i*) of length L/n gives

$$\dot{b}_i(t) = -\left(k_1 + \frac{v}{L/n}\right)b_i(t) + \frac{v}{L/n}b_{i-1}(t)$$

$$\dot{a}_i(t) = -\left(k_2 + \frac{v}{L/n}\right)a_i(t) + \frac{v}{L/n}a_{i-1}(t)$$

$$\dot{d}_i(t) = k_1 b_i(t) - \left(k_2 + \frac{v}{L/n}\right)d_i(t) + \frac{v}{L/n}d_{i-1}(t)$$

The transfer function representation of the *CSTR* model is

$$B_i(s) = \frac{\mu_c}{1 + T_c s} B_{i-1}(s)$$

$$A_i(s) = \frac{\mu'_c}{1 + T'_c s} A_{i-1}(s) \qquad (4\text{-}3\text{-}12)$$

$$D_i(s) = A_i(s) - \frac{k_1}{k_1 - k_2} B_i(s)$$

where

$$\mu_c = \frac{v}{v + k_1 L/n} (= \mu) \qquad T_c = \frac{L/n}{v\left(1 + k_1 \frac{L/n}{v}\right)}$$

$$\mu'_c = \frac{v}{v + k_2 L/n} (= \mu') \qquad T'_c = \frac{L/n}{v\left(1 + k_2 \frac{L/n}{v}\right)}$$

If the number *n* of reaches could be chosen such that

$$L/n = \frac{2D}{v}$$

then the following would be obtained

$$T_c = T \qquad T'_c = T'$$

i.e., the transfer function of the CSTR model would be the same as that of the approximate dispersion model. However, if $L/n > 2D/v$, as is usually the case, $T_c < T$ and $T'_c < T'$. Thus, it is necessary to add a *time delay* between the adjacent reaches to obtain the same wave propagation velocities (see Sec. 3-3) in both models. The transfer function representation of the *CSTR model with time delay* Δ is simply

4-3 APPROXIMATED STREETER-PHELPS DISPERSION MODELS

$$B_i(s) = \frac{\mu_c}{1 + T_c s} e^{-\Delta s} B_{i-1}(s)$$

$$A_i(s) = \frac{\mu'_c}{1 + T'_c s} e^{-\Delta s} A_{i-1}(s) \qquad (4\text{-}3\text{-}13)$$

$$D_i(s) = A_i(s) - \frac{k_1}{k_1 - k_2} B_i(s)$$

and in the time domain the model is described by the following differential equations (see Young and Beck, 1974)

$$\dot{b}_i(t) = -\left(k_1 + \frac{v}{L/n}\right) b_i(t) + \frac{v}{L/n} b_{i-1}(t - \Delta)$$

$$\dot{d}_i(t) = k_1 b_i(t) - \left(k_2 + \frac{v}{L/n}\right) d_i(t) + \frac{v}{L/n} d_{i-1}(t - \Delta) \qquad (4\text{-}3\text{-}14)$$

Choice of the pure time delay Δ as

$$\Delta = T - T_c$$

results in the same wave propagation velocities for the BOD peak in the CSTR model with time delay and in the approximate dispersion model. The CSTR model with time delay can be interpreted as a model describing a river constituted by a sequence of channels and pools, where the biochemical reactions take place only in the pools. Figure 4-3-3 shows the schematic diagram of a single reach of such a river.

If the channels are described by more complex transfer functions taking into account the BOD–DO reactions and the dispersion in the channel, the following

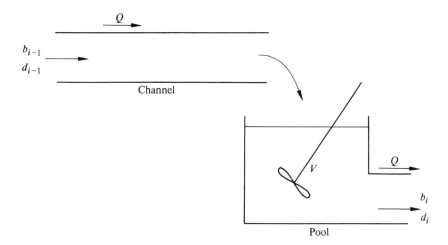

Figure 4-3-3 Schematic diagram of the CSTR model with time delay.

model is obtained

$$B_i(s) = \frac{\mu_c}{1 + T_c s} \Phi(s) B_{i-1}(s)$$

$$A_i(s) = \frac{\mu'_c}{1 + T_c s} \Psi(s) A_{i-1}(s) \qquad (4\text{-}3\text{-}15)$$

$$D_i(s) = A_i(s) - \frac{k_1}{k_1 - k_2} B_i(s)$$

where $\Phi(s)$ is the transfer function of the channel for BOD and $\Psi(s)$ is the one for the auxiliary variable. In the time domain the model (4-3-15) can be written as

$$\dot{b}_i(t) = -\left(k_1 + \frac{v}{L/n}\right) b_i(t) + \frac{v}{L/n} \int_0^t \phi(\tau) b_{i-1}(t - \tau) \, d\tau$$

$$\dot{d}_i(t) = k_1 b_i(t) - \left(k_2 + \frac{v}{L/n}\right) d_i(t) \qquad (4\text{-}3\text{-}16)$$

$$+ \frac{v}{L/n} \left[\int_0^t \psi(\tau) d_{i-1}(t - \tau) \, d\tau + \int_0^t \phi'(\tau) b_{i-1}(t - \tau) \, d\tau \right]$$

where

$$\phi'(\tau) = \frac{k_1}{k_1 - k_2} (\psi(\tau) - \phi(\tau))$$

and ϕ and ψ are the antitransforms of Φ and Ψ, i.e., the impulse responses of the channel. This is the continuous-time version of the *distributed-lag model* heuristically proposed by Tamura (1974). In this model the river is interpreted as a sequence of channels and pools, and the impulse responses of the channels representing biochemical reactions and distributed-time-delays due to the dispersion in the channel, should appear as shown in Fig. 4-3-4.

The BOD impulse responses of the dispersion model, the approximate dispersion model, the CSTR model, and the CSTR model with time delay are plotted for comparison in Fig. 4-3-5. For the dispersion model (4-3-10) the arrival time of the peak is not exactly an integer multiple of T as shown in Fig. (4-3-5a), but for $l \gg 3D/v$ this is true approximately. For the distributed-lag model it is possible

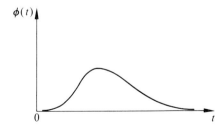

Figure 4-3-4 Impulse response for a channel of the distributed lag model.

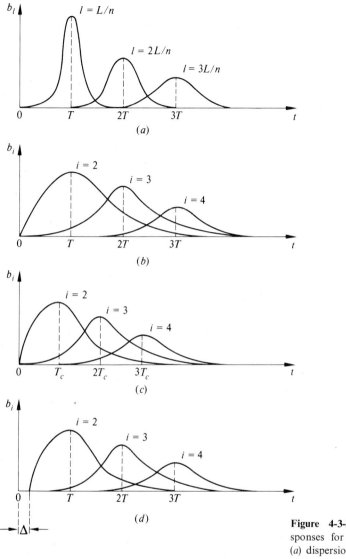

Figure 4-3-5 BOD impulse responses for three river reaches: (a) dispersion model (b) approximate dispersion model (c) CSTR model (d) CSTR model with time delay.

to obtain impulse responses which are as close as desired to those of the dispersion model by suitably selecting $\phi(\tau)$ and $\psi(\tau)$.

Finally, the general block diagram of all the models discussed in this section is shown in Fig. 4-3-6, where $m(s)$ is the BOD transfer function in each reach and $m'(s)$ is the transfer function for the auxiliary variable. The specific forms of $m(s)$ (gain and time constant) are summarized and compared in Table 4-3-1.

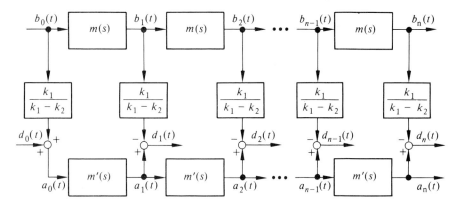

Figure 4-3-6 Block diagram of the n reach river model.

In conclusion, when the value of the dispersion coefficient D of a river is known, Dobbins' criterion can be used to decide whether to take dispersion into account or not. If the answer is positive, the plug flow model with modified velocities can be used as a simple and accurate steady state model, while models (4-3-11, 4-3-12, 4-3-13, 4-3-15) can be used for real time state estimation and control (see Sec. 5-4, 5-5 and 9-3, 9-4). When the value of the dispersion coefficient D is unknown, in general it is necessary to estimate the parameters of one of the models discussed, a procedure considered in Secs. 5-4 and 5-5.

Table 4-3-1 Comparison of approximated Streeter-Phelps models with dispersion (BOD equation)

Model	Transfer function $m(s)$	Gain $m(0)$	Time constant
Dispersion	$e^{\lambda_2(s)L/n}$	$e^{\lambda_2(0)L/n}$	
Approximate dispersion	$\dfrac{\mu}{1+Ts}$	$\mu = \dfrac{v}{v+k_1 L/n}$	$T = \dfrac{L/n}{v\left(1+\dfrac{2k_1 D}{v^2}\right)}$
CSTR with time delay	$\dfrac{\mu_c}{1+T_c s}e^{-s\Delta}$	$\mu_c = \dfrac{v}{v+k_1 L/n}$	$T_c = \dfrac{L/n}{v\left(1+k_1\dfrac{L/n}{v}\right)}$
Distributed lag	$\dfrac{\mu_c}{1+T_c s}\Phi(s)$	$\mu_c \Phi(0)$	$T_c = \dfrac{L/n}{v\left(1+k_1\dfrac{L/n}{v}\right)}$

4-4 AN ECOLOGICAL MODEL

As already mentioned in Sec. 3-5 (see Fig. 3-5-4), chemical models, like the Streeter-Phelps model, may be unsatisfactory due to their gross simplifications. These models are problematic, in particular, for investigations anticipating future changes of the wastewater load since the change in parameters cannot be predicted on a theoretical basis. On the other hand, detailed ecological models can avoid this problem, but the parameters involved are hard to estimate, because the necessary measurement effort is tremendous.

In the following, an ecological model is described which is simple enough to be identified with reasonable measurement effort and which is believed to be more realistic than the Streeter-Phelps model for a certain class of rivers (Stehfest, 1973). Beside the assumptions of vertical and lateral homogeneity (see Sec. 3-1), which are made throughout the main part of the book, the following assumptions have to be made for this model

1. river depth and velocity are high

2. longitudinal dispersion can be neglected

3. the river is heavily polluted

The first assumption implies that benthic variables can be neglected, for two reasons. First, sediments and benthic organisms can hardly develop because of the high stream velocity. Second, even if they do, their importance relative to the stream-borne organisms and pollutants will be small, because the water column above the river bottom is high. The latter argument is particularly applicable because the biochemical activity within bottom deposits is relatively small due to the slow material exchange. The insignificance of the benthic variables means that the model consists of equations of type (3-1-16) only (with nonvanishing v and D). If, in addition, the diffusion terms can be neglected (assumption 2) the model can be transformed into a system of ordinary differential equations (see Sec. 3-1). Assumption 2 is only made in order to simplify the following discussion, which is focused on the biochemical aspects. Little change would be required if longitudinal dispersion were to be taken into account in one of the ways described in Sec. 4-3. Assumption 3, in connection with assumption 1, further simplifies the model. The higher organisms of the food web can be neglected because they are fairly exacting, and because their reproduction time is usually comparable to the flow time (see Sec. 2-3). Nitrification and photosynthesis may be neglected for the same reasons. The development of photosynthesizing organisms is in addition hindered by the low mean light intensity, which is due to the great depth and the high turbidity. A river for which the preceding assumptions seem to be approximately fulfilled is the Rhine river in Germany; and in fact, this is the river for which the model was first developed (see Sec. 5-3).

On the basis of the simplifying assumptions the following state variables were selected:

w_1 = easily degradable pollutants

w_2 = slowly degradable pollutants

B = bacteria

P = protozoa

c = oxygen

All variables are defined as mass densities, the pollutants are measured by their chemical oxygen demand (see Sec. 3-5), and hence $w_1 + w_2$ is the biochemical oxygen demand of the pollutants. (The BOD of a water sample, however, also comprises the biochemical oxygen demand of the living matter.) The aggregation of all bacterial and protozoan species into two variables has already been discussed in Sec. 3-5. It is, in any case, more justified than dealing with all pollutants as one variable, because the differences between the pollutants with regard to degradability are certainly larger than the differences among the bacteria or protozoa with regard to growth rate or endogenous respiration.

The model equations chosen are:

$$\dot{w}_1 = -k_{11}g_1 B + L_1 \qquad (4\text{-}4\text{-}1a)$$

$$\dot{w}_2 = -k_{21}g_2 B + L_2 \qquad (4\text{-}4\text{-}1b)$$

$$\dot{B} = (g_1 + g_2 - k_{36})B - k_{37}g_3 P \qquad (4\text{-}4\text{-}1c)$$

$$\dot{P} = (g_3 - k_{43})P \qquad (4\text{-}4\text{-}1d)$$

$$\dot{c} = k_{51}(c_s - c) - (k_{52}g_1 + k_{53}g_2 + k_{54})B - (k_{55}g_3 + k_{56})P \qquad (4\text{-}4\text{-}1e)$$

with

$$g_1 = \frac{k_{31}w_1}{k_{32} + w_1} \qquad g_2 = \frac{k_{33}w_2}{k_{34} + w_2 + k_{35}w_1} \qquad g_3 = \frac{k_{41}B}{k_{42} + B}$$

All k_{ji}'s are positive parameters, while L_1 and L_2 are pollutant inputs. Obviously, the model is given with flow time as an independent variable (see Sec. 3-1). If it is assumed that there are no flows of material other than those shown in Fig. 4-4-1 (i.e., flows resulting from sedimentation or from interactions of protozoa and bacteria with higher levels in the food web are neglected) then, the three following relationships can be established between the parameters (see the analogous Eq. (3-5-20))

$$\frac{k_{54}}{k_{36}} = k_{11} - k_{52}$$

$$\frac{k_{54}}{k_{36}} = k_{21} - k_{53}$$

$$\frac{k_{56}}{k_{43}} = \frac{k_{54}}{k_{36}}k_{37} - k_{55}$$

However, if the parameters are considered to be independent these relationships may be used as a test for the estimation results (see Chapter 5). All relationships contained in Eq. (4-4-1) have already been discussed in Sec. 3-5. For the easily degradable pollutants the *Michaelis-Menten degradation kinetics* is assumed (Eq. (4-4-1a)), while for the degradation of the slowly degradable pollutants a *competitive inhibition* through the easily degradable pollutants is postulated. The latter assertion is justified by the fact that many enzymes which catalyze the degradation of slowly degradable matter are only formed after the more easily degradable substances have been used up (see Sec. 2-3). Of course, it is not mandatory to use the expression for competitive rather than allosteric inhibition. But it may be argued that for high concentrations of slowly degradable matter bacteria specialized on this matter become abundant, so that the inhibition can be overcome to a certain extent; this effect cannot be described by Eq. (3-5-22) for allosteric inhibition. The terms on the righthand side of Eq. (4-4-1c) stand for increase of bacterial mass by degradation activity, and loss of bacterial mass by endogenous respiration and protozoan predation. Equation (4-4-1d) describes the variation of the protozoan mass as the difference between growth due to digestion of bacteria and endogenous respiration. The latter term with an appropriate parameter value k_{43} may, in addition, approximately account for losses of protozoan mass through predation of higher organisms. Finally, the impacts of the processes just mentioned on the oxygen budget are listed in Eq. (4-4-1e) together with the term for physical reaeration. The structure of the model is shown in Fig. 4-4-1. The compartments correspond to the variables, and the arrows to the flows of material. The slanted lines indicate the surroundings of the river.

Model (4-4-1) contains quite a number of parameters which cannot be determined separately, either through experimentation or through theoretical considerations. The reaeration coefficient k_{51}, for example, cannot be determined

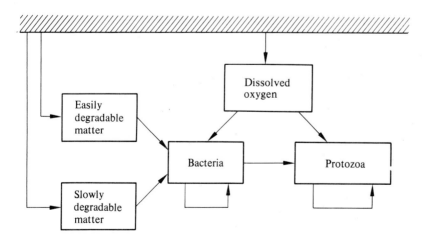

Figure 4-4-1 Structure of the ecological river quality model (4-4-1).

theoretically at sufficient accuracy, as discussed in Sec. 3-5. And it cannot be measured separately, (except for particular cases, see Churchill et al., 1962), because the reaeration process is always superimposed upon the biochemical oxygen consumption processes. Hence, the parameters have to be determined, in general, simultaneously from in situ measurements of the model variables. Various methods of solving this problem are discussed in Chapter 5. Even now, however, it should be intuitively clear that to determine the many parameters of Eq. (4-4-1) it is essential to have accurate measurements of as many variables as possible along the river. The least problematic variable in this respect is oxygen concentration, since it can easily be measured with high accuracy. For the measurement of protozoan mass, however, the only well established method is to determine microscopically number and size of the protozoa, which is rather expensive and not too accurate. For the bacterial mass a similar method could be used, while the traditional plate count (counting the number of colonies developing on a standardized culture medium) is too inaccurate. Another technique, which is not yet widely used, however, consists of measuring the total living biomass through a typical constituent of living matter, for instance, ATP (Jannasch, 1972), and then correcting for the protozoan mass. (Of course, it is also possible to use the observations of $P + B$ directly by defining in an appropriate way the output transformation matrix, see Sec. 1-2. The same applies to the corrections discussed below.) In measurements of the organic pollutants it is, in practice, impossible to differentiate between easily and slowly degradable pollutants. The total organic pollution, however, can be easily measured. One way is to measure the BOD of the water samples and to correct for the BOD of the living matter, so that measurements of $w_1 + w_2$ would be available for the parameter estimation. Another way is to measure the total COD and then correct for both living and undegradable matter (see Sec. 3-5). Quite often the major components of the undegradable organic pollution can be measured separately, e.g., chlorinated carbohydrates. The rest (or even all undegradable organic matter, see Sec. 5-3) may be considered as part of w_2. To summarize, observations of $(w_1 + w_2)$, B, P, and c can be made available at reasonable expense for parameter (and state) estimation of model (4-4-1). Whether this is sufficient or not for the determination of the parameters will be discussed in Sec. 5-3.

Far less can be proved analytically for model (4-4-1) than for the Streeter-Phelps model. But what can be proved is in accordance with what may be expected for an ecological river quality model. Thus, it can easily be shown that the values of w_1, w_2, B, and P can never become negative if the initial values are non-negative. The oxygen concentration may drop below zero (as with the Streeter-Phelps model), but this unrealistic result occurs only in extreme situations.

The *equilibria* of the model can also be easily calculated. In the case of no input ($L_1 = L_2 = 0$) there exists an infinity of non-isolated equilibrium points in the subspace given by $B = 0$, $P = 0$, $c = c_s$. These equilibria are stable, but not asymptotically, in the sense that after a small perturbation of the state, the system returns once more to an equilibrium, not necessarily the same one, though close to it. In other words, B, P, and $(c_s - c)$ tend to zero as τ goes to infinity, while certain amounts of w_1 and w_2 may be left over, depending on the initial values (and on the parameter values, of course). For constant, but non-zero input and realistic parameter values there are two isolated equilibrium points, one with all variables greater than zero, and one in the subspace given by $P = 0$. The assumption of realistic parameter values means, for example, that the maximum protozoan growth rate k_{41} is greater than the protozoan endogenous respiration rate k_{43}, which is certainly reasonable. The dependence of the equilibrium state upon the parameters can easily be determined, but it is not worth giving the formulae here, since they are rather lengthy. The procedure involves putting all derivatives equal to zero in Eq. (4-4-1) and then solving Eqs. (4-4-1d), (4-4-1a), (4-4-1b), (4-4-1c), and (4-4-1e) successively. For this the assumption $P \neq 0$ has to be made, so that the first one of the two isolated equilibrium points results. The other one may be calculated equally simply. It becomes clear during this procedure what the above mentioned realistic assumptions about the parameter values have to be, in order to obtain positive equilibrium values. The equilibrium point with $P \neq 0$ is an asymptotically stable one, as can be seen by inspection of the matrix of the system derived from Eq. (4-4-1) by linearization around the equilibrium. The eigenvalues of the matrix have strictly negative real parts (see Sec. 1-2). The imaginary parts are, in general, different from zero, so that the motions in the vicinity of the equilibrium are damped oscillations, as is typical for prey–predator relationships. Although it has not been proved, it is conjectured that all motions starting with w_1, w_2, B, and P greater than zero tend towards this equilibrium point. Figure 4-4-2 shows how the equilibrium point is approached from arbitrarily chosen initial conditions using realistic parameter values. If L_1 and L_2 are extremely high only B, P, and c reach stationary values, while w_1 and w_2 increase linearly for high τ values.

The *sensitivity* of system (4-4-1) to changes of initial and parameter values has been investigated extensively through numerical experimentation (see Sec. 4-1), but only a few general results have emerged from this study. No dramatically high sensitivities have been observed. Sensitivities show, in general, oscillations, as do the motions toward the equilibrium (see Fig. 4-4-2). It is surprising how far downstream disturbances of the initial values of w_1 and w_2 can be felt (for L_1 and L_2 different from zero the motions are asymptotically stable, which means that the difference between the perturbed and the unperturbed motion goes to zero as we go downstream). Probably the most interesting result is the sensitivity with respect to temperature: if the maximum specific growth rates and endogenous respiration rates are all changed simultaneously in correspondence with a certain temperature variation (see Sec. 3-5) changes of the state variables are remarkably smaller than in the case when only one of these parameters has been changed. Numerical examples of how the model solution for non-constant input varies if flow rate and temperature are changed will be given in Sec. 5-3, where model (4-4-1) is applied to a real case.

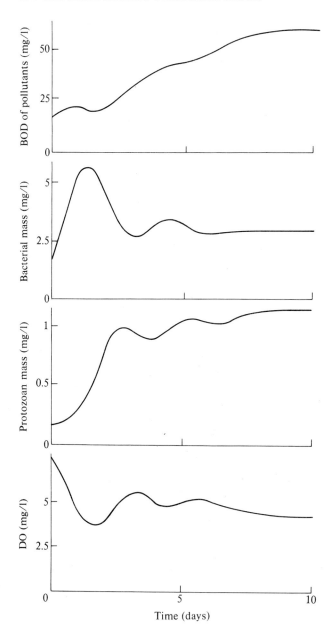

Figure 4-4-2 Typical solution of model (4-4-1) for constant, non-zero input.

4-5 OTHER ECOLOGICAL MODELS

As a representation of reality the ecological model described in the previous section must be considered to be very crude, and of restricted applicability. Many other ecological models can be found in the literature, most of which are much more complicated than the one described in Sec. 4-4. A few of them will be discussed in this section. A complete survey of all ecological river quality models would be beyond the scope of this book, because the material which would have to be presented would be too voluminous. (Since there is no clear distinction between river impoundments and lakes, the very broad field of eutrophication modeling would also have to be discussed.) None of the numerous models which contain ecological quantities, like algal or fish density, but not as dynamical state variables (see Sec. 1-2) are mentioned. In those models the ecological quantities usually do not depend on the pollutant input into the river, but appear as externally given functions in the oxygen balance equation (see, for example, O'Connor, 1967; Wolf, 1971; Willis et al., 1975). In effect, some of the modified Streeter-Phelps models presented in Sec. 4-2 are of this type (see, for instance, Eq. (4-2-2)).

An ecological model, which is even simpler than the one described in the preceding section, is the one by Gates et al., (1969), whose work was already mentioned in Sec. 3-5. Its state variables are the concentrations of organic pollutants, bacterial mass, and oxygen. The model can be derived from model (4-4-1) by removing all expressions related to slowly degradable pollutants, endogenous respiration, and protozoa. Because of its simplicity a formal (graphical) technique for estimating the parameters from measurements could be applied, but the uncertainty of the parameters seems to be high, due to the lack of measurements of the bacterial mass (see Stehfest, 1973, and Chapter 5). The model was used to describe both laboratory and field experiments with glucose as pollutant. This type of model has also been used to describe the nitrification process (Stratton and McCarty, 1967; O'Connor et al., 1976). Since nitrification is a two-stage process (see Sec. 2-3), in which the ammonium substrate is used in sequence by two different types of bacteria, it is necessary to couple two of these models appropriately.

An ecological model considerably more complex than the one in Sec. 4-4 was developed for the Delaware Estuary (USA) by Kelley (1975, 1976). The ecological relationships encompassed by this model are shown in Fig. 4-5-1, in analogy to Fig. 4-4-1 (cf. also Fig. 2-3-9). The mathematical description of the relationships is not worth giving here, especially since the single expressions used have essentially been discussed in Sec. 3-5 already. As one can see from Fig. 4-5-1, eutrophication is a major problem in the Delaware Estuary, while nitrification is considered to be less important. Another remarkable feature of the model is the intention to relate fish abundance to pollutants input. Since fish abundance is an important quality indicator, which also characterizes the recreational benefit of the river, it is very valuable to have it as a state variable of a river quality model. Another interesting aspect is the modeling of the mass transport phenomena in the estuary, which has to cope with tidal movements. Two versions of the model have been investigated: in the first, the river was viewed as a sequence of completely mixed tanks, similar to the CSTR model mentioned in Sec. 4-3. The river flow from reach to reach was averaged over the tidal cycles. As shown in Sec. 4-3, this model reproduces to a certain extent the dispersion phenomena. In the second

136 SOME PARTICULAR SELF-PURIFICATION MODELS

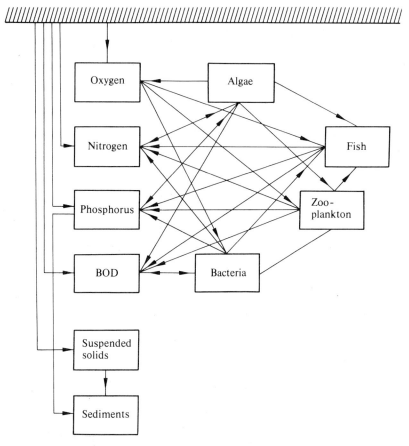

Figure 4-5-1 Structure of the Delaware estuary model.

version, diffusive exchange between the reaches was introduced additionally, mainly to account for tidal mixing. The second version proved to fit measured data better than the first one, but the differences are not dramatic. The model contains a huge number of parameters, so that in view of the scarce measurements for the dependent variables, application of a formal estimation technique would have been hopeless (see Sec. 5-1). Rather, the estimation has been done by playing around with the parameter values within reasonable limits until the model output fitted the measured values. In this way, satisfactory agreement with measurements of oxygen, BOD, phosphate, and nitrogen from one month could be obtained. No measurements were available for the other variables. The application of the model specified in this way to another month, where flow rate and temperature were considerably different from the month used for the parameter estimation, gave very poor agreement with the observations.

Another ecological model, which is of comparable complexity to the model just described, has been reported by Boes (1975). The parameters have been estimated in a similar manner, too. The model was used to predict BOD and oxygen concentrations in the Neckar river in West Germany, while dynamics of the other model variables were not discussed. The BOD and DO predictions have been compared with the corresponding predictions of two other, simpler models (Wolf, 1971; Abendt, 1975) and the differences have turned out to be fairly large.

REFERENCES

Section 4-1

Arbabi, M., Elzinga, J., and ReVelle, C. (1974). The Oxygen Sag Equation and a Linear Equation for the Critical Deficit. *Water Resour. Res.*, **10**, 921–929.
Di Toro, D. M. and O'Connor, D. J. (1968). The Distribution of Dissolved Oxygen in a Stream with Time Varying Velocity, *Water Resour. Res.*, **4**, 639–646.
Fair, G. M. and Geyer, J. C. (1965). *Water Supply and Waste-Water Disposal*. John Wiley, New York.
Liebman, J. C. (1965). The Optimal Allocation of Stream Dissolved Oxygen Resources. Ph.D. dissertation, Cornell Univ., Ithaca, N.Y.
Liebman, J. C. and Loucks, D. P. (1966). A Note on Oxygen Sag Equations. *J. WPCF*, **38**, 1963–1967.
Lin, S. H., Fan, L. T., and Hwang, C. L. (1973). Digital Simulation of the Effect of Thermal Discharge on Stream Water Quality. *Water Resour. Bull.*, **9**, 689–702.
Rinaldi, S. and Soncini-Sessa, R. (1976). Sensitivity Analysis of Streeter-Phelps Type Water Pollution Models (in Italian). *Ingegneria Ambientale*, **5**, 320–335.
Rinaldi, S. and Soncini-Sessa, R. (1978). Sensitivity Analysis of Generalized Streeter-Phelps Models. *Advances in Water Resources*, **1**, 141–146.
Stehfest, H. (1975). Decision Theoretical Remark on Sensitivity Analysis. Report IIASA RR-75-3, International Institute for Applied Systems Analysis, Laxenburg, Austria.
Streeter, H. W. and Phelps, E. B. (1925). A Study of the Pollution and Natural Purification of the Ohio River, vol. III, *Public Health Bulletin*, no. 146, United States Public Health Service, Reprinted by U.S. Department of Health, Education and Welfare, 1958.

Section 4-2

Braun, H. B. and Berthouex, P. M. (1970). Analysis of Lag Phase BOD Curves Using the Monod Equations. *Water Resour. Res.*, **6**, 838–844.
Camp, T. R. (1963). *Water and its Impurities*. Reinhold, New York.
Courchaine, R. J. (1963). The Significance of Nitrification in Stream Analysis—Effects on the Oxygen Balance. *Proc. of XVIII Industrial Waste Conference*, Purdue University, Lafayette, Indiana.
Di Toro, D. M. and O'Connor, D. J. (1968). The Distribution of Dissolved Oxygen in a Stream with Time Varying Velocity. *Water Resour. Res.*, **4**, 639–646.
Dobbins, W. E. (1964). BOD and Oxygen Relationships in Streams. *J. San. Eng. Div., Proc. ASCE*, **90**, 53–78.
Gameson, A. L. H. (1959). Some Aspects of the Carbon, Nitrogen and Sulphur Cycles in the Thames Estuary, Part II. In *Effects of Pollution on Living Material*. Institute of Biology, London.
Li, W. H. (1972). *Differential Equations of Hydraulic Transients and Groundwater Flow*. Prentice-Hall, Englewood Cliffs, N. J.
Li, W. H. and Kozlowski, M. E. (1974). DO-sag in Oscillating Flow. *J. Env. Eng. Div., Proc. ASCE*, **100**, 837–854.
Liebman, J. C. and Loucks, D. P. (1966). A Note on Oxygen Sag Equations. *J. WPCF*, **38**, 1963–1967.

O'Connor, D. J. (1967). The Temporal and Spatial Distribution of Dissolved Oxygen in Streams. *Water Resour. Res.*, **3**, 65–79.
O'Connor, D. J. and Di Toro, D. M. (1970). Photosynthesis and Oxygen Balance in Streams. *J. San. Eng. Div., Proc. ASCE*, **96**, 547–571.
Rinaldi, S. and Soncini-Sessa, R. (1974). Modelling River Pollution (in Italian). In *Ingegneria Sistemistica Ambientale*. C.L.U.P., Milan.
Shastry, J. S., Fan, L. T., and Erickson, L. E. (1975). Non Linear Parameter Estimation in Water Quality Modeling. *J. Env. Eng. Div., Proc. ASCE*, **99**, 315–331.
Thomas, H. A. (1948). Pollution Load Capacity of Streams. *Water and Sewage Works*, **95**, 405.
Thomas, H. A. (1953). Personal communication, cited in Woodward, (1953). Deoxygenation of Sewage: A Discussion. *Sewage and Industrial Wastes*, **25**, 518.
Westlake, D. F. (1968). A Model for Quantitative Studies of Photosynthesis by Higher Plants in Streams. *Air and Water Pollution Journal*, **10**, 883–896.
Young, J. C. and Clark, J. W. (1965). Second Order Equation for BOD. *J. San. Eng. Div., Proc. ASCE*, **91**, 43–58.

Section 4-3

Dobbins, W. E. (1964). BOD and Oxygen Relationships in Streams. *J. San. Eng. Div., Proc. ASCE*, **90**, 53–78.
Tamura, H. (1974). A Discrete Dynamic Model with Distributed Transport Delays and its Hierarchical Optimization for Preserving Stream Quality. *IEEE Trans.*, **SMC 4**, 424–431.
Young, P. and Beck, B. (1974). The Modelling and Control of Water Quality in a River System. *Automatica*, **10**, 455–468.

Section 4-4

Churchill, M. A., Elmore, H. L., and Buckingham, R. A. (1962). The Prediction of Stream Reaeration Rates. *Int. J. Air Wat. Pollution*, **6**, 467–504.
Jannasch, H. W. (1972). New Approaches to Assessment of Microbial Activity in Polluted Waters. In *Water Pollution Microbiology* (R. Mitchell, ed.). John Wiley, New York.
Stehfest, H. (1973). Mathematical Modelling of Self-Purification of Rivers (in German; English translation available as report IIASA PP-77-11, International Institute for Applied Systems Analysis, Laxenburg, Austria). Report KFK 1654 UF, Kernforschungszentrum Karlsruhe, Karlsruhe, W. Germany.

Section 4-5

Abendt, R. W. (1975). Models in Water Quality Planning. *Ecological Modelling*, **1**, 205–217.
Boes, M. (1975). Biocenotic Model for Simulating the Oxygen Balance of Rivers (in German). *GWF-Wasser/Abwasser*, **116**, 339–407.
Gates, W. E., Marlar, J. T., and Westfield, J. D. (1969). The Application of Bacterial Process Kinetics in Stream Simulation and Stream Analysis. *Water Research*, **3**, 663–686.
Kelley, R. A. (1975). The Deleware Estuary. In *Ecological Modelling in a Resource Management Framework* (C. S. Russell, ed.). Resources for the Future, Washington, D.C.
Kelley, R. A. (1976). Conceptual Ecological Model of the Delaware Estuary. In *Systems Analysis and Simulation in Ecology*, vol. IV (B. C. Patten, ed.). Academic Press, New York.
O'Connor, D. J. (1967). The Temporal and Spatial Distribution of Dissolved Oxygen in Streams. *Water Resour. Res.*, **3**, 65–79.
O'Connor, D. J., Thomann, R. V., and Di Toro, D. M. (1976). Ecological Models. In *Systems Approach to Water Management* (A. K. Biswas, ed.). McGraw-Hill, New York.

Stehfest, H. (1973). Mathematical Modelling of Self-Purification of Rivers (in German; English translation available as report IIASA PP-77-11, International Institute for Applied Systems Analysis, Laxenburg, Austria). Report KFK 1654 UF, Kernforschungszentrum Karlsruhe, Karlsruhe, W. Germany.

Stratton, F. E. and McCarty, P. L. (1967). Prediction of Nitrification Effects on the Dissolved Oxygen Balance of Streams. *Environ. Sci. Technol.*, **1**, 405–410.

Willis, R., Anderson, D. R., and Dracup, J. A. (1975). Steady-State Water Quality Modelling in Streams. *J. Env. Eng. Div., Proc. ASCE*, **101**, 245–258.

Wolf, D. (1971). Incorporation of Latest Findings into Oxygen Budget Calculations for Rivers (in German). *GWF-Wasser/Abwasser*, **112**, 200–203 and 250–254.

CHAPTER
FIVE

STATE AND PARAMETER ESTIMATION

5-1 GENERAL REMARKS

Problem Definition

The various models discussed previously are characterized by numerous parameters which cannot usually be determined theoretically or measured directly. Rather, they have to be determined on the basis of observations of the system output.

Confining ourselves, for the sake of simplicity, to the discrete time case, this means (see Eq. (1-2-20)) that the parameter vector $\boldsymbol{\theta} = [\theta_1\ \theta_2 \cdots \theta_q]^T$ in

$$\mathbf{x}(t+1) = \mathbf{f}(\mathbf{x}(t), \mathbf{u}(t), \boldsymbol{\theta}, t) \tag{5-1-1a}$$

$$\mathbf{y}(t) = \boldsymbol{\eta}(\mathbf{x}(t), \boldsymbol{\theta}, t) \tag{5-1-1b}$$

has to be determined on the basis of observations of the vector **y**. This problem is called *parameter estimation*. Its solution obviously requires the knowledge of the system state **x** (or something equivalent). Therefore, if the state is neither measurable directly nor known for other reasons, the problem of *state estimation*, which is an important problem not only in the context of parameter estimation but also in its own right (see, for example, Sec. 5-4), has to be considered.

As already mentioned in Sec. 1-3, there exists a close relationship between state and parameter estimation. Parameters θ_i ($i = 1, \ldots, q$) can always be considered as additional state variables for which equations of the type

$$\theta_i(t+1) = \theta_i(t)$$

hold. Hence, the problem of state and parameter estimation for a system (5-1-1) is equivalent to estimating the state of the system

$$\mathbf{z}(t+1) = \mathbf{f}'(\mathbf{z}(t), \mathbf{u}(t), t) \qquad (5\text{-}1\text{-}2a)$$

$$\mathbf{y}(t) = \boldsymbol{\eta}'(\mathbf{z}(t), t) \qquad (5\text{-}1\text{-}2b)$$

with

$$\mathbf{z} = \begin{bmatrix} \mathbf{x} \\ \boldsymbol{\theta} \end{bmatrix} \qquad \mathbf{f}'(\mathbf{z}, \mathbf{u}, t) = \begin{bmatrix} \mathbf{f}(\mathbf{x}, \mathbf{u}, \boldsymbol{\theta}, t) \\ \boldsymbol{\theta} \end{bmatrix}$$

$$\boldsymbol{\eta}'(\mathbf{z}, t) = \boldsymbol{\eta}(\mathbf{x}, \boldsymbol{\theta}, t)$$

In general, the new model is nonlinear, even if the original one was linear.

If the parameters are to be estimated alone, an input–output description of the system (see Sec. 1-1) can be used in order to eliminate the state equations. Then the parameters have to be estimated from

$$\boldsymbol{\theta}(t+1) = \boldsymbol{\theta}(t) \qquad (5\text{-}1\text{-}3a)$$

$$\mathbf{y}(t) = \boldsymbol{\psi}_{\mathbf{x}_0, t_0}\left(\mathbf{u}_{[t_0, t]}(\cdot), \boldsymbol{\theta}(t), t\right)$$

where $\boldsymbol{\psi}_{\mathbf{x}_0, t_0}$ is an input–output relationship. This formulation is again of the form (5-1-2). More generally, it can always be assumed in the following that a representation of type (5-1-2) in which the state $\mathbf{z}$ comprises exactly the quantities to be estimated has been found.

Unfortunately, estimation problems are, in practice, complicated by the fact that the system is randomly disturbed. The usual way of representing these disturbances is to add suitable noise terms to the model equations. Thus, for example, Eq. (5-1-2) is modified to give

$$\mathbf{z}(t+1) = \mathbf{f}'(\mathbf{z}(t), \mathbf{u}(t), t) + \mathbf{v}(t) \qquad (5\text{-}1\text{-}4a)$$

$$\mathbf{y}(t) = \boldsymbol{\eta}'(\mathbf{z}(t), t) + \mathbf{w}(t) \qquad (5\text{-}1\text{-}4b)$$

where $\mathbf{v}$ and $\mathbf{w}$ are called *process* and *measurement noise*, respectively. Equations (5-1-1) and (5-1-3) may be modified in a similar way. Of course, the addition of $\mathbf{v}$ and $\mathbf{w}$ is not the most general way of taking into account the random disturbances; but usually one does not know exactly the characteristics of the noise and so the most simple representation is chosen. If the noise terms are added in Eqs. (5-1-1)–(5-1-3), these equations are, in general, no longer equivalent, as they are in the deterministic case. However, since the usual way of taking noise into account in water pollution modeling is in one equation as arbitrary as in the other, the estimation problem is always assumed to be of type (5-1-4). The state estimation problem can be formulated precisely in the following way: given two sets $\mathbf{U}_\tau$ and $\mathbf{Y}_\tau$ of input and output observations of a system over the time interval $[0, \tau]$, and given a state model of the system, find a "suitable" *estimate* for the actual value of the state $\mathbf{z}$ at the prescribed time t_a. (For time-varying systems the observation interval ought to be defined more generally, but this minor loss of generality is not essential for what follows.) The argument t_α will be left out in the following if there is no danger of confusion. The "suitable" estimate of z is denoted by $\hat{z}$, the set of all possible z is denoted by Z, and the rule for calculating $\hat{z}$, i.e.,

$$\hat{z} = \hat{z}(U_\tau, Y_\tau) \qquad (5\text{-}1\text{-}5)$$

will be called the *estimator*. The notion "suitable" in the problem statement will be specified through the selection of the function $\hat{z}(\cdot, \cdot)$ appearing in Eq. (5-1-5).

The literature on the problem formulated just now is fairly well developed. More or less broad overviews of the area can be found in Åström and Eykhoff (1971), Eykhoff (1974), Isermann (1974), Unbehauen et al., (1974), and Strobel (1975). Many articles on particular topics are given in the proceedings of the IFAC Symposia on Identification and System Parameter Estimation.

The estimate $\hat{z}$ is a random variable because of the random disturbances. Its statistical properties (which are determined by the characteristics of v and w) are very important for selecting a suitable estimator. The most desirable ones are

Unbiasedness: $E[\hat{z}] = z$, where $E[\,\cdot\,]$ denotes expectation.
Consistency: $\lim_{\tau \to \infty} p(\|\hat{z} - z\| > \varepsilon) = 0 \;\forall\; \varepsilon > 0$, where $\|\cdot\|$ denotes norm.
Efficiency: $\operatorname{cov}[\hat{z}] = E[(\hat{z} - z)(\hat{z} - z)^T] \leq \operatorname{cov}[\hat{\zeta}]$ for all unbiased estimators $\hat{\zeta}$, where cov$[\,\cdot\,]$ denotes covariance. (The statement $A \leq B$, where A and B are matrices, means that $x^T A x \leq x^T B x$ for all vectors x.)

Selection of Estimators

The essential assumption for all statistical estimation techniques is that the *probability density function* (pdf) of the observations is known, apart from the unknown value z. This means, the function

$$p(Y_\tau | z)$$

is known, where the bar denotes *conditional upon*. (For the sake of notational simplicity the same symbol z is used for both the random variable and its actual value, and the input observations are omitted. In the applications shown in the following sections $u(t)$ is assumed to be known exactly, anyway.) For estimation the *a posteriori probability*

$$p(z | Y_\tau)$$

is of interest, i.e., the probability for the different z-values given certain observations Y_τ. The relationship between a priori and a posteriori probabilities is given by *Bayes' rule* (see, for instance, Raiffa, 1968):

$$p(z | Y_\tau) = \frac{p(Y_\tau | z) p(z)}{\int_z p(Y_\tau | z) p(z) \, dz} \qquad (5\text{-}1\text{-}6)$$

The *a priori probability* $p(\mathbf{z})$ for $\mathbf{z}$ might reflect very vague, subjective knowledge. Since the integration in the denominator of Eq. (5-1-6) extends over all $\mathbf{z}$, the denominator equals $p(\mathbf{Y}_\tau)$. A natural choice for an estimator $\hat{\mathbf{z}}$ would be the solution of the maximization problem

$$\max_{\mathbf{z}} \left[p(\mathbf{z} | \mathbf{Y}_\tau) \right] \qquad (5\text{-}1\text{-}7)$$

i.e., $\hat{\mathbf{z}}$ is defined as that value of $\mathbf{z}$ which maximizes the a posteriori probability given by Eq. (5-1-6). This kind of estimator is called *Bayes' estimator*, because of the use of Bayes' formula. It can still be refined, if one knows the costs associated with a deviation of the estimate from the true value of $\mathbf{z}$. If $\mathscr{C}(\mathbf{z}', \mathbf{z})$ indicates the costs to be incurred if $\mathbf{z}'$ is used instead of the true value $\mathbf{z}$, it is reasonable to use as an estimate that value of $\mathbf{z}'$ which minimizes the expected costs, i.e., to use the solution of the minimization problem

$$\min_{\mathbf{z}'} \int \mathscr{C}(\mathbf{z}', \mathbf{z}) p(\mathbf{z} | \mathbf{Y}_\tau) \, d\mathbf{z} \qquad (5\text{-}1\text{-}8)$$

In general, $\mathscr{C}(\mathbf{z}', \mathbf{z})$ is known only after the model has been applied to the problem causing the modeling effort (see Sec. 1-3 and 10-1). Hence, if the estimator is given by expression (5-1-8) no use is made of the *separation hypothesis* mentioned in Sec. 1-3.

In most cases, however, even less a priori knowledge than required by Eq. (5-1-6) is available. If nothing is known about $p(\mathbf{z})$, the assumption usually made is

$$p(\mathbf{z}) = \text{const.}$$

for all $\mathbf{z} \in Z$. Then Eq. (5-1-7) yields the same $\hat{\mathbf{z}}$ as

$$\max_{\mathbf{z}} \left[p(\mathbf{Y}_\tau | \mathbf{z}) \right] \qquad (5\text{-}1\text{-}9)$$

The function $p(\mathbf{Y}_\tau | \cdot)$ (which is not a pdf!) is the so-called *likelihood function*; corresponding to this, the estimate from Eq. (5-1-9) is called *maximum likelihood estimate*. It gives the value of $\mathbf{z}$ for which the observed values $\mathbf{Y}_\tau$ have maximum probability.

The statistical properties of the maximum likelihood estimator have been extensively studied. If there is no process noise and certain other general assumptions are fulfilled the maximum likelihood estimator can be shown to be (Eykhoff, 1974) asymptotically unbiased (i.e., $E[\hat{\mathbf{z}}] \to \mathbf{z}$ as $\tau \to \infty$), consistent, and asymptotically efficient. Moreover, it is asymptotically normally distributed and the covariance matrix is given by

$$\text{cov}[\hat{\mathbf{z}}] = - \left[E \left[\frac{\partial^2 \ln p(\mathbf{Y}_\tau | \mathbf{z})}{\partial \mathbf{z} \, \partial \mathbf{z}^T} \bigg|_{\mathbf{z}=\hat{\mathbf{z}}} \right] \right]^{-1} \qquad (5\text{-}1\text{-}10)$$

The matrix within the brackets on the righthand side of Eq. (5-1-10) is called *information matrix* (see Sec. 5-3).

If also the conditional probability density $p(\mathbf{Y}_\tau|\mathbf{z})$ is unknown it is usually assumed that it can be characterized completely by its expectation and covariance and that only the expectation depends on $\mathbf{z}(t_a)$. In other words, a $(\tau+1)p$-variate Gaussian (normal) distribution is assumed for $p(\mathbf{Y}_\tau|\mathbf{z})$:

$$p(\mathbf{Y}_\tau|\mathbf{z}) = C \cdot \exp(-\tfrac{1}{2}(\mathbf{Y}_\tau - E[\mathbf{Y}_\tau])^T (\text{cov}[\mathbf{Y}_\tau])^{-1}(\mathbf{Y}_\tau - E[\mathbf{Y}_\tau])) \quad (5\text{-}1\text{-}11)$$

where C is a normalizing constant and $\mathbf{Y}_\tau$ has to be interpreted as the vector

$$\mathbf{Y}_\tau = [\mathbf{y}(0)^T \; \mathbf{y}(1)^T \ldots \mathbf{y}(\tau)^T]^T$$

If the system (5-1-4) is linear and both $\mathbf{v}(t)$ and $\mathbf{w}(t)$ are uncorrelated ("white") Gaussian disturbances, the observations $\mathbf{Y}_\tau$ are exactly a sample from such a distribution (see Sec. 5-4). Since the logarithm is a monotonic function, maximization of $p(\mathbf{Y}_\tau|\mathbf{z})$ from Eq. (5-1-11) with respect to $\mathbf{z}$ yields the same estimate $\hat{\mathbf{z}}$ as

$$\min_{\mathbf{z}} (\mathbf{Y}_\tau - E[\mathbf{Y}_\tau])^T (\text{cov}[\mathbf{Y}_\tau])^{-1} (\mathbf{Y}_\tau - E[\mathbf{Y}_\tau]) \quad (5\text{-}1\text{-}12)$$

The rule (5-1-12) for calculating the estimate $\hat{\mathbf{z}}$ is called *Markov estimator*. For reasons which will become obvious in the following, it is occasionally also called *generalized least-squares estimator*.

Often even the knowledge about cov $[\mathbf{Y}_\tau]$ is defective. Then one may assume that the covariance matrix is diagonal and that the diagonal elements (variances) for each component of $\mathbf{y}$ are the same for all t. This particular covariance matrix is obtained if there is no process noise in system (5-1-4) and $\mathbf{w}(t)$ is a white Gaussian noise with covariance matrix

$$\text{cov}[\mathbf{w}] = \begin{bmatrix} \sigma_1^2 & 0 & \ldots & 0 \\ 0 & \sigma_2^2 & \ldots & 0 \\ \vdots & \vdots & & \vdots \\ 0 & 0 & \ldots & \sigma_p^2 \end{bmatrix} \quad (5\text{-}1\text{-}13)$$

Under this simplifying assumption maximization of $p(\mathbf{Y}_\tau|\mathbf{z})$ becomes equivalent to

$$\min_{\mathbf{z}} \sum_{i=1}^{p} \frac{1}{\sigma_i^2} \sum_{t=0}^{\tau} (y_i(t) - E[y_i(t)])^2 \quad (5\text{-}1\text{-}14)$$

i.e. equivalent to minimizing the sum of the squared deviations of $y_i(t)$ from the undisturbed output, where the squares are weighted by $1/\sigma_i^2$. If nothing is known about the σ_i^2 the most natural choice is to assume that they are all equal. An estimator which minimizes the sum of squared deviations between actual measurements and the output of a model is called *least-squares estimator*. It is the most widely used type of estimator.

Maximum likelihood, Markov, and least-squares estimator all have been derived as simpler and simpler special cases of the Bayes' estimator. As already indicated they are usually applied not because the assumptions made for their derivation are fulfilled, but because the knowledge about the noise characteristics is not sufficient, or a more sophisticated estimate is too complicated to be determined. In these cases the estimator has, in general, poorer statistical properties than a more sophisticated one. However, quite often it is possible to transform the output y(t) through a filter into a signal which has exactly the characteristics required for the derivation of a simpler estimator (Åström and Eykhoff, 1971).

Numerical Techniques

Once an expression either for $p(\mathbf{z}|\mathbf{Y}_\tau)$ or $p(\mathbf{Y}_\tau|\mathbf{z})$ has been derived the problem of determining the maximum over z remains. In general, this is a difficult task for which iterative numerical techniques have usually to be used (see e.g., Eykhoff, 1974). If, however, the function to be maximized is quadratic in the components of z, and z is unconstrained the maximization can easily be done by putting all derivatives of the function with respect to z_i equal to zero, which yields a system of linear algebraic equations. This way of solving the maximization problem is feasible with the least-squares estimator, if $E[y_i(t)]$ in Eq. (5-1-14) is a linear function of $\mathbf{z}(t_\alpha)$.

In any case, there are two strategies for numerically calculating the estimate. They might be called *accumulative* and *recursive estimation*. Accumulative procedures are designed for calculating the estimate from a fixed set of observations, while recursive schemes are suited for estimates which are to be updated as new observations become available. Hence, accumulative procedures are appropriate for off-line use and in particular for estimation of parameters which are supposed not to vary in time. Recursive methods are used for on-line estimation, where usually t_α increases at each recursion in accordance with the sampling interval. Each recursion may be interpreted as an application of Bayes' rule (5-1-6) for combining the information on $\mathbf{z}(t_\alpha)$ which has been extracted from the previous measurements (a priori probability) with the information on $\mathbf{z}(t_\alpha)$ obtained by the new measurement. To start a recursive estimation scheme one can either use an estimation result from an accumulative scheme or a subjective guess with an appropriately large variance. Recursive schemes are used, for example, for real-time state estimation in feedback control problems (see Sec. 9-1) or for estimation of parameters which can be expected to drift. They may also be used as a tool for determining the structure of a model: observation of how a parameter estimate changes under repeated updating may yield hints for improving the model structure, or confirm the structure proposed (see Sec. 5-5).

A critical problem for all estimation techniques is the problem of "*identifiability*." It is not at all obvious a priori whether the value of $\hat{\mathbf{z}}(t_\alpha)$ can be determined uniquely from the available observations. If we try, for instance, to estimate too many parameters, which means that the dimension of z is high, there will be many completely different combinations of parameter values with about the same a posteriori probability, or in other words, $[\operatorname{cov}[\hat{\mathbf{z}}]]^{-1}$ becomes singular (see also Eq. (5-1-10)). This means, only certain combinations of the originally proposed parameters can be estimated uniquely. The result may be physically meaningless parameter values or difficulties in the numerical calculations. Similar things may happen with too short observation sequences or too large noise variances. Very little can be said on a theoretical basis, and even the definition of *"identifiability"* (in probabilistic terms) is problematic. Therefore, numerical experimentation is often the only way to solve the problem of identifiability (see, for example, Mehra and Tyler, 1973, and Sec. 5-3). In the case of linear systems without noise, conditions under which $z(t_\alpha)$ can be determined uniquely (conditions for *"observability"*) can be specified theoretically; they are discussed in Sec. 5-4.

Classification of the Applications

In the following sections four examples of estimation techniques, which have been applied to river quality models, are given in detail; two are of the accumulative and two of the recursive type. Both accumulative techniques use the least-squares criterion. In Sec. 5-2 the parameters of a modified Streeter-Phelps model are estimated by a direct search in the parameter space. In Sec. 5-3 state and parameters are estimated simultaneously for the ecological model described in Sec. 4-4 by solving the estimation problem for a sequence of linear approximations to the original model.

The recursive technique described in Sec. 5-4 calculates a Markov estimate for the state of a linear system at time $t_\alpha = \tau$. The algorithm, which is known as *Kalman filter*, is also obtained if the estimation result for $z(\tau - 1)$ and the new observation $y(\tau)$ are to be combined in such a way that

$$E\left[(z(\tau) - \hat{z}(\tau))^T (z(\tau) - \hat{z}(\tau))\right]$$

is minimal.

The Kalman filter is also used in Sec. 5-4 for estimating solely the parameters of a linear system. As already mentioned, system (5-1-4) need not be linear even if the parameters appeared linearly in the original model. A linear system may be obtained, however, by applying the following trick. The single input–single output linear state model is transformed into a model of type (1-2-17) (in discrete time)

$$y(t) + a_1 y(t-1) + \cdots + a_n y(t-n) = b_1 u(t-1) + b_2 u(t-2) + \cdots + b_n u(t-n) \quad (5\text{-}1\text{-}15)$$

following the procedure indicated in Sec. 1-2. (In general, however, the parameters a_i and b_i are not identical with the original parameters but with certain combinations of them.) If the state vector z is now defined as

$$z = [a_1 \cdots a_n \; b_1 \cdots b_n]^T$$

and the output matrix

$$\mathbf{h}^T(t) = [-y(t-1) \cdots -y(t-n) \; u(t-1) \cdots u(t-n)]$$

Eq. (5-1-15) can be put into the form

$$z(t+1) = z(t) \quad (5\text{-}1\text{-}16\text{a})$$

$$y(t) = \mathbf{h}^T(t)z(t) \quad (5\text{-}1\text{-}16\text{b})$$

System (5-1-16) is a time-varying linear system and if the random disturbances, which are not included in Eq. (5-1-16), have the appropriate characteristics, the Kalman filter may be applied (Åström and Eykhoff, 1971). This technique could not have been applied in a straightforward manner to the problem treated in Sec. 5-2, since the observations were not regularly spaced. A discrete time model like (5-1-15) would be time-varying in this case and therefore Eq. (5-1-16a) could not be used.

In Sec. 5-5 a recursive technique for state estimation in nonlinear models is described and applied to combined state and parameter estimation. It is an extension of the Kalman filtering technique described in Sec. 5-4. Finally, in Sec. 5-6 a few other applications of estimation techniques to water quality models are reviewed.

5-2 NONLINEAR PARAMETER ESTIMATION OF STREETER-PHELPS MODELS

The aim of this section is to describe a very simple least-squares estimator for the parameters of linear models such as those described in Secs. 4-1 and 4-2 (see Rinaldi et al., 1976). The estimation scheme is nonlinear and accumulative and requires that steady state BOD and/or DO values have been sampled in a certain number of points along the river. For the sake of clarity, reference to the particular case of the Bormida river (Italy), to which the estimation scheme has been applied, will be made.

The stretch of the river which has been considered is 68 km long and shown in Fig. 5-2-1; the numbers $(0, 1, \ldots, 6)$ indicate the points in which measurements of temperature, BOD, and DO were taken. Because of the very high BOD load discharged by a factory in point 0 the downstream dissolved oxygen concentration was very low and some control action had to be taken in order to satisfy some required steady state stream standards. The installation of a wastewater treatment plant and/or the use of the upstream reservoir (see Fig. 5-2-1) for low flow augmentation were the two feasible alternatives open to the local authority. In order to make the decision on a rational basis a mathematical model was needed which could allow the computation of the steady state dissolved oxygen concentration at any point of the stretch under different hydrological, thermal, and biological conditions. Since algae had been observed in the final part of the stretch and sedimentation was supposed to take place in all the stretch (see Marchetti and Provini, 1969) a modified Streeter-Phelps model of the Dobbins kind (see Eq. (4-2-1)) was selected for further investigation.

Moreover, dispersion was neglected together with some minor distributed BOD load so that the model was given by

$$\frac{\partial b}{\partial t} + v \frac{\partial b}{\partial l} = -\left[k_1(T) + (k_3(v(Q))/A(Q))\right]b \qquad (5\text{-}2\text{-}1a)$$

$$\frac{\partial c}{\partial t} + v \frac{\partial c}{\partial l} = -k_1(T)b + \frac{k_2(T, Q)}{H(Q)}(c_s(T) - c) + \frac{k_4}{A(Q)} \qquad (5\text{-}2\text{-}1b)$$

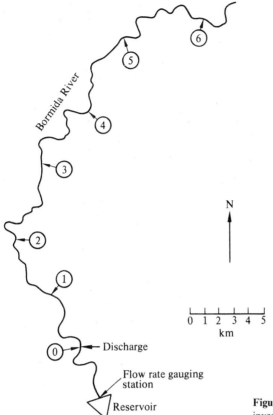

Figure 5-2-1 The reach of the Bormida river investigated, and locations of point (0) and measurement stations (1–6).

where, as usual, T is the water temperature, A the cross-sectional area, v the average stream velocity ($v = Q/A$) and $H(Q)$ is the mean river depth as a function of flow rate Q. Notice that the parameters are defined differently than in model (4-2-1, 4-2-2) in order to make clear the dependence on T and Q. The selection of a particular model among this class is, strictly speaking, a functional problem since the functions $k_1(T)$, $k_2(T, Q)$, $k_3(v)$, and k_4 (deoxygenation, reoxygenation, suspended BOD sedimentation, and photosynthetic oxygen production rate) are unknown. Moreover, the functions $v(Q)$, $A(Q)$, and $H(Q)$ are also unknown in the majority of the applications.

Thus, a more convenient set up must be obtained, and this is possible if a particular isothermical regime, characterized by constant BOD load and constant flow rate Q, is considered. Under these assumptions the daily average BOD and DO concentrations satisfy the following differential equations

5-2 NONLINEAR PARAMETER ESTIMATION OF STREETER-PHELPS MODELS

$$\frac{db}{dl} = -K_1 b \qquad (5\text{-}2\text{-}2a)$$

$$\frac{dc}{dl} = -K_2 b + K_3(c_s - c) + K_4 \qquad (5\text{-}2\text{-}2b)$$

where the functions K_h, $h = 1,\ldots,4$, depend upon the two independent variables Q and T, i.e.,

$$K_1(T, Q) = k_1(T)/v(Q) + k_3(v(Q))/Q \qquad (5\text{-}2\text{-}3a)$$

$$K_2(T, Q) = k_1(T)/v(Q) \qquad (5\text{-}2\text{-}3b)$$

$$K_3(T, Q) = k_2(T, Q)/H(Q)v(Q) \qquad (5\text{-}2\text{-}3c)$$

$$K_4(Q) = k_4/Q \qquad (5\text{-}2\text{-}3d)$$

In order to further simplify the estimation problem the structure of these unknown functions is specified in terms of a finite dimensional parameter vector $\theta = (\theta_1,\ldots,\theta_q)$, so that the functional estimation problem is reduced to a much more simple finite dimensional parameter estimation problem (determination of the optimal value of θ). Thus, from now on, the four functions $K_h(\theta, T, Q)$, $h = 1,\ldots,4$ are assumed to be given.

The solution of Eq. (5-2-2) is well known and is given by (see Sec. 4-2)

$$b(l, K_1, b_0) = b_0 e^{-K_1 l} \qquad (5\text{-}2\text{-}4a)$$

$$c(l, K_1,\ldots, K_4, b_0, c_0) = c_s + K_4/K_3 - [c_s + (K_4/K_3) - c_0] e^{-K_3 l}$$
$$+ [K_2 b_0/(K_1 - K_3)][e^{-K_1 l} - e^{-K_3 l}] \qquad (5\text{-}2\text{-}4b)$$

where b_0 and c_0 are the concentrations at the upstream end of the stretch ($l = 0$).

Consider n different steady state regimes characterized by n pairs (Q^i, T^i), $i = 1,\ldots,n$ of flow rate and water temperature and assume that the daily average concentrations of BOD and DO have been measured for each one of the above regimes at the initial point ($l = 0$) and at r fixed points (stations) $j = 1,\ldots,r$ along the river. Thus, a set of initial conditions

$$(b_0^i, c_0^i) \qquad i = 1,\ldots,n$$

and a set of measurements along the river

$$(b_j^i, c_j^i) \qquad i = 1,\ldots,n \qquad j = 1,\ldots,r$$

are available together with the distance l_j, $j = 1,\ldots,r$, of each station from the origin of the stretch. The square of the deviations for each station and for each regime can therefore be defined as

$$\varepsilon_b^{ji} = [b(l_j, K_1, b_0^i) - b_j^i]^2$$

$$\varepsilon_c^{ji} = [c(l_j, K_1,\ldots, K_4, b_0^i, c_0^i) - c_j^i]^2$$

so that the square error J^i associated with the i-th regime is given by

$$J^i = \sum_{j=1}^{r} [\lambda \varepsilon_b^{ji} + (1-\lambda)\varepsilon_c^{ji}] \qquad 0 \leq \lambda \leq 1 \qquad (5\text{-}2\text{-}5)$$

and the total square error of J is

$$J = \sum_{i=1}^{n} J^i$$

Thus the least-squares estimation problem can be formulated as follows:

Problem A
   Assume that the structure of the functions K_h is known, i.e., let $K_h = K_h(\theta, T, Q), h = 1,\ldots,4$, where θ is a q-dimensional vector of parameters. Then determine the value of θ which minimizes the total error J.

This problem could be solved by applying a suitable searching algorithm in the q-dimensional vector space of the parameters. Unfortunately, reasonable values for q are quite high (about 10–12) which is probably the reason why this problem has never been dealt with in the literature.

A suboptimal solution of Problem A can be obtained by solving the following two problems in series:

Problem B-1
   For each regime $i = 1,\ldots,n$, determine the four-dimensional parameter vector $\mathbf{K}^i = (K_1^i, K_2^i, K_3^i, K_4^i)$ which minimizes the error J^i.

Problem B-2
   Determine the parameter vector θ which solves the following regression problem

$$\min_{\theta} \sum_{i=1}^{n} \sum_{h=1}^{4} (K_h(\theta, T^i, Q^i) - K_h^i)^2$$

Problems B-1 and B-2 are, in general, simpler to solve than Problem A and the dimensionality of Problem B-1 can be reduced by taking advantage of the particular structure of the model. Moreover, if data and estimates of Problem B-1 are highly correlated for all regimes the solution of Problem B-2 may be expected to be similar to that of Problem A. In the following, these correlations are computed for the Bormida river. They turn out to be very high, from which the conclusion may be drawn that Problem B is, in general, a meaningful problem. This result also justifies the fact that Problem B-1 is the only one which has been dealt with until now in the literature.

Solution of Problem B-1

Problem B-1 is a four-dimensional problem since four parameters must be determined. Nevertheless, the dimensionality of the problem can be reduced to two. In fact, from Eqs. (5-2-4) and (5-2-5) it follows that J^i is quadratic in K_2 and K_4, so that the stationarity conditions

$$\frac{\partial J^i}{\partial K_2} = 0, \qquad \frac{\partial J^i}{\partial K_4} = 0 \qquad (5\text{-}2\text{-}6)$$

are linear in the same parameters. It is therefore possible to obtain two functions from Eq. (5-2-6)

$$K_2 = K_2(K_1, K_3), \qquad K_4 = K_4(K_1, K_3)$$

which can be substituted into the expression for J^i. Thus J^i becomes a function only of the two parameters K_1, K_3 and a searching method in the parameter space (K_1, K_3) must therefore be used to solve Problem B-1. Since nonlinear searching algorithms are described in some detail in Sec. 6-2 the reader is referred to that section for this point.

The two limit cases $\lambda = 1$ and $\lambda = 0$, which correspond to weighting only BOD or DO errors in the performance, are of particular interest. The first case ($\lambda = 1$) implies that the estimation of the parameters can in fact be performed in two separate and very simple steps, while the second case ($\lambda = 0$) requires only DO data to be collected, which is a definite advantage considering the effort needed for measuring BOD.

Two-step estimation ($\lambda = 1$) For $\lambda = 1$ Eq. (5-2-5) gives

$$J_b^i = \sum_{j=1}^{r} \varepsilon_b^{ji} = \sum_{j=1}^{r} [b(l_j, K_1^i, b_0^i) - b_j^i]^2 \qquad (5\text{-}2\text{-}7)$$

from which the estimate K_1^i can be obtained by applying a one-dimensional search algorithm (see Sec. 6-2). The three remaining parameters can be obtained by minimizing the DO error

$$\sum_{j=1}^{r} \varepsilon_c^{ji} = \sum_{j=1}^{r} [c(l_j, K_1^i, K_2^i, K_3^i, K_4^i, b_0^i, c_0^i) - c_j^i]^2$$

where K_1^i is the value determined in the first step of the procedure (minimization of (5-2-7)). As mentioned above, the two parameters K_2^i and K_4^i can be eliminated, thus reducing the problem to a simple one-dimensional search with respect to K_3^i. In conclusion, the four parameters $(K_1^i, K_2^i, K_3^i, K_4^i)$ can be estimated by means of two successive one-dimensional search procedures. Because of its simplicity this is the estimation scheme which has been used for the Bormida river case.

Estimation from DO measurements ($\lambda = 0$) For $\lambda = 0$ the objective function (5-2-5) becomes

$$J_c^i = \sum_{j=1}^{r} [c(l_j, K_1^i, K_2^i, K_3^i, K_4^i, b_0^i, c_0^i) - c_j^i]^2$$

from which it is still possible to derive estimates of the four parameters by means of a two-dimensional search.

It must be noticed that one BOD measurement is used in the estimation procedure. If this measurement is not available only three parameters can be identified. In fact, if only DO data are available the analysis must be based on the second order differential equation derived from Eq. (5-2-2) (see Eq. (1-2-17))

$$\frac{d^2c}{dl^2} = -(K_1 + K_3)\frac{dc}{dl} + K_1 K_3 (c_s - c) + K_1 K_4 \qquad (5\text{-}2\text{-}8)$$

in which the parameter K_2 does not appear. From this equation, the solution $\tilde{c}_j^i$ can be obtained in each point l_j, $j = 2, \ldots, r$, as a function of the two measurements c_0^i and c_1^i, i.e.,

$$\tilde{c}_j^i = \tilde{c}(l_j, K_1, K_3, K_4, c_0^i, c_1^i)$$

This expression can be used to define the objective function

$$J_c^i = \sum_{j=2}^{r} [\tilde{c}_j^i - c_j^i]^2$$

which must be minimized in order to obtain the three best estimates (K_1^i, K_3^i, K_4^i) of the parameters appearing in Eq. (5-2-8). Again this minimization can be performed in a two-dimensional vector space since $\tilde{c}_j^i$ is linear in K_4.

The conclusion is that the deoxygenation rate K_2 cannot be identified from DO measurements: at least one BOD measurement (e.g., b_0^i) is necessary to estimate this parameter. The only trivial exception is when sedimentation effects are a priori known to be negligible since in this case $K_1 = K_2$.

Estimation Results

Returning to the problem of the Bormida river, the BOD, DO, and water temperature were measured in six stations and at the wastewater outlet at monthly intervals for four years, while the flow rate Q was continuously recorded at an upstream station. Thus, 48 sets of data were available from which fifteen regimes corresponding to roughly stationary hydrological and thermal conditions were selected in order to apply the algorithms described above (see Table 5-2-1).

The flow rate Q^i was taken as the average value of the flow rate during the three days immediately preceding the moment in which the data of the i-th regime were collected. The temperature T^i was chosen as the average of the measurements obtained in the six stations at the same time as BOD and DO were sampled. Since these measurements were not taken at the same time of the day this average value T^i is not very significant. Finally, since continuous records of BOD and DO

5-2 NONLINEAR PARAMETER ESTIMATION OF STREETER-PHELPS MODELS

Table 5-2-1 The data of the 13 regimes used for parameter estimation and of the two regimes used for the model validation (the first row of each regime refers to BOD (mg/l), the second one to DO (mg/l)).

Station number Distance (km)	0 0.00	1 1.75	2 4.20	3 14.00	4 25.00	5 40.00	6 68.00	Flow rate (10^3 m³/day)	Water Temperature (°C)	
									Average	Range
Regime										
1	196.0	180.0	200.0	118.0	64.0	38.0	10.0	55	17.5	4.2
	2.5	3.0	0.0	4.5	5.5	6.5	9.0			
2	149.0	118.0	120.0	92.0	72.0	58.0	24.0	60	9.0	5.1
	4.0	4.5	3.0	5.5	9.0	9.5	9.5			
3	222.0	220.0	162.0	126.0	110.0	66.0	40.0	125	0.5	5.0
	5.0	0.0	1.0	3.0	5.0	6.5	10.5			
4	116.0	105.0	105.0	84.0	70.0	44.0	38.0	100	19.0	3.0
	3.0	3.5	2.0	5.0	5.5	6.0	7.5			
5	155.0	160.0	125.0	78.0	46.0	18.0	14.0	75	18.0	3.2
	1.5	0.0	1.5	3.5	4.5	5.5	7.0			
6	129.0	150.0	125.0	86.0	70.0	46.0	20.0	80	17.0	3.3
	3.0	3.5	2.0	5.0	6.0	6.0	6.5			
7	86.0	70.0	68.0	56.0	50.0	34.0	24.0	225	5.0	2.5
	8.0	0.0	2.0	6.0	7.0	9.5	12.0			
8	171.0	160.0	145.0	72.0	68.0	30.0	16.0	100	25.0	3.7
	1.5	0.0	0.0	1.2	2.2	3.6	5.8			
9	205.0	200.0	200.0	104.0	98.0	60.0	58.0	55	10.0	8.9
	1.0	0.0	0.0	4.0	6.0	6.0	7.0			
10	101.0	100.0	90.0	70.0	68.0	58.0	22.0	200	1.8	3.5
	7.0	5.0	4.0	4.0	8.0	9.0	9.0			
11	84.0	80.0	80.0	60.0	50.0	36.0	24.0	250	3.5	2.4
	8.0	3.0	6.0	8.0	10.0	10.5	11.0			
12	163.0	150.0	135.0	100.0	85.0	62.0	50.0	125	11.8	2.4
	4.5	0.0	0.5	4.0	5.0	6.0	8.0			
13	74.0	80.0	70.0	60.0	44.0	46.0	22.0	200	16.0	2.5
	5.5	5.5	3.0	6.0	7.0	7.5	8.0			
14	106.0	90.0	85.0	70.0	55.0	40.0	20.0	200	11.5	5.5
	7.5	3.0	3.0	6.0	7.0	9.0	9.5			
15	90.0	75.0	80.0	40.0	30.0	20.0	12.0	150	16.0	6.0
	6.0	5.5	2.5	5.0	7.0	8.5	9.0			

were not available, the samples were assumed to be equal to the average daily values.

Using these very rough assumptions concerning the data, Problem B-1 has been solved for different values of λ for the first thirteen regimes of Table 5-2-1, the main result being that sedimentation and photosynthesis could reasonably be neglected, i.e., $K_1 \cong K_2$ and $K_4 \cong 0$.

Then, the estimation of the two parameters (K_1, K_3) of the Streeter-Phelps model was carried out again for the same thirteen regimes by means of the

154 STATE AND PARAMETER ESTIMATION

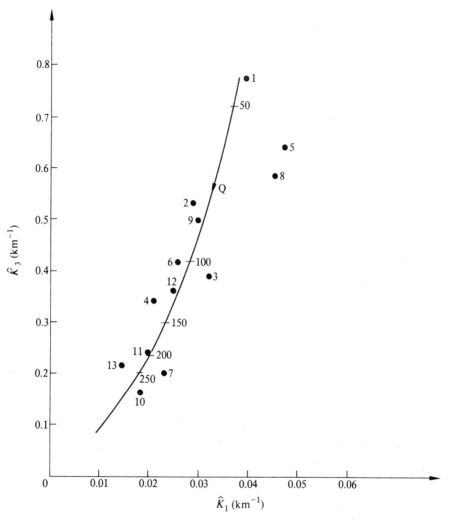

Figure 5-2-2 Estimates of parameters K_1 and K_3 for the 13 considered regimes and curve (5-2-10).

two-step procedure outlined above. The estimates of the parameters are shown in Fig. 5-2-2 where each number refers to the corresponding regime of Table 5-2-1. The correlation coefficient between estimated and measured BOD turned out to be greater than 0.95 for all regimes (average value 0.98), while for the DO concentrations the minimum correlation coefficient was 0.75 (average value 0.90). These high correlation coefficients justify the interest in Problem B-1, since the solution of Problem B-2 can now be expected to be similar to that of Problem A. Unfortunately, as could be predicted from the uncertainties of T^i, the regression problem B-2 did not give satisfactory results as far as the dependence of the parameters (K_1^i, K_3^i) upon T^i is concerned. Nevertheless, disregarding the

5-2 NONLINEAR PARAMETER ESTIMATION OF STREETER-PHELPS MODELS

dependence upon T^i and letting

$$K_1 = \theta_1 Q^{\theta_2} \qquad K_3 = \theta_3 Q^{\theta_4}$$

the solution of problem B-2 gives

$$K_1 = 0.2 Q^{-0.43} \qquad K_3 = 16.4 Q^{-0.8} \qquad (5\text{-}2\text{-}9)$$

with correlation coefficients $r = 0.68$ and $r = 0.89$ respectively. Then, by eliminating Q in the two preceding expressions the following relationship is obtained

$$K_3 = 327.5 K_1^{1.86} \qquad (5\text{-}2\text{-}10)$$

which is represented in Fig. 5-2-2, while the ratio K_3/K_1 is given by

$$K_3/K_1 = 82.0 Q^{-0.37} \qquad (5\text{-}2\text{-}11)$$

This relationship can be proved to be approximately satisfied even if the temperature is taken into account. In fact, consider Eq. (5-2-3) with $k_3 = k_4 = 0$ and, disregarding the dependence of k_2 upon turbulence (see, for example, Metcalf and Eddy, 1972), let (see Eq. (3-5-9))

$$k_1(T) \propto \alpha^{(T-20°)} \qquad k_2(T, Q) \propto \beta^{(T-20°)}$$

where α and β are suitable parameters and $\propto$ denotes proportionality.

Then, from Eq. (5-2-3)

$$K_3/K_1 \propto (\alpha/\beta)^{(T-20°)}/H(Q) \qquad (5\text{-}2\text{-}12)$$

Since (α/β) is approximately unity (see, for instance, Metcalf and Eddy, 1972) the ratio K_3/K_1 should only depend upon Q, as it does in fact in Eq. (5-2-11). Moreover, from Eqs. (5-2-11) and (5-2-12) one can derive

$$Q \propto H^{2.78}$$

which is convex, as a stage-discharge relationship must be (see Sec. 3-3).

The result of the parameter estimation phase is that the Bormida river is described by a Streeter-Phelps model with the parameters K_1 and K_3 dependent on the flow rate Q as indicated by Eq. (5-2-9). These equations have been used to validate the model by means of the last two regimes of Table 5-2-1. The results of the validation turned out to be satisfactory and are shown in Fig. 5-2-3.

The problem of the sensitivity of the estimates with respect to the data of the first measurement point and with respect to the BOD data is now briefly discussed.

As is well known, BOD and DO data collected in the vicinity of a discharge point are often not very useful, since the waste and the river are not yet completely mixed at that point (with the measurement of the Bormida river this was not the case). Therefore, it is of interest to know if the parameters (K_1, K_3) can still be estimated without using the data of the first measurement point. The results obtained from the Bormida using only the data of stations $1,\ldots,6$ are highly consistent with the preceding ones (see Fig. 5-2-4). Therefore, it is possible to state that in doubtful cases it would be better to omit unreliable initial data,

156 STATE AND PARAMETER ESTIMATION

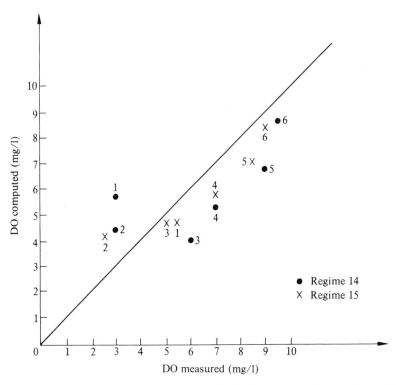

Figure 5-2-3 Measured and computed values of DO for the two regimes used to validate the model.

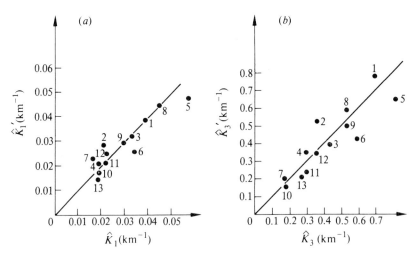

Figure 5-2-4 Effect of the first measurement point on: (*a*) estimate of deoxygenation rate, (*b*) estimate of reoxygenation rate (the primed parameters are estimated without data of the first section).

even if this entails the new initial point being approximately on the minimum of the DO sag curve (as it is in many of our regimes).

Finally, the parameters (K_1, K_3) have also been estimated by means of DO measurements only, as indicated on page 152. The estimates which have been obtained are relatively consistent for the decay rate K_1 as shown in Fig. 5-2-5, while the estimates of K_3 were found to be, in general, very inconsistent with the preceding ones. This fact can easily be explained, since it follows from Eq. (5-2-8) with $K_1 \ll K_3$ (as it is in the case under discussion) that the final part of the DO sag can be approximated by

$$c(l) = c_s - (c_s - c_0) e^{-K_1 l}$$

which depends only upon the deoxygenation rate K_1. Thus, if station 2 is not located around the minimum of the DO curve a good estimate for K_1, and a poor estimate for K_3, may be expected, and this is indeed what happened for the majority of the thirteen regimes. On the other hand, for the two regimes in which station 2 was around the point of the minimum DO (see regimes 10 and 13) highly consistent estimates for K_1 (see Fig. 5-2-5) and quite satisfactory estimates for K_3 were obtained. In conclusion, if only DO measurements are available the two parameters of the Streeter-Phelps model can still be estimated provided that at least three data points are in the critical part of the DO curve.

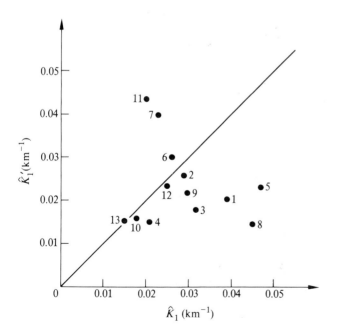

Figure 5-2-5 Estimates of decay rate K_1 with BOD and DO measurements (K_1) and with DO measurements only (K'_1).

5-3 QUASILINEARIZATION TECHNIQUE WITH APPLICATION TO THE RHINE RIVER

Another application of an accumulative estimation technique to a real case is the identification of two water quality models for a major part of the *Rhine river* in Germany (Stehfest, 1973). The Rhine river, which is depicted in Fig. 5-3-1, is

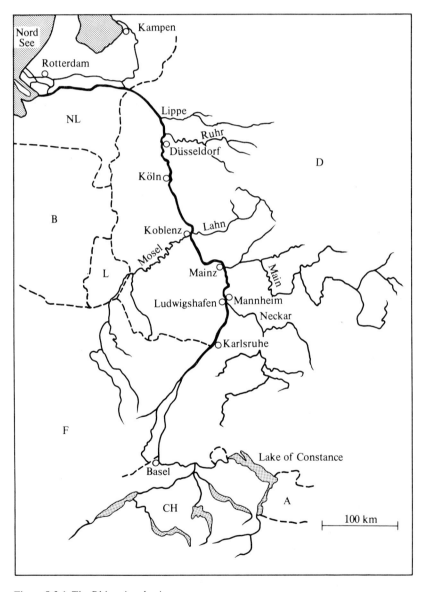

Figure 5-3-1 The Rhine river basin.

one of the most heavily polluted, large rivers in the world. At the same time, its water is extensively used for drinking water production, and also for other purposes such as navigation, recreation, etc. In view of this situation, the need for a river quality model is obvious. In order to be able to estimate the level of model complexity necessary to obtain a realistic river management tool, two models were used: the ecological model described in Sec. 4-4 and a Streeter-Phelps model. The part of the river modeled extends from the cities of Mannheim and Ludwigshafen to the Dutch–German border. Upstream from Mannheim/Ludwigshafen the river pollution is considerably less severe, and downstream from the Dutch–German border the estuary part, which has to be modeled in a different way, begins.

Quasilinearization Technique

The technique used for the estimation part of the system identification (see Sec. 1-3) is the quasilinearization technique for solving nonlinear *multi-point boundary value problems*, described in detail by Bellman and Kalaba (1965) and Lee (1968) (see also Eykhoff, 1974). Looking at the estimation problem as a boundary value problem means that the parameters are treated as additional state variables, as described in Sec. 5-1. Then the initial values of the augmented state are determined such that the sum of the squared differences between the measurements and the output of the model is minimal. Confining ourselves to time-invariant systems and linear observation equations, the problem can be formulated as follows (see Eq. (5-1-14))

$$\min_{\mathbf{x}(0)} J(\mathbf{x}(0)) = \min_{\mathbf{x}(0)} \left[\sum_{i=1}^{p} \frac{1}{2\sigma_i^2} \sum_{k=0}^{K_i} (y_{ik} - \mathbf{h}_i^T \mathbf{x}(t_{ik}))^2 \right] \quad (5\text{-}3\text{-}1a)$$

subject to

$$\dot{\mathbf{x}} = \mathbf{f}(\mathbf{x}) \quad (5\text{-}3\text{-}1b)$$

$$\mathbf{y} = \mathbf{H}\mathbf{x} + \mathbf{w} \quad (5\text{-}3\text{-}1c)$$

where

σ_i^2 = variance of w_i
t_{ik} = time at which the k-th measurement of y_i was taken
y_{ik} = measurement of y_i at time t_{ik}
$\mathbf{h}_i^T$ = i-th row of the output transformation matrix $\mathbf{H}$
$\mathbf{w}$ = measurement noise

The input **u** has been omitted from Eq. (5-3-1b) for notational simplicity. As explained in Sec. 5-1, the least-square criterion used here is optimal in the maximum likelihood sense only if there is no process noise. But it is a reasonable substitute for the maximum likelihood criterion if the knowledge on the noise characteristics is not sufficient.

The method of quasilinearization for solving problem (5-3-1) consists of calculating better and better approximations $\mathbf{x}^j(0)$ to $\mathbf{x}(0)$ by solving iteratively a modified version of problem (5-3-1) obtained by substituting Eq. (5-3-1b) with its linearized form

$$\dot{\mathbf{x}}^j = \mathbf{f}(\mathbf{x}^{j-1}) + \mathbf{F}^{j-1}(t)(\mathbf{x}^j - \mathbf{x}^{j-1}) \tag{5-3-2}$$

where $\mathbf{F}^j(t)$ is the *Jacobian matrix* of $\mathbf{f}$ (see Sec. 1-2) evaluated at $\mathbf{x}^j(t)$, i.e.,

$$\mathbf{F}^j(t) = \left[\frac{\partial \mathbf{f}}{\partial \mathbf{x}}\right]_{\mathbf{x}^j}$$

The initial solution $\mathbf{x}^0(t)$ may be the solution of Eq. (5-3-1b) with an initial guess for $\mathbf{x}(0)$.

The optimal solution of the problem defined by Eqs. (5-3-1a) and (5-3-1c) (with $\mathbf{x}^j$ instead of $\mathbf{x}$) and Eq. (5-3-2) can be determined relatively easily, since $\mathbf{x}^j(t)$ is linear in the initial values $\mathbf{x}^j(0)$. The solution of Eq. (5-3-2) can be written in the form (see Eq. (1-2-4))

$$\mathbf{x}^j(t) = \mathbf{\Phi}^j(t)\mathbf{x}^j(0) + \mathbf{a}^j(t) \tag{5-3-3}$$

where $\mathbf{\Phi}^j$ is the transition matrix of system (5-3-2) and is the solution of

$$\dot{\mathbf{\Phi}}^j = \mathbf{F}^j(t)\mathbf{\Phi}^j \tag{5-3-4}$$

with initial condition

$$\mathbf{\Phi}^j(0) = \mathbf{I}$$

and $\mathbf{a}^j(t)$ is that solution of Eq. (5-3-2) which satisfies $\mathbf{x}^j(0) = \mathbf{0}$. The minimization in Eq. (5-3-1a) can easily be performed by putting all derivatives with respect to $x_r^j(0)$, $r = 1, \ldots, n$, equal to zero and solving the resulting linear algebraic system (see Sec. 5-1). The linear algebraic system is

$$\frac{\partial}{\partial x_r^j(0)} J(\mathbf{x}(0)) = \sum_{i=1}^{p} \frac{1}{\sigma_i^2} \sum_{k=0}^{K_i} \left[\mathbf{h}_i^T(\mathbf{\Phi}^j(t_{ik})\mathbf{x}^j(0) + \mathbf{a}^j(t_{ik})) - y_{ik}\right]\mathbf{h}_i^T \boldsymbol{\phi}_r^j = 0$$

$$r = 1, \ldots, n \tag{5-3-5}$$

where $\boldsymbol{\phi}_r^j$ denotes the r-th column of $\mathbf{\Phi}^j$. It can be solved easily using one of the standard techniques.

The main advantage of the quasilinearization technique is its fast convergence. However, whether or not the series of the $\mathbf{x}^j(0)$ converges normally cannot be ascertained from the outset, but has to be decided through numerical experimentation. If it converges, then it is certain that at least a local minimum of $J(\mathbf{x}(0))$ has been reached. But if the procedure does not converge, then it is not possible to infer that the parameters and the state of the system are not identifiable (see Sec. 5-1). Usually, inspection of $\mathbf{x}^j(0)$ and $J(\mathbf{x}^j(0))$ *for different j* yields some indication of whether there are identifiability problems. If, for instance, $\mathbf{x}^j(0)$ varies greatly over j while the value of $J(\mathbf{x}^j(0))$ remains almost constant, then it is probable that the estimate is not unequivocal. The same is the case if the coefficient matrix of system (5-3-5) becomes singular, since this means that there is a whole variety of solutions of Eq. (5-3-5). (The coefficients matrix of Eq. (5-3-5) is the information matrix of the linearized estimation problem, see Eq. (5-1-10).) In general, it is fair to say that application of any nonlinear estimation technique, and hence also of quasilinearization, is an art, at least to a certain extent. It is always advisable experimenting with various initial guesses and, if possible, with different estimation techniques also in order to get an estimate in which one has confidence. In practice, convergence can always be forced by treating a sufficient number of components of the initial guess $\mathbf{x}^0(0)$ as measurements. This can be quite reasonable since often the numerical value of a parameter is roughly known. One can then pretend to have a measurement of that value and change the matrix $\mathbf{H}$ in Eqs. (5-3-1a) and (5-3-1c) accordingly (see page 167). The confidence in that guess can be expressed by selecting appropriately the value of the corresponding variance.

Several time series of observations which have been obtained from the same system under different circumstances may be used for the estimation; steady state observations along a river for different flow rates and/or temperatures can be used, for instance. In this case the variables which are affected by a change of circumstances, have simply to be split into as many variables as there are observation series.

When determining $\mathbf{x}^j(\cdot)$ from Eq. (5-3-2) the preceding approximation $\mathbf{x}^{j-1}(\cdot)$ has to be available for the entire range of t. This can be achieved by storing $\mathbf{x}^{j-1}(\cdot)$ as a sufficiently dense table function, though with large systems or long observation periods the storage requirement may become prohibitive. Integration at each iteration of all previously used equations of type (5-3-2) may be made instead, thereby trading storage for computing time.

The quasilinearization technique may be modified in the following way: for the evaluation of $\mathbf{f}$ and $\mathbf{F}$ in Eq. (5-3-2), the solution of the original system (5-3-1b) may be used instead of $\mathbf{x}^{j-1}(t)$, which is defined by the previous optimal values $\mathbf{x}^{j-1}(0)$ as initial values. This technique is known as *Gauss-Newton algorithm* (see Mataušek and Milovanović, 1973). In this case the columns of $\mathbf{\Phi}^j$ in Eq. (5-3-4) are the sensitivity vectors of the original system with respect to the initial values, and the nominal solution is defined by $\mathbf{x}(0) = \mathbf{x}^{j-1}(0)$ (see Sec. 4-1). Finally, it should be mentioned that the quasilinearization technique can also be applied to distributed parameter models.

The Models

As mentioned above, the initial states and the parameters of two models, the Streeter-Phelps model and the ecological model described in Sec. 4-4, were to be estimated for a section of the Rhine river. Dispersion was not taken into account because of the relatively high velocity (see Sec. 4-3). Both models were augmented by an additional equation for the concentration w_3 of nondegradable pollutants, which simply describes the accumulation of those pollutants. Since the amount of nondegradable waste discharged into the river is assumed to be known (although its definition is problematic, see Sec. 3-5) these equations are not at all involved in the estimation procedure. In the final results, however, this waste component is always included (see, for example, Fig. 5-3-5), since later on (Sec. 8-4) it must be taken into account. In principle, it would be possible to dispense with the nondegradable waste fraction as a separate variable and include it instead in the slowly degradable fraction. What is said on identifiability of the models in the following would essentially remain true in this case, and the fit to the measured data would deteriorate only slightly. The models were also augmented by an additive constant in the oxygen equation, to account for biogenic aeration (see Fig. 5-3-5).

The measurements on which the estimation had to be based were total COD, bacterial mass density, protozoan mass density, and oxygen concentration (see Sec. 4-4). They are very fragmentary and in several cases they even had to be derived from other measurements (Stehfest, 1973). The total COD measurements, for instance, had to be derived from measurements of the dissolved COD and measurements of particulate organic matter, the COD of the latter having to be estimated.

The wastewater discharged into the river was assumed to contain dead matter only. The total amount and the proportion which is nondegradable was assumed to be known, while the ratio between easily and slowly degradable components in the ecological model, which is not measurable, was subject to the estimation procedure. Figure 5-3-2 depicts the assumptions on the BOD of pollutants

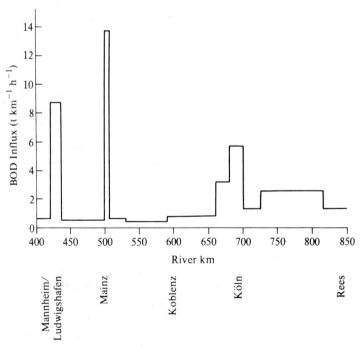

Figure 5-3-2 Estimated discharge of degradable COD into the Rhine river between Mannheim/Ludwigshafen and the Dutch–German border.

entering the river per river-km and hour. The river section between Mannheim/ Ludwigshafen and the Dutch–German border was divided into 12 reaches within which the discharge was assumed to be equally distributed along the river. In Fig. 5-3-2 the major pollution sources along the Rhine river are clearly recognizable: the cities of Mannheim and Ludwigshafen at the mouth of the heavily polluted Neckar river, with large chemical industries and pulp mills; the cities of Mainz and Wiesbaden at the mouth of the Main river which is extremely polluted by chemical industries and pulp mills; the area of Bonn, Cologne, and Leverkusen also having many chemical industries; and finally the heavily industrialized Ruhr district. The pollutants brought in by the affluents are included in Fig. 5-3-2. The composition of the wastewater, i.e., the proportions between easily, slowly, and nondegradable components, are assumed to be constant everywhere, which is quite a rough approximation. The total amounts of COD production in Fig. 5-3-2 are also relatively rough estimates; they refer to 1970. On the assumption that all discharges in Fig. 5-3-2 are due to the domestic, industrial, and agricultural activity of 35 million inhabitants (which is realistic), a production of approximately 600 g COD/(capita · day) results. This seems to be fairly high, in particular, in view of the fact that the values of Fig. 5-3-2 are lower than the actual production, since part of the COD produced has already been removed by treatment plants and by self-purification in the affluents. In Löffler-Ertel and Reichert (1975) considerably lower estimates are given, but the authors emphasize the great uncertainties in the data. In Roberts and Kreijci (1975) a daily TOC production (see Sec. 3-5), for the Glatt valley in Switzerland, of 200 g per capita was estimated for the year 2000, which corresponds roughly to 600 g COD. The main industries in this region are metal working and machine building industries, which produce less organic pollution than the chemical and pulp industries, which determine the pollutional situation of the Rhine river.

The quantities characterizing the hydrology, namely, the lateral inflow q and velocity v, were also assumed to be given and constant within the reaches. Figure 5-3-3 shows some of the hydrologic characteristics of the Rhine river.

The two models, written in flow time, are as follows:

Streeter-Phelps model

$$\dot{b} = -k_{11}b + \frac{v}{Q}(L - qb) \tag{5-3-6a}$$

$$\dot{w}_3 = \frac{v}{Q}(\alpha L - qw_3) \tag{5-3-6b}$$

$$\dot{c} = (k_{21} + k_{22}v)(c_s - c) - k_{11}b + \frac{v}{Q}(L_c - qc) + \beta \tag{5-3-6c}$$

$$\dot{Q} = qv \tag{5-3-6d}$$

$$b(0) = k_{01}, w_3(0) = w_{30}, c(0) = k_{03}, Q(0) = Q_0$$

164 STATE AND PARAMETER ESTIMATION

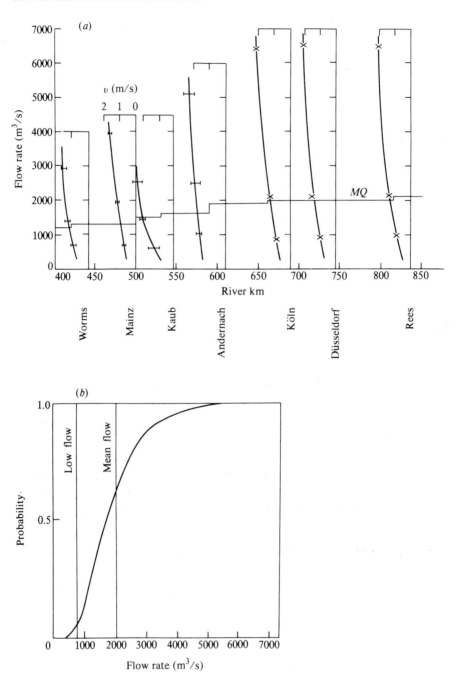

Figure 5-3-3 Hydrologic characteristics of the Rhine river:
(a) mean flow rate MQ (approximated by a step function) and relationship between v and Q in selected points
(b) distribution function of flow rate at Köln.

Ecological model

$$\dot{w}_1 = -k_{11}g_1 B + \frac{v}{Q}(k_{12}L - qw_1) \tag{5-3-7a}$$

$$\dot{w}_2 = -k_{21}g_2 B + \frac{v}{Q}((1 - k_{12})L - qw_2) \tag{5-3-7b}$$

$$\dot{w}_3 = \frac{v}{Q}(\alpha L - qw_3) \tag{5-3-7c}$$

$$\dot{B} = (g_1 + g_2 - k'_{46})B - k'_{47}g_3 P + \frac{v}{Q}(L_B - qB) \tag{5-3-7d}$$

$$\dot{P} = (g_3 - k_{53})P + \frac{v}{Q}(L_P - qP) \tag{5-3-7e}$$

$$\dot{c} = (k'_{61} + k_{62}v)(c_s - c) - (k_{63}g_1 + k_{64}g_2 + k'_{65}k'_{46})B$$
$$\quad - (k'_{66}g_3 + k'_{67}k_{53})P + \beta + \frac{v}{Q}(L_c - qc) \tag{5-3-7f}$$

$$\dot{Q} = qv \tag{5-3-7g}$$

where

$$g_1 = \frac{k'_{41}w_1}{k_{42} + w_1} \quad g_2 = \frac{k_{43}w_1}{k'_{44} + w_2 + k'_{45}w_1} \quad g_3 = \frac{k'_{51}B}{k'_{52} + B}$$

$$w_1(0) = k_{01}, w_2(0) = k_{02}, w_3(0) = w_{30}, B(0) = k_{04}, P(0) = k_{05},$$
$$c(0) = k_{06}, Q(0) = Q_0$$

All k_{ij}'s are quantities which are subject to optimal estimation. The primes on some of the k_{ij}'s will be explained later. The quantities L, L_B, L_P, and L_c denote the distributed loads (in flow time) of pollutants, bacterial mass, protozoan mass, and dissolved oxygen, respectively, which are added to the river as components of all kinds of affluxes. All these functions are assumed to be reachwise constant. The quantity α gives the ratio between the COD of the nondegradable pollutants and L, so that $(1 + \alpha)L$ is the total COD distributed load of the pollutants. The constant β in Eq. (5-3-7f) accounts for biogenic aeration and its numerical value was derived from measurements during summertime.

The composition of the pollution input is defined by the values of the parameters α and k_{12}. For all reaches which do not receive major tributaries these values were assumed to be the same, i.e., the assumption is made that the wastewater has the same composition everywhere and is directly discharged into the Rhine river. For reaches 2, 4, and 7, where major tributaries flow in, the values of α and k_{12} were modified empirically for the effects of self-purification of the tributaries (k_{12} becomes smaller, while α increases). Similarly, the values of L_c for those three reaches were chosen according to the quality of the tributaries.

In all other cases, the admixture terms (i.e., the last terms in Eqs. (5-3-6c), (5-3-7d)–(5-3-7f)) were assumed to vanish, which means that any lateral inflow has the same concentration of bacterial mass, protozoan mass, and oxygen as the receiving river.

Estimation Results

Figure 5-3-4 shows the result of the estimation for the Streeter-Phelps model. The quasilinearization technique converged to a unique set of initial values k_{ij} over a wide range of initial guesses. Flow rate Q and temperature T were chosen to be the same as for the ecological model. This was done in order to be able to consistently compare the two models (see Sec. 8-4). For the same reason the BOD measurements used were the sum of the $(w_1 + w_2)$ values used for the ecological model and the chemical oxygen demand of bacterial and protozoan mass; the oxygen observations used are exactly the same as for the ecological model. The parameter estimates turned out to be

$$\hat{k}_{11} = 0.045 \text{ h}^{-1}, \quad \hat{k}_{21} = 0.28 \text{ h}^{-1}, \quad \hat{k}_{22} \simeq 0$$

The negligible value of k_{22} means that the physical reaeration rate does not depend significantly upon the velocity variations along the river. The value of k_{21} seems to be somewhat high, which is consistent with the statement made above that the estimates for the input of pollutants are a little high. But it is still within the range marked by numerous publications about reaeration rates of comparable rivers (see, for example, Negulescu and Rojanski, 1969; Bansal, 1973). There is also very dense shipping traffic on the Rhine river, which considerably enhances physical reaeration. The fit of the measured data is not very good, and a recursive estimation (see Sec. 5-5) would clearly have shown variations of the parameter values along the river.

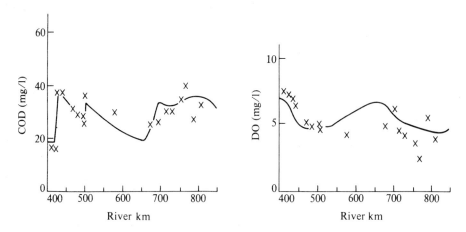

Figure 5-3-4 Description of the self-purification of the Rhine river by the Streeter-Phelps model.

An unambiguous estimation for the ecological model was possible only if the initial guesses for numerous parameters were treated as observations, because the number and accuracy of the observations was not sufficient. The optimal model output together with the measurements is shown in Fig. 5-3-5. Under the conditions selected for that figure ($Q = 1.25MQ$, $T = 20\,°C$), a relatively large number of accurate measurements were taken along a characteristic line in the upper part of the river section. For other Q and T values even fewer measurements were available, and therefore the use of several measurement series (see p. 161) would not have removed the identifiability problems. The ecological model obviously fits the observations considerably better than the Streeter-Phelps model. The validation tests described below also indicate that the ecological model is more adequate than the Streeter-Phelps model.

The dashed oxygen sag curve in Fig. 5-3-5 shows that the influence of biogenic aeration on the oxygen balance is relatively small in this part of the Rhine river. The reasons for this have already been discussed in Sec. 4-4. One of them is the great depth of the river: because of this the phytoplankton receives on the average very little sunlight; this effect is particularly important because of the high turbidity of the Rhine river. Other reasons for the relative unimportance of biogenic aeration are high velocity and high turbulence of the river. High velocity simply means that the time for the development of a dense phytoplankton population is short. Finally, high turbulence is unfavorable for algal reproduction.

If the measurements from a system are not sufficient to obtain state and parameter estimates under satisfactory conditions, as in the case described, the estimation technique may be used as a tool for planning additional measurements (see Bellman et al., 1966). Numerical experimentation may show which variables have to be measured, where, and with what accuracy in order to obtain an unambiguous state and parameter estimate. The quasilinearization technique was applied to the ecological model for this purpose: under the assumption of reasonably dense and accurate measurements of $w_1 + w_2$, B, P, and c the estimation of the initial values of all variables and of the parameters for model (5-3-7) was attempted. Parameters k_{62} and β, all input and admixture terms, and the equation for nondegradable pollutants were left out. The measurements were generated on a computer using a river quality model much more complex than the one given by Eq. (5-3-7). This model, which is described in Stehfest (1973), contains 30 different pollutants, all of them having different degradation kinetics (mutual inhibitions according to Eqs. (3-5-21) and (3-5-22), purely additive degradation, formation of exoenzymes, see Sec. 3-5), as well as two protozoa types with different metabolic dynamics. The kinetic parameters were generated within realistic ranges by a random number generator. It turned out that the estimation was not possible in a unique way if all parameters were left completely free. But if, for instance, a priori estimates for the primed parameters in Eq. (5-3-7) were considered as measurements, the estimation technique converged, even with very high variances of the parameter guesses and very noisy measurements. In selecting the parameters, for which approximate values are to be prescribed, it is necessary to question whether reasonable guesses for the parameters concerned

168 STATE AND PARAMETER ESTIMATION

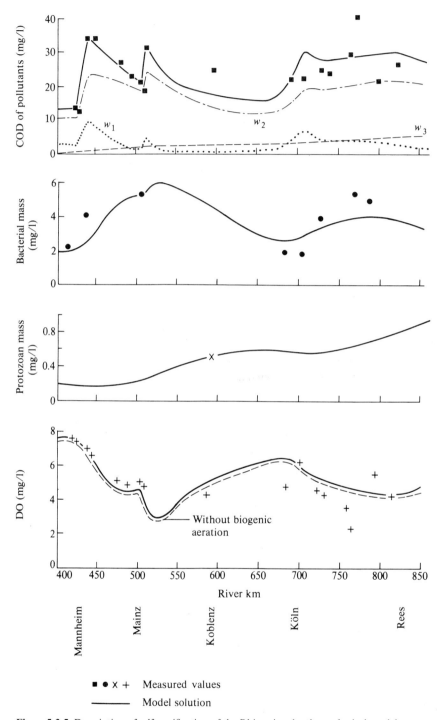

Figure 5-3-5 Description of self-purification of the Rhine river by the ecological model.

are available. It is better, for example, to use an approximate value for k_{41}, rather than for k_{42}, because k_{41} would not be expected to be very much smaller than the largest known growth rate of bacteria at the given temperature, while for a complex nutrient mixture, very little can be predicted about k_{42}. Kinetic parameters for the interaction between bacteria and protozoa which have been found in laboratory experiments can also be used as measured values for the estimation; but it is desirable to have the parameter k_{53} totally free, since the term $k_{53}P$ is to account approximately for the unknown influence of higher order links of the food chain.

Figure 5-3-6 shows the optimal model solution for noise-free measurements (*a*), and measurements distributed by a reasonable noise (*b*). In Fig. 5-3-6a the initial solution $\mathbf{x}^0(t)$ is given, which was also used for Fig. 5-3-6b. It can be seen that the estimation is possible, even if the initial estimate is quite wrong. It should be emphasized again that the ratio between w_1 and w_2 is determined through the estimation procedure on the basis of measurements of $w_1 + w_2$, B, P, and c only.

Validation Tests

As explained in Sec. 1-3 model validation is an indispensable step for building any model: good fit to an observation series alone is not yet very meaningful, in particular if there are many parameters to be adjusted. It is always necessary to check if the optimal model is able to reproduce measurements which were taken independently of the ones used for estimation.

Figure 5-3-7 gives an example: in Fig. 5-3-7a that solution of model (5-3-7) is given which best fits the measurements generated by the complex river model. In Fig. 5-3-7b both models with the same parameter values but with completely different initial values for the variables are checked against each other. Since model (5-3-7) describes well the "observations" also under changed initial conditions, it may be concluded that the simple model (5-3-7) is a good approximation of the complex one.

Some validation tests could also be carried out for the ecological model as applied to the Rhine river, although the measurements are so sparse (see also Stehfest, 1978). Figure 5-3-8 shows, for instance, the changes of the model behavior if temperature is lowered from 20 °C to 10 °C. Two most remarkable changes have been confirmed by real measurements: the COD increases more in the upstream part than in the lower Rhine; and bacterial mass density is almost constant from river km 550 to 700 at $T = 10\,°C$, while at $T = 20\,°C$ there was a decline along this river section. The latter effect has already been mentioned and explained in Sec. 2-3.

Figure 5-3-9 shows the changes in the model behavior if the flow rate decreases from $1.25MQ$ to $0.77MQ$. The consequences of this decrease are governed by two effects: the dilution ratio for the discharged pollutants is changed, and the flow times between the pollution sources are changed. Both effects result in an increase of the pollutants removal over a fixed river section. This can be seen from the figure and has also been observed in practice. The

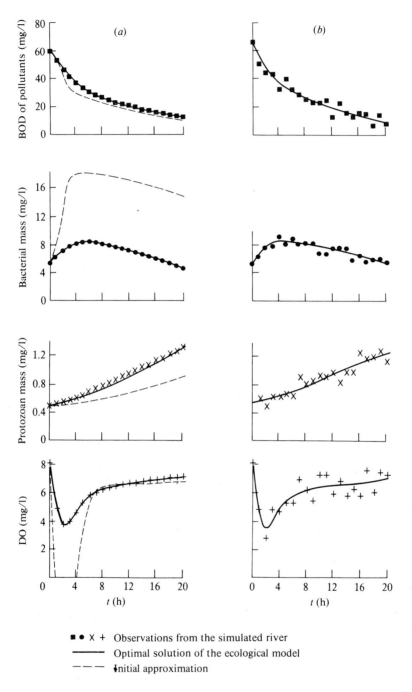

■ ● × + Observations from the simulated river
——— Optimal solution of the ecological model
— — — Initial approximation

Figure 5-3-6 State and parameter estimation for the ecological model, using synthetic data: (a) noise-free observations (b) noise-corrupted observations.

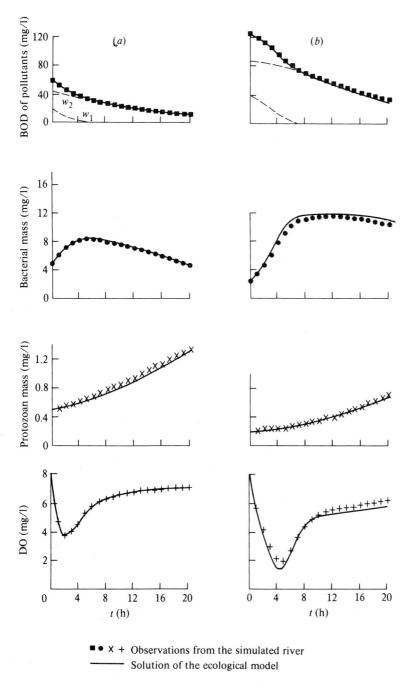

Figure 5-3-7 Model validation for the ecological model using synthetic data:
(a) estimation result
(b) comparison between observations and model solution under changed initial values.

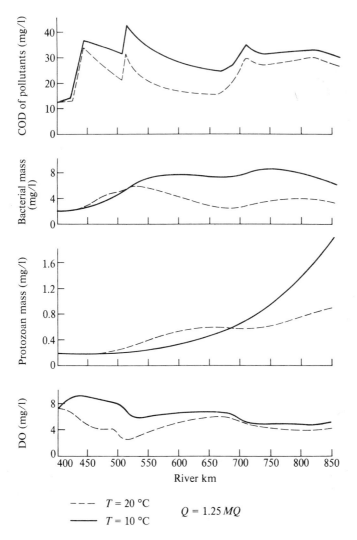

Figure 5-3-8 Changes in the self-purification behavior of the Rhine river when temperature is lowered.

serious deterioration of the oxygen conditions shown by Fig. 5-3-9 also corresponds to real observations.

Finally, Fig. 5-3-10 gives the model behavior in the case that 50 percent of the easily degradable component of the wastewater is removed before discharge. Although this means that as much as about 25 percent of the total COD had been removed, the COD concentration did not decrease considerably anywhere, and in certain parts even increased. The reason for this is the decrease of the growth rate of the bacteria relative to the protozoan consumption rate and to endogenous respiration. This result might be considered as a kind of validation, since a reduction of the easily degradable components is achieved by biological treatment plants. The fact that the quality of the Rhine river has not improved during the last years in spite of remarkable efforts at building biological treatment plants could possibly be attributed to this effect.

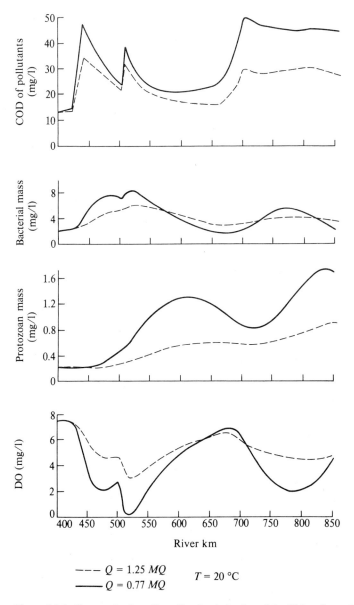

Figure 5-3-9 Changes in the self-purification behavior of the Rhine river when flow rate is lowered.

174 STATE AND PARAMETER ESTIMATION

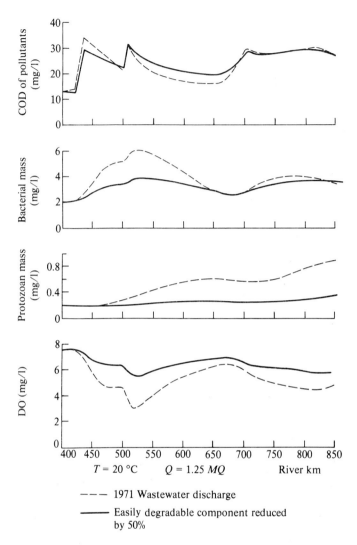

Figure 5-3-10 Changes in the self-purification behavior of the Rhine river due to removal of part of the easily degradable component from sewage.

5-4 KALMAN FILTERING FOR PARAMETER AND STATE ESTIMATION

In this section, examples of parameter and state estimation by means of linear recursive schemes are shown. These schemes do not require the knowledge of the whole string of past input–output data at each step, since the new estimate is found by updating the old one only on the basis of the new input–output observation.

The method applied in this section is the *Kalman filtering* technique, which is

5-4 KALMAN FILTERING FOR PARAMETER AND STATE ESTIMATION

a linear recursive scheme well suited for real time *state estimation*. For this reason, the main results of state estimation theory are now described both in the deterministic case (Luenberger state reconstructor) and in the stochastic case (Kalman filter).

State Reconstructor and Kalman Filter

Consider a discrete, time-invariant, linear system of the form

$$\mathbf{x}(t+1) = \mathbf{F}\mathbf{x}(t) + \mathbf{G}\mathbf{u}(t) \tag{5-4-1a}$$

$$\mathbf{y}(t) = \mathbf{H}\mathbf{x}(t) \tag{5-4-1b}$$

where t is an integer, $\mathbf{u}$, $\mathbf{x}$, and $\mathbf{y}$ are the input, state, and output vectors of dimension m, n, and p, respectively, and $\mathbf{F}$, $\mathbf{G}$, and $\mathbf{H}$ are constant matrices of suitable order.

Our problem is to look at the possibility of determining $\mathbf{x}(t)$ from the input–output measurements on the interval $[0, t]$, i.e., from the enlarged vectors

$$\mathbf{u}_{t-1} = [\mathbf{u}(0)^T \quad \mathbf{u}(1)^T \cdots \mathbf{u}(t-1)^T]^T$$

$$\mathbf{y}_t = [\mathbf{y}(0)^T \quad \mathbf{y}(1)^T \cdots \mathbf{y}(t)^T]^T$$

In order to solve this problem, consider first a slight modification of it, namely the possibility of computing the initial state $\mathbf{x}(0) = \mathbf{x}_0$ from $\mathbf{u}_{t-1}$ and $\mathbf{y}_t$. By using Eq. (5-4-1) recursively t times we obtain

$$\mathbf{y}(0) = \mathbf{H}\mathbf{x}_0$$

$$\mathbf{y}(1) = \mathbf{H}\mathbf{F}\mathbf{x}_0 + \mathbf{H}\mathbf{G}\mathbf{u}(0)$$

$$\mathbf{y}(2) = \mathbf{H}\mathbf{F}^2\mathbf{x}_0 + \mathbf{H}\mathbf{F}\mathbf{G}\mathbf{u}(0) + \mathbf{H}\mathbf{G}\mathbf{u}(1)$$

$$\vdots$$

$$\mathbf{y}(t) = \mathbf{H}\mathbf{F}^t\mathbf{x}_0 + \sum_{i=0}^{t-1} \mathbf{H}\mathbf{F}^{t-1-i}\mathbf{G}\mathbf{u}(i)$$

which can be written in the compact form

$$\mathbf{y}_t = \mathbf{O}^T(t)\mathbf{x}_0 + \mathbf{f}(\mathbf{u}_{t-1}) \tag{5-4-2}$$

where the matrix $\mathbf{O}(t)$ is given by

$$\mathbf{O}(t) = [\mathbf{H}^T \quad \mathbf{F}^T\mathbf{H}^T \quad (\mathbf{F}^T)^2\mathbf{H}^T \cdots (\mathbf{F}^T)^t\mathbf{H}^T] \tag{5-4-3}$$

and

$$\mathbf{f}(\mathbf{u}_{t-1}) = \begin{bmatrix} 0 \\ \mathbf{H}\mathbf{G}\mathbf{u}(0) \\ \vdots \\ \sum_{i=0}^{t-1} \mathbf{H}\mathbf{F}^{t-1-i}\mathbf{G}\mathbf{u}(i) \end{bmatrix}$$

The initial state x_0 can be uniquely determined if and only if the rank of the matrix $O(t)$ is equal to n (the dimension of the system), since from Eq. (5-4-2) it follows

$$[O(t)O^T(t)]x_0 = O(t)(y_t - f(u_{t-1})) \tag{5-4-4}$$

and the matrix $[O(t)O^T(t)]$ is invertible if the rank of $O(t)$ is equal to n. A well known result of linear algebra says that the rank of the matrix $O(t)$ given by Eq. (5-4-3) is constant for $t \geq n - 1$; therefore, the first important conclusion is as follows: if x_0 cannot be determined for $t = n - 1$, then it cannot be determined for $t \geq n$. In other words, if the initial state is to be determined from input–output observations, there is no reason to store more than the first n observations. Moreover, if we consider the matrix $O(t)$ for $t = n - 1$, called *observability matrix* and denoted by O,

$$O = [H^T \quad F^T H^T \cdots (F^T)^{n-1} H^T] \tag{5-4-5}$$

and we assume that it is full rank, from Eq. (5-4-4) we obtain

$$x_0 = (OO^T)^{-1} O^T (y_{n-1} - f(u_{n-2})) \tag{5-4-6}$$

which is the general expression for the computation of the initial state (notice that the same expression can be used for any instant of time after shifting the indexes). In conclusion, we have proved that the initial state can be determined from input–output measurements if and only if the observability matrix (5-4-5) is full rank and this is why a system satisfying this condition is said to be *completely observable*.

Once the initial state x_0 has been determined the state at any time t can be computed by means of Eq. (5-4-1a). Nevertheless, this scheme for the determination of $x(t)$ is not very satisfactory since it implies the need to store a relatively large number of data. A recursive scheme would be much more convenient even at the price of some approximation in the evaluation of $x(t)$. The most widely known recursive scheme available today (*Luenberger state reconstructor*) is represented in block diagram form in Fig. 5-4-1. The upper part of the diagram is simply a symbolic representation of the system (see Eq. (5-4-1)), while the lower part is the state reconstructor which is an exact copy of the system, in general simulated on a digital computer. The input $u(t)$ of the system is applied also to the reconstructor and its state vector $\hat{x}(t)$ is an estimate of the state of the system. If $x(t)$ is equal to its estimate $\hat{x}(t)$, the output $y(t)$ is also equal to its estimate $\hat{y}(t)$, while if $\hat{x}(t) \neq x(t)$ in general $y(t) \neq \hat{y}(t)$ and the difference $(\hat{y}(t) - y(t))$ is used as an input of the reconstructor in order to improve the next estimate $\hat{x}(t+1)$.

The equations describing the state reconstructor are

$$\hat{x}(t+1) = F\hat{x}(t) + Gu(t) - L(\hat{y}(t) - y(t))$$

$$\hat{y}(t) = H\hat{x}(t)$$

and they clearly show the recursive nature of the scheme since the new estimate $\hat{x}(t+1)$ is obtained from the last one $(\hat{x}(t))$ and from the last pair of input–output

5-4 KALMAN FILTERING FOR PARAMETER AND STATE ESTIMATION

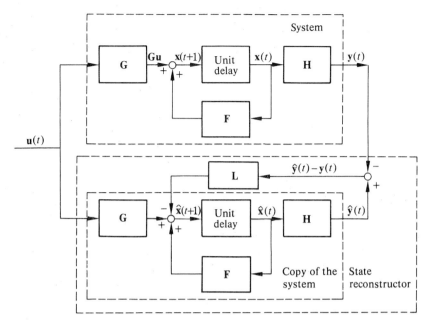

Figure 5-4-1 The structure of the Luenberger state reconstructor.

observations $(\mathbf{u}(t), \mathbf{y}(t))$. The matrix $\mathbf{L}$ defining the reconstructor is an $n \times p$ constant matrix which has to be determined such as to obtain a satisfactory behavior of the estimation procedure. If the *estimation error* is defined as follows

$$\varepsilon(t) = \hat{\mathbf{x}}(t) - \mathbf{x}(t)$$

we obtain

$$\varepsilon(t+1) = \hat{\mathbf{x}}(t+1) - \mathbf{x}(t+1)$$
$$= \mathbf{F}\hat{\mathbf{x}}(t) + \mathbf{G}\mathbf{u}(t) - \mathbf{L}(\mathbf{H}\hat{\mathbf{x}}(t) - \mathbf{H}\mathbf{x}(t)) - \mathbf{F}\mathbf{x}(t) - \mathbf{G}\mathbf{u}(t)$$
$$= (\mathbf{F} - \mathbf{L}\mathbf{H})\varepsilon(t)$$

which says that the error $\varepsilon(t)$ does not depend upon the input $\mathbf{u}(t)$. If the initial estimate $\hat{\mathbf{x}}(0)$ is different from $\mathbf{x}(0)$, as it usually is, then the estimation error goes to zero for t going to infinity if the matrix $(\mathbf{F} - \mathbf{L}\mathbf{H})$ is asymptotically stable, i.e., if its eigenvalues are in module strictly less than one. Thus, a reasonable criterion in the design of a state reconstructor is to select $\mathbf{L}$ in such a way that the eigenvalues of the matrix $(\mathbf{F} - \mathbf{L}\mathbf{H})$ are sufficiently small in module. One of the main results in state reconstruction theory (see Luenberger, 1971) is that the eigenvalues of $(\mathbf{F} - \mathbf{L}\mathbf{H})$ can be fixed at will by suitably selecting $\mathbf{L}$ if and only if the system is completely observable. Thus, if the system is completely observable the Luenberger state reconstructor can be considered to be very attractive since its rate of convergence can be fixed at will. The procedure for determining the matrix $\mathbf{L}$ is trivial for single-output systems ($p = 1$), while it is relatively complex for multi-output systems ($p > 1$). Since applications of the state reconstructor are not given in this book, the algorithm for the determination of **L** is not described and the interested reader is referred to the previously mentioned paper by Luenberger.

A more realistic description of the problem of state estimation takes into account the presence of uncertainties at least in the form of additive noise. Thus, the system is described by the following equations

$$\mathbf{x}(t+1) = \mathbf{F}\mathbf{x}(t) + \mathbf{G}\mathbf{u}(t) + \mathbf{v}(t)$$

$$\mathbf{y}(t) = \mathbf{H}\mathbf{x}(t) + \mathbf{w}(t)$$

where $\mathbf{v}(t)$ and $\mathbf{w}(t)$ are *process* and *measurement noise*. If the noise is relatively small, the Luenberger state reconstructor can still be used, but for considerable noise an alternative approach for the estimation of $\mathbf{x}(t)$ must be followed. Among the different criteria proposed in the past the technique known as *Kalman filtering* is certainly the most popular one. The idea is to find the estimate $\hat{\mathbf{x}}(t)$ which minimizes (see also Sec. 5-1)

$$E\left[\left(\hat{\mathbf{x}}(t) - \mathbf{x}(t)\right)^T \left(\hat{\mathbf{x}}(t) - \mathbf{x}(t)\right)\right] = E\|\varepsilon(t)\|^2$$

where E stands for *expectation* and $\|\cdot\|$ for norm. It can be shown (see, for example, Kalman, 1960; Luenberger, 1969; and Kwakernaak and Sivan, 1972) that if the system is completely observable and the noise vectors $\mathbf{v}(t)$ and $\mathbf{w}(t)$ are suitable stochastic processes the solution to this problem exists and can be given a very simple recursive form. The stochastic processes $\mathbf{v}(t)$ and $\mathbf{w}(t)$ are assumed to be stationary, Gaussian (normal), and uncorrelated in time (white), so that they are completely described by

$$E[\mathbf{v}(t)] = \mathbf{0} \qquad E[\mathbf{v}(t)\mathbf{v}(\tau)^T] = \mathbf{V}_v \delta_\tau^t$$
$$E[\mathbf{w}(t)] = \mathbf{0} \qquad E[\mathbf{w}(t)\mathbf{w}(\tau)^T] = \mathbf{V}_w \delta_\tau^t$$

where δ_τ^t is the Kronecker delta ($\delta_\tau^t = 0$ for $\tau \neq t$, and $\delta_\tau^t = 1$ for $\tau = t$), and the *covariance matrix* $\mathbf{V}_v$ of the process noise is *positive semidefinite* ($\mathbf{v}^T \mathbf{V}_v \mathbf{v} \geq 0$ for all $\mathbf{v}$), while the covariance matrix $\mathbf{V}_w$ of the measurement noise is *positive definite* ($\mathbf{w}^T \mathbf{V}_w \mathbf{w} > 0$ for all $\mathbf{w} \neq \mathbf{0}$), which corresponds to saying that there are no linear combinations of the components of the output vector $\mathbf{y}$ which are insensitive to noise. Moreover, the initial state vector $\mathbf{x}(0)$ is assumed to be normally distributed with mean $\bar{\mathbf{x}}_0$ and covariance matrix $\mathbf{V}_\varepsilon$

$$E[\mathbf{x}(0)] = \bar{\mathbf{x}}_0 \qquad E\left[\left(\mathbf{x}(0) - \bar{\mathbf{x}}_0\right)\left(\mathbf{x}(0) - \bar{\mathbf{x}}_0\right)^T\right] = \mathbf{V}_\varepsilon$$

and $\mathbf{v}(t)$, $\mathbf{w}(t)$, and $\mathbf{x}(0)$ are assumed to be independent, i.e.,

$$E[\mathbf{v}(t)\mathbf{w}(\tau)^T] = \mathbf{0} \qquad E[\mathbf{v}(t)\mathbf{x}(0)^T] = \mathbf{0} \qquad E[\mathbf{w}(t)\mathbf{x}(0)^T] = \mathbf{0}$$

Under these conditions the optimal estimate $\hat{\mathbf{x}}(t)$ can be obtained in a recursive way by means of the following linear scheme (Kalman filter)

$$\hat{\mathbf{x}}(t+1) = \mathbf{F}\hat{\mathbf{x}}(t) + \mathbf{G}\mathbf{u}(t) - \mathbf{L}(t+1)\{\mathbf{H}[\mathbf{F}\hat{\mathbf{x}}(t) + \mathbf{G}\mathbf{u}(t)] - \mathbf{y}(t+1)\} \quad (5\text{-}4\text{-}7)$$

with initial estimate
$$\hat{x}(0) = \bar{x}_0$$
This scheme is very similar to the one shown in Fig. 5-4-1, the main difference being that the matrix **L** of the state reconstructor has been substituted by a time-varying matrix **L**(t), often called the *gain matrix of the filter*.

The gain matrix **L**(t) can be computed by using the following recursive algorithm

$$V_\varepsilon^a(t+1) = FV_\varepsilon(t)F^T + V_v \qquad V_\varepsilon(0) = V_\varepsilon$$
$$L(t+1) = V_\varepsilon^a(t+1)H^T[HV_\varepsilon^a(t+1)H^T + V_w]^{-1} \qquad (5\text{-}4\text{-}8)$$
$$V_\varepsilon(t+1) = V_\varepsilon^a(t+1) - L(t+1)HV_\varepsilon^a(t+1)$$

which is the discrete form of the well known *Riccati equation*. The matrix $V_\varepsilon(t)$ appearing in Eq. (5-4-8) is the covariance matrix of the estimation error

$$V_\varepsilon(t) = E[\varepsilon(t)\varepsilon(t)^T]$$

and represents, for this reason, a measure of the reliability of the estimate, and the matrix $V_\varepsilon^a(t)$ is the a priori covariance matrix of the estimation error. In Eq. (5-4-7) the term $[F\hat{x}(t) + Gu(t)]$ represents the best estimate of the state at time $(t + 1)$ that one can make at time t on the basis of the estimate $\hat{x}(t)$. For this reason, it is called state prediction and denoted by $\hat{x}(t+1|t)$. Similarly, the term $H[F\hat{x}(t) + Gu(t)]$ is the output prediction (forecast) and can be written as $\hat{y}(t+1|t)$, so that Eq. (5-4-7) can be given the form

$$\hat{x}(t+1) = \hat{x}(t+1|t) - L(t+1)[\hat{y}(t+1|t) - y(t+1)] \qquad (5\text{-}4\text{-}9)$$

which shows the intimate structure of the Kalman filter. As t increases the gain matrix **L**(t) tends to a constant matrix (the solution of the algebraic Riccati equation) and the resulting filter is time-invariant and asymptotically stable as the Luenberger state reconstructor.

All these results can easily be extended to the case of time-varying systems and this is actually the case which will be considered in the following. Also the normality assumption for $v(t)$ and $w(t)$ may be dropped, in which case the Kalman filter provides the optimal *linear* estimate.

Problem Statement and Discretization of Distributed-Lag Model

The following discussion is divided into two parts; in the first part recursive parameter estimation is dealt with, while in the second the state is estimated under the assumption that all the parameters are known. The first part is mainly based on Tamura (1976), and the second part on Tamura and Ueno (1975a, b).

For the parameter estimation discrete-time BOD and DO measurements are assumed to be available every Δt days, while in the state estimation scheme, unsteady state BOD values are estimated by using discrete-time DO measurements only.

The river quality model used in this section is the distributed-lag model described in Sec. 4-3 and a heavily polluted river, namely the Yomo river, Amagasaki, Japan, is considered for testing the estimation scheme. Since only steady state data of this river were available, realistic dynamic data were generated by computer simulation, and these "synthetic" data were used later on for the estimation study. Moreover, only discrete-time measurement are assumed to be available, so that it is necessary to first discretize the distributed-lag model of Sec. 4-3 which is of the form

$$\dot{b}_i = -\left(k_{1i} + \frac{Q_i}{V_i}\right)b_i + \frac{Q_{i-1}}{V_i}b^l_{i-1}(t) + \frac{Q_{Ei}}{V_i}e_i(t)$$

$$\dot{d}_i = k_{1i}b_i - \left(k_{2i} + \frac{Q_i}{V_i}\right)d_i + \frac{Q_{i-1}}{V_i}d^l_{i-1}(t) + \frac{Q_{Ei}}{V_i}p_i + \frac{r_i}{V_i}$$

(5-4-10a)

where

$$b^l_{i-1}(t) = \int_0^t \phi_{i-1}(\tau)b_{i-1}(t-\tau)\,d\tau$$

$$d^l_{i-1}(t) = \int_0^t \phi'_{i-1}(\tau)b_{i-1}(t-\tau)\,d\tau + \int_0^t \psi_{i-1}(\tau)d_{i-1}(t-\tau)\,d\tau$$

(5-4-10b)

$$i = 1, 2, \ldots, N \quad b_0(t), d_0(t) \text{ given}$$

In Eq. (5-4-10a) $e_i(t)$, p_i, and Q_{Ei} are BOD, DO deficit, and flow rate of effluent discharge, and r_i is the net addition of DO deficit to the i-th reach by the combined effects of photosynthesis and respiration of bottom sludge. By integrating Eq. (5-4-10a) from $t\Delta t$ to $(t+1)\Delta t$ and writing, for simplicity, $b_i(t)$ and $d_i(t)$ instead of $b_i(t\Delta t)$ and $d_i(t\Delta t)$ we obtain

$$b_i(t+1) = e^{-\left(k_{1i}+\frac{Q_i}{V_i}\right)\Delta t} b_i(t) + \frac{Q_{i-1}\Delta t}{V_i} b^l_{i-1}(t) + \frac{Q_{Ei}\Delta t}{V_i} e_i(t)$$

$$d_i(t+1) = \frac{k_{1i}}{k_{1i}-k_{2i}}\left(e^{-\left(k_{2i}+\frac{Q_i}{V_i}\right)\Delta t} - e^{-\left(k_{1i}+\frac{Q_i}{V_i}\right)\Delta t}\right)b_i(t)$$

$$+ e^{-\left(k_{2i}+\frac{Q_i}{V_i}\right)\Delta t} d_i(t) + \frac{Q_{i-1}\Delta t}{V_i} d^l_{i-1}(t)$$

$$+ \frac{Q_{Ei}\Delta t}{V_i} p_i + \frac{r_i\Delta t}{V_i}$$

(5-4-11)

$$i = 1, 2, \ldots, N \quad t = 0, 1, \ldots, T-1$$

Consistently, the discrete version of Eq. (5-4-10b) can be written as

$$b^l_{i-1}(t) = \sum_{j=0}^{\theta} \phi_{i-1}(j)b_{i-1}(t-j)$$

$$d^l_{i-1}(t) = \sum_{j=0}^{\theta} \phi'_{i-1}(j)b_{i-1}(t-j) + \sum_{j=0}^{\theta} \psi_{i-1}(j)d_{i-1}(t-j)$$

(5-4-12)

5-4 KALMAN FILTERING FOR PARAMETER AND STATE ESTIMATION

By defining the following three parameters

$$\alpha_i = e^{-\left(k_{1i} + \frac{Q_i}{V_i}\right)\Delta t}$$

$$\alpha'_i = \frac{k_{1i}}{k_{1i} - k_{2i}} \left(e^{-\left(k_{2i} + \frac{Q_i}{V_i}\right)\Delta t} - e^{-\left(k_{1i} + \frac{Q_i}{V_i}\right)\Delta t} \right)$$

$$\beta_i = e^{-\left(k_{2i} + \frac{Q_i}{V_i}\right)\Delta t}$$

Eq. (5-4-11) and (5-4-12) can be aggregated as follows

$$b_i(t+1) = \alpha_i b_i(t) + \frac{Q_{i-1}}{V_i} \sum_{j=0}^{\theta} \phi_{i-1}(j) b_{i-1}(t-j) + \frac{Q_{Ei}}{V_i} e_i(t) \quad (5\text{-}4\text{-}13a)$$

$$d_i(t+1) = \alpha'_i b_i(t) + \beta_i d_i(t) + \frac{Q_{i-1}}{V_i} \sum_{j=0}^{\theta} \phi'_{i-1}(j) b_{i-1}(t-j)$$

$$+ \frac{Q_{i-1}}{V_i} \sum_{j=0}^{\theta} \psi_{i-1}(j) d_{i-1}(t-j) + \frac{Q_{Ei}}{V_i} p_i(t) + \frac{r_i}{V_i} \quad (5\text{-}4\text{-}13b)$$

$$i = 1, 2, \ldots, N \qquad t = 0, 1, \ldots, T-1$$

where Q_i, Q_{i-1}, Q_{Ei} and r_i stand for $Q_i \Delta t$, $Q_{i-1} \Delta t$, $Q_{Ei} \Delta t$ and $r_i \Delta t$, respectively. The boundary conditions $b_0(t)$, $d_0(t)$, and the initial conditions are assumed to be given. Equation (5-4-13) is the *discrete distributed-lag model* which is used in the following, both for parameter and state estimation.

Parameter Estimation

The interest to estimate the unknown parameters in a recursive way, instead of using a one shot procedure, arises from the following reasons.

(i) The rate of convergence of the estimation can be followed in real time so that the procedure can be stopped when the estimates do not vary significantly from one step to the other.
(ii) It is possible to estimate slowly varying parameters.

For simplicity in notation, the following parameters are defined ($j = 0, 1, \ldots, \theta$)

$$\tilde{\phi}_{i-1}(j) = \frac{Q_{i-1}}{V_i} \phi_{i-1}(j) \quad (5\text{-}4\text{-}14a)$$

$$\tilde{\phi}'_{i-1}(j) = \frac{Q_{i-1}}{V_i} \phi'_{i-1}(j) \quad (5\text{-}4\text{-}14b)$$

$$\tilde{\psi}_{i-1}(j) = \frac{Q_{i-1}}{V_i} \psi_{i-1}(j) \tag{5-4-14c}$$

$$W_i = 1/V_i$$

$$\pi_i = \frac{Q_{Ei}}{V_i} p_i + \frac{r_i}{V_i}$$

where the DO deficit concentration p_i of the effluent discharge or of the tributary flow is assumed to be time-invariant. Thus, the parameters to be estimated are

$$\alpha_i, W_i, \tilde{\phi}_{i-1}(j) \qquad (j = 0, 1, \ldots, \theta)$$

as far as the BOD equation is concerned (Eq. (5-4-13a)), and

$$\alpha'_i, \beta_i, \pi_i, \tilde{\phi}'_{i-1}(j), \tilde{\psi}_{i-1}(j) \qquad (j = 0, 1, \ldots, \theta)$$

as for the DO deficit equation (Eq. (5-4-13b)).

Estimation of $\alpha_i, W_i, \tilde{\phi}_{i-1}(j)$ (BOD equation) Suppose that the BOD measurements in each reach are given by

$$b_i^*(t) = b_i(t) + \delta b_i(t) \tag{5-4-15a}$$

and that the BOD measurements in the effluent discharge or in the tributary flow are given by

$$E_i^*(t) = Q_{Ei} e_i(t) + \delta E_i(t) \tag{5-4-15b}$$

where the measurement noises δb_i and δE_i are assumed to be stationary white noises with zero mean and known variance.

By substituting Eqs. (5-4-14) and (5-4-15) into Eq. (5-4-13a) we obtain

$$b_i^*(t+1) - \delta b_i(t+1) = \alpha_i(b_i^*(t) - \delta b_i(t)) + \sum_{j=0}^{\theta} \tilde{\phi}_{i-1}(j)(b_{i-1}^*(t-j) - \delta b_{i-1}(t-j))$$

$$+ W_i(E_i^*(t) - \delta E_i(t)) \tag{5-4-16}$$

Then, define the parameter vector $\mathbf{x}(t)$ as (see Sec. 5-1)

$$\mathbf{x}(t) = [\alpha_i \quad \tilde{\phi}_{i-1}(0) \quad \tilde{\phi}_{i-1}(1) \cdots \tilde{\phi}_{i-1}(\theta) \quad W_i]^T \tag{5-4-17a}$$

and let

$$y(t) = b_i^*(t+1) \tag{5-4-17b}$$

$$\mathbf{h}^T(t) = [b_i^*(t) \quad b_{i-1}^*(t) \cdots b_{i-1}^*(t-\theta) \quad E_i^*(t)] \tag{5-4-17c}$$

$$w(t) = \delta b_i(t+1) - \alpha_i \delta b_i(t) - \sum_{j=0}^{\theta} \tilde{\phi}_{i-1}(j) \delta b_{i-1}(t-j) - W_i \delta E_i(t) \tag{5-4-17d}$$

Since all the parameters are assumed to be time-invariant, we can write

$$\mathbf{x}(t+1) = \mathbf{x}(t) \tag{5-4-18a}$$

and from Eqs. (5-4-16) and (5-4-17) we obtain

$$y(t) = \mathbf{h}^T(t)\mathbf{x}(t) + w(t) \quad (5\text{-}4\text{-}18b)$$

where the expectation and variance of $w(t)$ are given by

$$E[w(t)] = 0 \quad (5\text{-}4\text{-}19a)$$

$$V_w = \text{var}\left[\delta b_i(t+1)\right] + \alpha_i^2 \text{ var}\left[\delta b_i(t)\right] + \sum_{j=0}^{\theta} \tilde{\phi}_{i-1}^2(j) \text{ var}\left[\delta b_{i-1}(t-j)\right]$$
$$+ W_i^2 \text{ var}\left[\delta E_i(t)\right] \quad (5\text{-}4\text{-}19b)$$

Equation (5-4-18) represents a particular linear dynamical system with measurement noise so that the Kalman filtering technique can be applied to it, thus obtaining the estimate $\hat{\mathbf{x}}(t)$ of the parameters in a recursive way. The statistics of the observation noise $w(t)$ can be obtained at each step of the procedure by substituting the parameters $\alpha_i, W_i, \tilde{\phi}_{i-1}$ in Eq. (5-4-19b) by their estimates $\hat{\alpha}_i, \hat{W}_i, \tilde{\tilde{\phi}}_{i-1}(j)$.

Strictly speaking, since the observation noise $w(t)$ is not a white noise (see Eq. (5-4-17d)) and the parameters $\alpha_i, W_i,$ and $\tilde{\phi}_{i-1}(j)$ have been approximated by their estimates in Eq. (5-4-19b), the use of the Kalman filtering technique is not theoretically justified. However, extensive simulation showed that the estimates of the parameters successfully converge to the true values.

Estimation of $\alpha'_i, \beta_i, \pi_i, \tilde{\phi}'_{i-1}(j), \tilde{\psi}_{i-1}(j)$ **(DO equation)** For this case both BOD and DO deficit measurements are required in each reach,

$$b_i^*(t) = b_i(t) + \delta b_i(t) \qquad d_i^*(t) = d_i(t) + \delta d_i(t) \quad (5\text{-}4\text{-}20)$$

where δb_i and δd_i represent stationary white noises with zero mean and known variance.

By substituting Eqs. (5-4-14) and (5-4-20) into Eq. (5-4-13b), we obtain

$$d_i^*(t+1) - \delta d_i(t+1) = \alpha'_i(b_i^*(t) - \delta b_i(t)) + \beta_i(d_i^*(t) - \delta d_i(t))$$
$$+ \sum_{j=0}^{\theta} \tilde{\phi}'_{i-1}(j)(b_{i-1}^*(t-j) - \delta b_{i-1}(t-j))$$
$$+ \sum_{j=0}^{\theta} \tilde{\psi}_{i-1}(j)(d_{i-1}^*(t-j) - \delta d_{i-1}(t-j)) + \pi_i$$

$$(5\text{-}4\text{-}21)$$

and again if the parameter vector $\mathbf{x}(t)$ is defined as

$$\mathbf{x}(t) = \begin{bmatrix} \alpha'_i & \beta_i & \tilde{\phi}'_{i-1}(0) \cdots \tilde{\phi}'_{i-1}(\theta) & \tilde{\psi}_{i-1}(0) \cdots \tilde{\psi}_{i-1}(\theta) & \pi_i \end{bmatrix}^T$$

184 STATE AND PARAMETER ESTIMATION

and

$$y(t) = d_i^*(t+1)$$

$$\mathbf{h}^T(t) = [b_i^*(t) \quad d_i^*(t) \quad b_{i-1}^*(t) \cdots b_{i-1}^*(t-\theta) \quad d_{i-1}^*(t) \cdots d_{i-1}^*(t-\theta) \quad 1]$$

$$w(t) = \delta d_i(t+1) - \alpha_i' \delta b_i(t) - \beta_i \delta d_i(t) - \sum_{j=0}^{\theta} \tilde{\phi}_{i-1}'(j) \delta b_{i-1}(t-j)$$

$$- \sum_{j=0}^{\theta} \tilde{\psi}_{i-1}(j) \delta d_{i-1}(t-j)$$

we obtain

$$\mathbf{x}(t+1) = \mathbf{x}(t) \tag{5-4-22a}$$

$$y(t) = \mathbf{h}^T(t)\mathbf{x}(t) + w(t) \tag{5-4-22b}$$

with

$$E[w(t)] = 0$$

$$V_w = \text{var}\,[\delta d_i(t+1)] + \alpha_i'^2 \,\text{var}\,[\delta b_i(t)] + \beta_i^2 \,\text{var}\,[\delta d_i(t)]$$

$$+ \sum_{j=0}^{\theta} \tilde{\phi}_{i-1}'^2(j) \,\text{var}\,[\delta b_{i-1}(t-j)]$$

$$+ \sum_{j=0}^{\theta} \tilde{\psi}_{i-1}^2(j) \,\text{var}\,[\delta d_{i-1}(t-j)]$$

Equation (5-4-22) is of the same form as Eq. (5-4-18) so that the Kalman filtering technique can be applied to it as in the case for the BOD equation.

A stretch of 4 km of the Yomo river is now considered to test the filter

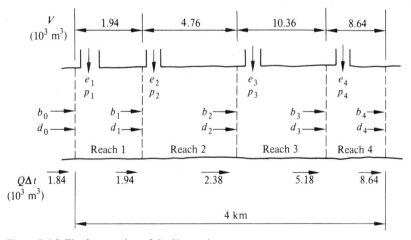

Figure 5-4-2 The four reaches of the Yomo river.

algorithm. The stretch is divided into 4 reaches as shown in Fig. 5-4-2. The values of the parameters k_1 and k_2 for all the four reaches are

$$k_{1i} = 0.53 \text{ days}^{-1} \qquad k_{2i} = 0.10 \text{ days}^{-1}$$

while the values of the other parameters are listed in Table 5-4-1, where $\Delta t = \frac{1}{8}$ days. For simplicity, the unknown parameters to be estimated were assumed to be only the BOD distributed-lag coefficients $\phi_{i-1}(j)$ $(j = 0, 1, \ldots, \theta)$. Their true values are shown in Table 5-4-2 where $\theta = 3$.

Table 5-4-1 Parameter values in the Yomo river

Reach	Steady state b_i (mg/l)	d_i (mg/l)	$Q_i \Delta t$ (10^3 m³)	V_i (10^3 m³)	$Q_{Ei} \Delta t$ (10^3 m³)	e_i (mg/l)	π_i (mg/l)
0	12.0	6.13	1.84				
1	11.4	6.13	1.94	1.94	0.10	15.3	0.02
2	17.6	7.79	2.38	4.76	0.44	57.8	1.24
3	80.0	9.50	5.18	10.36	2.80	151.0	4.32
4	49.5	9.10	8.64	8.64	3.46	11.7	3.20

Table 5-4-2 True values of distributed-lag coefficients

i	$\phi_{i-1}(0)$	$\phi_{i-1}(1)$	$\phi_{i-1}(2)$	$\phi_{i-1}(3)$
1	1.00	0	0	0
2	0.60	0.25	0.15	0
3	0.20	0.50	0.20	0.10
4	0.20	0.50	0.20	0.10

As an example, a set of computational results for the last reach ($i = 4$) is shown in Fig. 5-4-3. The recursive technique outlined above gave satisfactory results. These results were obtained by putting the variance of measurement noise equal to 0.1 (i.e., var $[\delta b_4(t)] = $ var $[\delta b_3(t)] = 0.1$), the initial estimates of the 4 parameters equal to (3.0, 3.0, 3.0, 3.0) and the variance of the initial estimation error equal to (10.0, 10.0, 10.0, 10.0). The parameter estimates do converge to the true values, (0.2, 0.5, 0.2, 0.1) within 50 steps of the recursive estimation, and the values of the diagonal elements of the error covariance matrix $\mathbf{V}_\varepsilon(t)$ are shown in Table 5-4-3 for $t = 10, 20, \ldots, 50$.

186 STATE AND PARAMETER ESTIMATION

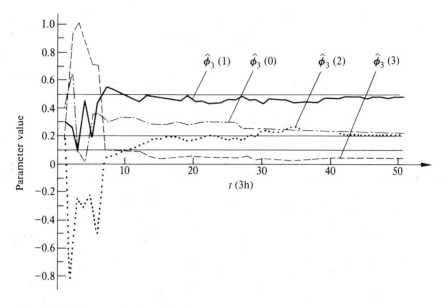

Figure 5-4-3 Computational results for time-invariant parameter estimation by Kalman filtering technique.

So far the parameters have been assumed to be time-invariant. As seen in Table 5-4-3, the values of the error covariance matrix are converging to zero as the recursive estimation process develops in time, and consequently the filter gain matrix is also converging to zero (see Eq. (5-4-8) with $V_v = 0$). Suppose that for some reason after a sufficiently long period of time (i.e., after we had obtained convergence) the parameters start to vary in time. Then, the parameter estimates will not efficiently keep track of the variations of the parameters because of the small values of the error covariance matrix and of the filter gain matrix. Of course, if the dynamics of the parameters were to be known, it could be incorporated into the system equation (5-4-18a) and/or (5-4-22a), and the parameter estimates would satisfactorily track the true values. However, in reality, it is difficult to obtain information about the dynamics of the parameters and therefore other solutions must be found.

Table 5-4-3 The diagonal elements of the error covariance matrix $V_\varepsilon(t)$ for $t = 10, 20, \ldots, 50$

	$t = 10$	$t = 20$	$t = 30$	$t = 40$	$t = 50$
$[V_\varepsilon(t)]_{11}$	0.0113	0.0039	0.0022	0.0015	0.0014
$[V_\varepsilon(t)]_{22}$	0.0224	0.0111	0.0061	0.0041	0.0037
$[V_\varepsilon(t)]_{33}$	0.0253	0.0111	0.0060	0.0041	0.0037
$[V_\varepsilon(t)]_{44}$	0.0138	0.0039	0.0021	0.0016	0.0013

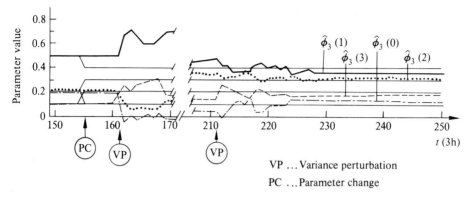

Figure 5-4-4 Computational results for drifted parameter estimation by variance perturbation method.

One way to estimate drifting parameters is to perturb periodically the values of the error covariance matrix in the Kalman filtering procedure. This method, called the *variance perturbation method*, gives artificial periodic errors to the parameter estimates, and to the filter gain matrix accordingly. If in the previous example the parameters vary from (0.2, 0.5, 0.2, 0.1) to (0.1, 0.4, 0.3, 0.2) at time 155 the result of the estimation by means of the variance perturbation method is that shown in Fig. 5-4-4. In this example the variance is perturbed every 50 steps, that is, the diagonal elements of the error covariance matrix are reset to 0.2 and the off diagonal elements to zero every 50 steps.

Although the variance perturbation method is based on a heuristic argument and there is no theoretical guarantee of convergence, the numerical experience shows that the method is quite effective for estimating time-varying parameters without using the a priori knowledge of the dynamics of the parameters.

State Estimation

Assuming that all the parameters of the model are known, a method of estimating both BOD and DO concentration in rivers in real time by using only DO measurement data is developed.

First, Eq. (5-4-13) with noise reads in vector form as follows

$$\mathbf{x}(t+1) = \mathbf{F}_0 \mathbf{x}(t) + \mathbf{F}_1 \mathbf{x}(t-1) + \cdots + \mathbf{F}_\theta \mathbf{x}(t-\theta) + \mathbf{u}(t) + \mathbf{v}(t) \quad (5\text{-}4\text{-}23)$$

where

$$\mathbf{x}(t) = [b_1(t) \ d_1(t) \cdots b_N(t) \ d_N(t)]^T$$

$$\mathbf{F}_0 = \begin{bmatrix} F_{011} & 0 & \cdots & 0 & 0 \\ F_{021} & F_{022} & \cdots & 0 & 0 \\ 0 & F_{032} & \cdots & 0 & 0 \\ \vdots & \vdots & & \vdots & \vdots \\ 0 & 0 & \cdots & F_{0,N,N-1} & F_{0NN} \end{bmatrix}$$

$$\mathbf{F}_j = \begin{bmatrix} 0 & 0 & \cdots & 0 & 0 \\ \mathbf{F}_{j21} & 0 & \cdots & 0 & 0 \\ 0 & \mathbf{F}_{j32} & \cdots & 0 & 0 \\ \vdots & \vdots & & \vdots & \vdots \\ 0 & 0 & \cdots & \mathbf{F}_{j,N,N-1} & 0 \end{bmatrix} \qquad j = 1,\ldots,\theta$$

$$\mathbf{F}_{0ii} = \begin{bmatrix} \alpha_i & 0 \\ \alpha_i' & \beta_i \end{bmatrix} \qquad i = 1,\ldots,N$$

$$\mathbf{F}_{j,i,i-1} = \frac{Q_{i-1}}{V_i} \begin{bmatrix} \phi_{i-1}(j) & 0 \\ \phi_{i-1}'(j) & \psi_{i-1}(j) \end{bmatrix} \qquad \begin{array}{l} i = 2,\ldots,N \\ j = 0,1,\ldots,\theta \end{array}$$

$$\mathbf{u}(t) = [\mathbf{u}_1(t)^T \quad \mathbf{u}_2(t)^T \cdots \mathbf{u}_N(t)^T]^T$$

$$\mathbf{u}_1(t) = \left[\frac{Q_0}{V_1} b_0(t) + \frac{Q_{E1}}{V_1} e_1(t) \quad \frac{Q_0}{V_1} d_0(t) + \frac{Q_{E1}}{V_1} p_1 + \frac{r_1}{V_1} \right]^T$$

$$\mathbf{u}_i(t) = \left[\frac{Q_{Ei}}{V_i} e_i(t) \quad \frac{Q_{Ei}}{V_i} p_i + \frac{r_i}{V_i} \right]^T \qquad i = 2,3,\ldots,N$$

$\mathbf{v}(t) = 2N$ dimensional white noise with zero mean and covariance matrix $\mathbf{V}_v$.

Equation (5-4-23) can be thought of as a *general discrete-time distributed-lag model*. Since only DO is measured at one point in each reach, the observation equation is given by

$$d_i^*(t) = d_i(t) + \delta d_i(t) \qquad i = 1,2,\ldots,N \qquad (5\text{-}4\text{-}24)$$

where $d_i^*(t)$ is the measured DO data in reach i, and $\delta d_i(t)$ is a stationary white measurement noise with zero mean and variance $\operatorname{var}[\delta d_i(t)]$. The vector form of Eq. (5-4-24) is

$$\mathbf{y}(t) = \mathbf{H}\mathbf{x}(t) + \mathbf{w}(t) \qquad (5\text{-}4\text{-}25)$$

where

$$\mathbf{y}(t) = [d_1^*(t) \quad d_2^*(t) \cdots d_N^*(t)]^T$$

$$\mathbf{H} = \begin{bmatrix} 0 & 1 & 0 & 0 & \cdots & 0 & 0 \\ 0 & 0 & 0 & 1 & \cdots & 0 & 0 \\ \vdots & \vdots & \vdots & \vdots & & \vdots & \vdots \\ 0 & 0 & 0 & 0 & \cdots & 0 & 1 \end{bmatrix} \qquad (N \times 2N \text{ matrix})$$

$$\mathbf{w}(t) = [\delta d_1(t) \quad \delta d_2(t) \cdots \delta d_N(t)]^T$$

The covariance matrix for $\mathbf{w}(t)$ is given by

$$\mathbf{V}_w = \begin{bmatrix} \operatorname{var}[\delta d_1(t)] & 0 & \cdots & 0 \\ 0 & \operatorname{var}[\delta d_2(t)] & \cdots & 0 \\ \vdots & \vdots & & \vdots \\ 0 & 0 & \cdots & \operatorname{var}[\delta d_N(t)] \end{bmatrix}$$

5-4 KALMAN FILTERING FOR PARAMETER AND STATE ESTIMATION 189

The state estimation problem can then be stated as follows: given the process equation (5-4-23), the initial conditions, the deterministic input $\mathbf{u}(t)$, and the measurement equation (5-4-25), find the minimum variance estimates $\hat{\mathbf{x}}(t)$ of $\mathbf{x}(t)$.

Three different state estimation techniques are presented, and compared from the point of view of computational efficiency.

Kalman filter Equations (5-4-23) and (5-4-25) can be converted to an equivalent first-order system by using the *augmented state* $\mathbf{x}^*(t)$ as follows

$$\mathbf{x}^*(t+1) = \mathbf{F}^*\mathbf{x}^*(t) + \mathbf{G}^*\mathbf{u}(t) + \mathbf{v}^*(t)$$
$$\mathbf{y}(t) = \mathbf{H}^*\mathbf{x}^*(t) + \mathbf{w}(t) \tag{5-4-26}$$

where

$$\mathbf{x}^*(t) = [\mathbf{x}(t)^T \quad \mathbf{x}(t-1)^T \cdots \mathbf{x}(t-\theta)^T]^T$$
$$\mathbf{v}^*(t) = [\mathbf{v}(t)^T \quad \mathbf{0}^T \cdots \mathbf{0}^T \quad]^T$$

$$\mathbf{F}^* = \begin{bmatrix} \mathbf{F}_0 & \mathbf{F}_1 & \cdots & \mathbf{F}_{\theta-1} & \mathbf{F}_\theta \\ \mathbf{I} & \mathbf{0} & \cdots & \mathbf{0} & \mathbf{0} \\ \mathbf{0} & \mathbf{I} & \cdots & \mathbf{0} & \mathbf{0} \\ \vdots & \vdots & & \vdots & \vdots \\ \mathbf{0} & \mathbf{0} & \cdots & \mathbf{I} & \mathbf{0} \end{bmatrix} \quad \mathbf{G}^* = \begin{bmatrix} \mathbf{I} \\ \mathbf{0} \\ \mathbf{0} \\ \vdots \\ \mathbf{0} \end{bmatrix}$$

$$\mathbf{H}^* = [\mathbf{H} \quad \mathbf{0} \cdots \mathbf{0} \quad \mathbf{0}]$$

Kalman filtering technique can be directly applied to Eq. (5-4-26). Since the dimension of the new state variable $\mathbf{x}^*(t)$ is $2N(\theta+1)$, it is clear that for large N and/or θ, the application of the Kalman filtering technique leads to an excessive computational burden, both in computer storage and computation time.

Suboptimal recursive filter (SMART) The suboptimal recursive filter (SMART), which will now be described, can be directly applied to a generalized distributed-lag model of the kind described by Eqs. (5-4-23) and (5-4-25).

The SMART algorithm which is a direct extension of Eqs. (5-4-7) and (5-4-8) is as follows:

$$\hat{\mathbf{x}}(t+1) = \sum_{j=0}^{\theta} \mathbf{F}_j \hat{\mathbf{x}}(t-j) + \mathbf{u}(t)$$
$$- \mathbf{L}(t+1)\left[\mathbf{H}\left\{\sum_{j=0}^{\theta} \mathbf{F}_j \hat{\mathbf{x}}(t-j) + \mathbf{u}(t)\right\} - \mathbf{y}(t+1)\right] \tag{5-4-27}$$

where $\hat{\mathbf{x}}(t)$ is given for $t = -\theta, -\theta+1, \ldots, 0$, and the gain matrix $\mathbf{L}(t)$ is obtained recursively by (compare with Eq. (5-4-8))

$$\mathbf{V}_\varepsilon^a(t+1) = \sum_{j=0}^{\theta} \mathbf{F}_j \mathbf{V}_\varepsilon(t-j) \mathbf{F}_j^T + \mathbf{V}_\mathbf{v}$$
$$\mathbf{V}_\varepsilon(t) = \text{given for } t = -\theta, -\theta+1, \ldots, 0 \tag{5-4-28}$$
$$\mathbf{L}(t+1) = \mathbf{V}_\varepsilon^a(t+1)\mathbf{H}^T[\mathbf{H}\mathbf{V}_\varepsilon^a(t+1)\mathbf{H}^T + \mathbf{V}_\mathbf{w}]^{-1}$$
$$\mathbf{V}_\varepsilon(t+1) = \mathbf{V}_\varepsilon^a(t+1) - \mathbf{L}(t+1)\mathbf{H}\mathbf{V}_\varepsilon^a(t+1)$$

This recursive scheme can be intuitively understood, since in Eq. (5-4-27) the term

$$\left\{ \sum_{j=0}^{\theta} \mathbf{F}_j \hat{\mathbf{x}}(t-j) + \mathbf{u}(t) \right\}$$

represents the best estimate $\hat{\mathbf{x}}(t+1 \mid t, t-1, \ldots, t-\theta)$ of the state at time $(t+1)$ that one can make at time t on the basis of the estimates $\hat{\mathbf{x}}(t), \hat{\mathbf{x}}(t-1), \ldots, \hat{\mathbf{x}}(t-\theta)$, while the term

$$\mathbf{H} \left\{ \sum_{j=0}^{\theta} \mathbf{F}_j \hat{\mathbf{x}}(t-j) + \mathbf{u}(t) \right\}$$

is $\hat{\mathbf{y}}(t+1 \mid t, t-1, \ldots, t-\theta)$. Then Eq. (5-4-27) can be written as

$$\hat{\mathbf{x}}(t+1) = \hat{\mathbf{x}}(t+1 \mid t, t-1, \ldots, t-\theta)$$
$$- \mathbf{L}(t+1)[\hat{\mathbf{y}}(t+1 \mid t, t-1, \ldots, t-\theta) - \mathbf{y}(t+1)]$$

which has exactly the same form as Eq. (5-4-9).

A comparison of the SMART with the Kalman filter applied to Eq. (5-4-26), leads to the conclusion that the error covariance matrices $\mathbf{V}_\varepsilon(t), \mathbf{V}_\varepsilon(t-1), \ldots, \mathbf{V}_\varepsilon(t-\theta)$ of the SMART correspond to the diagonal blocks of $\mathbf{V}_{\varepsilon^*}(t)$ in the Kalman filter, where

$$\varepsilon^*(t) = \hat{\mathbf{x}}^*(t) - \mathbf{x}^*(t)$$

That is, the SMART is a Kalman filter in which the off diagonal blocks of $\mathbf{V}_{\varepsilon^*}(t)$ have been forced to be zero for all t: for this reason the SMART is only a suboptimal recursive filter.

Recursive filter in time and space (RFTS) Taking advantage of the special structure of the $\mathbf{F}_j$ matrices, a very simple *recursive filter in time and space* (RFTS) can be constructed. The continuous version of this filter was first proposed by Singh (1973, 1975), while the extension to the discrete case with distributed-lag was done by Tamura and Ueno (1975b).

First, Eqs. (5-4-23) and (5-4-25) are divided into N groups as follows. For $i = 1$ we write

$$\mathbf{x}_1(t+1) = \mathbf{F}_{011}\mathbf{x}_1(t) + \mathbf{u}_1(t) + \mathbf{v}_1(t)$$
$$y_1(t) = \mathbf{h}_1^T \mathbf{x}_1(t) + w_1(t) \tag{5-4-29}$$

and for $i = 2, 3, \ldots, N$

$$\mathbf{x}_i(t+1) = \mathbf{F}_{0ii}\mathbf{x}_i(t) + \sum_{j=0}^{\theta} \mathbf{F}_{j,i,i-1}\mathbf{x}_{i-1}(t-j) + \mathbf{u}_i(t) + \mathbf{v}_i(t)$$
$$y_i(t) = \mathbf{h}_i^T \mathbf{x}_i(t) + w_i(t) \tag{5-4-30}$$

where for $i = 1, \ldots, N$

$$\mathbf{x}_i(t) = [b_i(t) \quad d_i(t)]^T$$

$$y_i(t) = d_i^*(t)$$

$$\mathbf{h}_i^T = [0 \quad 1]$$

$\mathbf{v}_i(t)$ = two-dimensional white noise with zero mean and covariance matrix $\mathbf{V}_{v_i}(t)$

$w_i(t) = \delta d_i(t)$

The RFTS algorithm is as follows: apply the Kalman filter first to Eq. (5-4-29), then recursively, for $i = 2, 3, \ldots, N$, to

$$\mathbf{x}_i(t+1) = \mathbf{F}_{0ii}\mathbf{x}_i(t) + \sum_{j=0}^{\theta} \mathbf{F}_{j,i,i-1}\hat{\mathbf{x}}_{i-1}(t-j) + \mathbf{u}_i(t) + \mathbf{v}_i^*(t) \quad (5\text{-}4\text{-}31)$$

$$y_i(t) = \mathbf{h}_i^T \mathbf{x}_i(t) + w_i(t)$$

where

$$\mathbf{v}_i^*(t) = \mathbf{v}_i(t) + \sum_{j=0}^{\theta} \mathbf{F}_{j,i,i-1}\varepsilon_{i-1}(t-j)$$

$$E[\mathbf{v}_i^*(t)] = \mathbf{0} \quad (5\text{-}4\text{-}32)$$

$$\text{cov}[\mathbf{v}_i^*(t)] = \mathbf{V}_{v_i} + \sum_{j=0}^{\theta} \mathbf{F}_{j,i,i-1}\mathbf{V}_{\varepsilon_{i-1}}(t-j)\mathbf{F}_{j,i,i-1}^T$$

As is apparent from Eq. (5-4-32), the process noise $\mathbf{v}_i^*(t)$ in Eq. (5-4-31) is *not* stationary. Therefore, when the Kalman filter is applied to Eq. (5-4-31) $\mathbf{V}_v$ in Eq. (5-4-8) must be replaced by $\mathbf{V}_v(t)$.

The mechanism of this filter is very simple. It starts from the first reach and estimates BOD and DO in the reach. Then, the results of the estimation are feedforwarded to the adjacent downstream reach where a new state estimation is accomplished by applying the usual Kalman filtering technique, and this operation is repeated until all reaches have been worked out.

Comparison The three different filters described above were applied to the same example (Yomo river) used for parameter estimation, with

$$\mathbf{V}_{v_i}(t) = \begin{bmatrix} 0.10 & 0 \\ 0 & 0.01 \end{bmatrix} \quad \text{var}[\delta d_i(t)] = 0.1$$

The results are shown in Fig. 5-4-5.

Even when starting from a crude initial estimate of BOD and DO, the three filters perform satisfactorily, and no remarkable difference between the Kalman filter and the two suboptimal filters, SMART and RFTS, can be found. Note that the estimates are following the sudden changes of the true states in each reach. Convergence of the state estimates for the measured variable, DO, is very quick, while, as expected, the convergence for the unmeasured variable,

192 STATE AND PARAMETER ESTIMATION

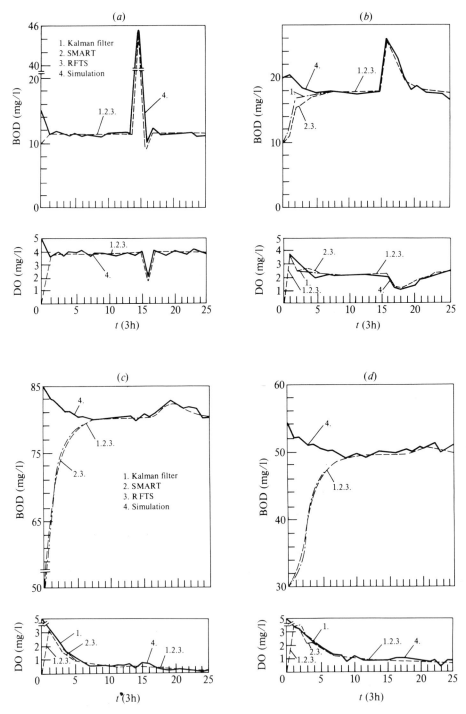

Figure 5-4-5 Computational results for state estimation: (a) reach 1 (b) reach 2 (c) reach 3 (d) reach 4.

BOD, is slower than for DO. The convergence for the BOD in the downstream reaches is slower than in the upstream reaches, since in order to estimate the BOD in a downstream reach we need the state estimates of the BOD in the upstream reaches. The computation times for the three different filters were compared and the SMART took about $\frac{1}{10}$ of the Kalman filter, while the RFTS took about $\frac{1}{100}$ of the Kalman filter. Therefore, the RFTS seems to be the best filter at least for the case in which there are not too many reaches.

5-5 EXTENDED KALMAN FILTERING TECHNIQUE AND MODEL DISCRIMINATION

The objective of this section is to show examples of simultaneous recursive estimation of states and parameters in river quality models, and to illustrate how these recursive schemes can be used for determining the appropriate model structure (*model discrimination*) (see Sec. 5-1). The examples shown are from Beck and Young (1976) (see also Lettenmaier and Burges, 1976).

The interest in solving the simultaneous estimation problem, instead of estimating separately parameters and states as in Sec. 5-4, arises if

(i) parameters are to be estimated on the basis of very noisy measurements of BOD and DO
(ii) the state has to be estimated for a system having time-varying parameters

For solving the problem the parameters are considered as additional state variables, as described in Sec. 5-1. In this way a pure state estimation problem, which is, in general, nonlinear, is obtained even if the original model was linear.

Extended Kalman Filter

The estimation scheme used to solve the nonlinear state estimation problem is the *extended Kalman filtering* technique, which is a modification of the linear estimation technique described in the previous section.

Consider a *nonlinear continuous time system* of the form

$$\dot{\mathbf{x}}(t) = \mathbf{f}(\mathbf{x}(t), \mathbf{u}(t)) + \mathbf{v}(t) \qquad t \geq 0 \qquad (5\text{-}5\text{-}1a)$$

with *discrete measurement equation*

$$\mathbf{y}(t) = \mathbf{H}\mathbf{x}(t) + \mathbf{w}(t) \qquad t = 1, 2, \ldots, \qquad (5\text{-}5\text{-}1b)$$

where **v** and **w** are white and stationary noises with zero mean and known variances.

As for the linear problem solved in Sec. 5-4, the estimate $\hat{\mathbf{x}}(t+1)$, $t = 1, 2, \ldots$, is calculated by adding to the prediction based on $\hat{\mathbf{x}}(t)$ a correction proportional to the difference between the actual observation $\mathbf{y}(t+1)$ and the output prediction $\hat{\mathbf{y}}(t+1|t)$, i.e.,

$$\hat{\mathbf{x}}(t+1) = \hat{\mathbf{x}}(t+1|t) - \mathbf{L}(t+1)[\hat{\mathbf{y}}(t+1|t) - \mathbf{y}(t+1)] \qquad (5\text{-}5\text{-}2)$$

In Eq. (5-5-2) the state prediction $\hat{\mathbf{x}}(t+1|t)$ is obtained by integrating Eq. (5-5-1a) from t to $t+1$ with initial value $\hat{\mathbf{x}}(t)$, and with input fixed at its latest measured value $\mathbf{u}(t)$, i.e.,

$$\hat{\mathbf{x}}(t+1|t) = \hat{\mathbf{x}}(t) + \int_t^{t+1} \mathbf{f}(\tilde{\mathbf{x}}(\tau), \mathbf{u}(t)) \, d\tau$$

where

$$\dot{\tilde{\mathbf{x}}}(\tau) = \mathbf{f}(\tilde{\mathbf{x}}(\tau), \mathbf{u}(t)) \qquad t \leq \tau \leq t+1$$

$$\tilde{\mathbf{x}}(t) = \hat{\mathbf{x}}(t)$$

The output prediction is simply obtained by

$$\hat{\mathbf{y}}(t+1|t) = \mathbf{H}\hat{\mathbf{x}}(t+1|t)$$

Also the gain matrix $\mathbf{L}(t+1)$ is computed in the same fashion as in Sec. 5-4. Since, however, Eq. (5-5-1a) is nonlinear, the matrix $\mathbf{F}$ needed for the calculation of $\mathbf{L}(t+1)$ (see Eq. (5-4-8)) has to be taken from a system which is derived from Eq. (5-5-1) through linearization around $\hat{\mathbf{x}}(t)$ (see Sec. 1-2) and discretization. Hence, $\mathbf{F}$, which is now a function of t, is calculated according to

$$\mathbf{F}(t) = e^{\left[\frac{\partial \mathbf{f}}{\partial \mathbf{x}}\right]_{\hat{\mathbf{x}}(t), \mathbf{u}(t)}}$$

where $[\partial \mathbf{f}/\partial \mathbf{x}]_{\hat{\mathbf{x}}(t), \mathbf{u}(t)}$ is the Jacobian matrix of $\mathbf{f}(\mathbf{x}(t), \mathbf{u}(t))$ (see Sec. 1-2). The gain matrix $\mathbf{L}(t+1)$ of the extended Kalman filter is now calculated (as in Eq. (5-4-8)) according to

$$\mathbf{V}_\varepsilon^a(t+1) = \mathbf{F}(t)\mathbf{V}_\varepsilon(t)\mathbf{F}(t)^T + \mathbf{V}_\mathbf{v}$$

$$\mathbf{L}(t+1) = \mathbf{V}_\varepsilon^a(t+1)\mathbf{H}^T\left[\mathbf{H}\mathbf{V}_\varepsilon^a(t+1)\mathbf{H}^T + \mathbf{V}_\mathbf{w}\right]^{-1}$$

$$\mathbf{V}_\varepsilon(t+1) = \mathbf{V}_\varepsilon^a(t+1) - \mathbf{L}(t+1)\mathbf{H}\mathbf{V}_\varepsilon^a(t+1)$$

where $\mathbf{V}_\varepsilon(t)$ is the covariance matrix of the estimation error and $\mathbf{V}_\varepsilon^a(t)$ is the a priori covariance matrix of the estimation error. The initial error covariance matrix $\mathbf{V}_\varepsilon(0)$ and $\hat{\mathbf{x}}(0)$ are assumed to be given. In the extended Kalman filter the linearization around the latest state estimate $\hat{\mathbf{x}}(t)$ and the discretization must be done at each time step, and consequently $\mathbf{F}(t)$ must be evaluated at each step of the algorithm.

Since the extended Kalman filter is a linear approximation to a *nonlinear filter*, there is no theoretical guarantee of convergence for the estimates, nor is the matrix $\mathbf{V}_\varepsilon(t)$ an accurate indication of the true covariance matrix of the estimation error. On the other hand, nonlinear filters are so complicated that one is obliged to use approximate filters like the extended Kalman filter or others (see Jazwinski, 1970). As for the Kalman filter, the choice of the initial estimate $\hat{\mathbf{x}}(0)$, the initial covariance matrix $\mathbf{V}_\varepsilon(0)$, and the matrices $\mathbf{V}_\mathbf{v}$ and $\mathbf{V}_\mathbf{w}$ is usually subjective and the estimate is sensitive to this choice.

Model Description

The model chosen to demonstrate the application of the nonlinear recursive state estimation technique is the CSTR model described in Sec. 4-3. Since this is only an illustrative example the discussion is confined to a single river reach. The data used were collected at a reach of the Cam river in England which is depicted in Fig. 5-5-1. The field data were those obtained for the eighty-day period 6 June–25 August, 1972 (Beck and Young 1975). Samples of BOD_5 and DO were taken daily at the upstream and downstream boundaries of the reach, volumetric flow rate measurements were taken at Bottisham, and river water temperature $T(°C)$ was observed at both Bottisham and Bait's Bite. The volume V of the reach of the river is assumed to be constant and equal to 150×10^3 m³, and the flow rate $Q(t)$ observed was between 100 and 200×10^3 m³/day. The following discussion is based on once-daily sample observations of all the variables, which means t in Eq. (5-5-1b) is in days.

The model considered is the following (see Sec. 4-3):

$$\dot{b}(t) = -\left(k_1 + \frac{Q(t)}{V}\right)b(t) + \frac{Q(t)}{V}b_i(t) + L(t) + v_b(t)$$

$$\dot{c}(t) = -k_1 b(t) - \left(k_2 + \frac{Q(t)}{V}\right)c(t) + \frac{Q(t)}{V}c_i(t)$$

$$+ k_2 c_s(t) + D(t) + v_c(t) \qquad t \geq 0$$

(5-5-3)

where b_i and c_i refer to the water entering the reach from upstream, $v_b(t)$ and $v_c(t)$ are the process noises on BOD and DO dynamics, which are assumed to be white and stationary with zero mean and given variance, and

$L(t)$ = rate of addition of BOD to the reach by local surface runoff.

$D(t)$ = net rate of addition of DO to the reach through photosynthetic/respiratory activity of plants and decomposition of mud deposits.

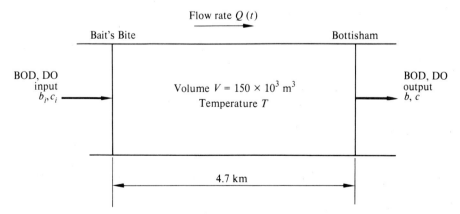

Figure 5-5-1 Reach of the Cam river used for estimation.

The particular form of L and D is subject to the model discrimination process and will be discussed below. Since the temperature variations during the measurement period were small, only the temperature dependence of c_s was taken into account (see Eq. (3-5-8)). Moreover, $Q(t)$ is assumed to be a known function of t.

The observation $b^*(t)$ and $c^*(t)$ of $b(t)$ and $c(t)$ are assumed to be given by

$$b^*(t) = b(t) + \delta b(t)$$
$$c^*(t) = c(t) + \delta c(t) \qquad (5\text{-}5\text{-}4)$$

where $\delta b(t)$ and $\delta c(t)$ are white and stationary measurement noises with zero mean and known variance.

Model Validation and Discrimination

Two different cases for L and D are now analyzed

Case 1 $L(t) = 0$, $D(t) = 0$. In this case, model (5-5-3) is of the Streeter-Phelps type, and the terms in Eq. (5-5-1) read

$$\mathbf{x}(t) = \begin{bmatrix} b(t) & c(t) & k_1 & k_2 \end{bmatrix}^T$$

$$\mathbf{u}(t) = \begin{bmatrix} b_i(t) & c_i(t) & c_s(t) \end{bmatrix}^T$$

$$\mathbf{f}(\mathbf{x}(t), \mathbf{u}(t)) = \begin{bmatrix} -\left(k_1 + \dfrac{Q(t)}{V}\right) b(t) + \dfrac{Q(t)}{V} b_i(t) \\ -k_1 b(t) - \left(k_2 + \dfrac{Q(t)}{V}\right) c(t) + \dfrac{Q(t)}{V} c_i(t) + k_2 c_s(t) \\ 0 \\ 0 \end{bmatrix}$$

$$\mathbf{y}(t) = \begin{bmatrix} b^*(t) & c^*(t) \end{bmatrix}^T$$

$$\mathbf{H} = \begin{bmatrix} 1 & 0 & 0 & 0 \\ 0 & 1 & 0 & 0 \end{bmatrix}$$

$$\mathbf{v}(t) = \begin{bmatrix} v_b(t) & v_c(t) & 0 & 0 \end{bmatrix}^T$$

$$\mathbf{w}(t) = \begin{bmatrix} \delta b(t) & \delta c(t) \end{bmatrix}^T$$

and the Jacobian matrix is

$$\left[\dfrac{\delta \mathbf{f}}{\delta \mathbf{x}}\right]_{\hat{\mathbf{x}}(t), \mathbf{u}(t)} = \begin{bmatrix} -\left(\hat{k}_1(t) + \dfrac{Q(t)}{V}\right) & 0 & -\hat{b}(t) & 0 \\ -\hat{k}_1(t) & -\left(\hat{k}_2(t) + \dfrac{Q(t)}{V}\right) & -\hat{b}(t) & (c_s(t) - \hat{c}(t)) \\ 0 & 0 & 0 & 0 \\ 0 & 0 & 0 & 0 \end{bmatrix}$$

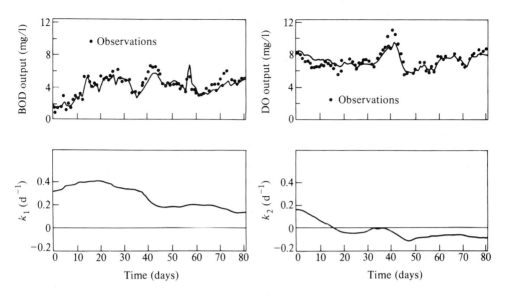

Figure 5-5-2 State and parameter estimation results for Case 1 (after Beck and Young, 1976).

The parameter and state estimation results for this case are shown in Fig. 5-5-2. The drift of both parameters k_1 and k_2 clearly demonstrates that the model chosen is not adequate, in particular, because k_2 assumes even negative values. Similarly, the state estimation results are not satisfactory, since the estimates during some subperiods are obviously biased.

It is therefore justified to relax the conditions $L = 0$ and $D = 0$ as is done in the following case.

Case 2 Random walk process for $L(t)$ and $D(t)$. Model (5-5-3) is now a CSTR model based on Dobbins Model, where $L(t)$ and $D(t)$ are assumed to follow a continuous *random walk* process, i.e.,

$$\dot{L}(t) = v_L(t)$$
$$\dot{D}(t) = v_D(t) \qquad (5\text{-}5\text{-}5)$$

where $v_L(t)$ and $v_D(t)$ are stationary white noises of known (guessed) variance. The parameters k_1 and k_2 are assumed to be undisturbed, as for Case 1. Therefore $\mathbf{x}(t)$ and $\mathbf{v}(t)$ in Eq. (5-5-1) now read

$$\mathbf{x}(t) = \begin{bmatrix} b(t) & c(t) & k_1 & k_2 & L(t) & D(t) \end{bmatrix}^T$$
$$\mathbf{v}(t) = \begin{bmatrix} v_b(t) & v_c(t) & 0 & 0 & v_L(t) & v_D(t) \end{bmatrix}^T$$

The estimation results for this case are shown in Fig. 5-5-3. The drift of k_1 and k_2 and the deviations are much smaller now. However, this is not surprising, since all discrepancies are taken up now by $L(t)$ and $D(t)$, which are allowed to change randomly and for which relatively large initial error covariances were assumed. Hence, Case 2 is hardly more useful for direct application than Case 1.

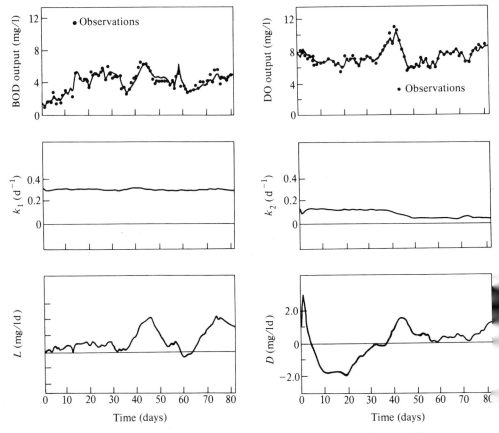

Figure 5-5-3 State and parameter estimation results for Case 2 (after Beck and Young, 1976).

However, what has been achieved through the estimation for Case 2 is that inspection of the random walks of $L(t)$ and $D(t)$ may yield suggestions for establishing relationships between $L(t)$ and $D(t)$ on the one hand, and external factors causing the seemingly random walks on the other. Once such a relationship has been discovered, it can be incorporated into the model and state and parameters of the modified model again can be estimated. In the case of Fig. 5-5-3 a relationship between sunlight and $L(t)$ and $D(t)$ has been discovered, and several attempts to model it have been undertaken and proved to give better state and parameter estimates (Fawcett et al., 1974; Beck, 1975; Beck and Young, 1976).

5-6 OTHER APPLICATIONS

It is certainly not an easy task to survey the applications and the areas of future research in the field of state and parameter estimation of river pollution without being too long and diffuse. In fact, both the number of models and the number of estimation methods are considerable, so that the number of combinations (model-method) which, strictly speaking, qualify an estimation problem becomes tremendously large, even if we consider that some combinations do not make sense. Of course, a survey of all these possibilities does not exist and would not even be of great help, since the actual need is for general guidelines for the selection of the model and of the method. Although such general guidelines may not emerge clearly from the preceding exposition, the reader has had the opportunity to become familiar with the main problems and difficulties of state and parameter estimation.

What is presently available in the literature is a relatively large number of applications or case studies in which, unfortunately, the attention is a priori focused on a particular estimation scheme, the selection of which is never qualified. Only a few exceptions to this attitude can be found, among which are the works of Shastry et al., (1973), and a series of works on the Cam river by Beck (1974, 1975, and 1976) and Young and Whitehead (1975). In the work by Shastry et al., (1973) a general nonlinear estimation procedure is used to estimate the parameters of many different models on the basis of BOD and DO data. The study shows how parameter estimation can be used in order to discriminate among different models and to what extent calibrated models can be considered reliable objects.

In the series of works on the Cam river (see also Sec. 5-5) the same problem, namely model discrimination, is dealt with from a more methodological point of view. Some rational criteria are mathematically defined and used to compare the reliability of different models. Moreover, the discussion is expanded in the direction of the methods (in particular the effectiveness of different estimation schemes), among which the maximum likelihood method and the *Instrumental Variable method* are shown.

More abstract and sophisticated studies can be found in the literature about some particular problems in state and parameter estimation. These works are mainly concerned with linear models and the results have not yet been applied to real cases. For this reason, they must be interpreted as first contributions in some new direction of research and certainly not as the final word on the argument. One of the most interesting investigations of this kind (see Sec. 5-2, 5-3) is related to the possibility of avoiding BOD measurements which, as already pointed out, are time consuming, very unreliable and quite costly. The simplest problem in this context relates to the possibility of estimating BOD values starting from DO measurements only, under the assumption that a model describing the behavior of the river is available. As we know from the preceding sections, this is a state estimation problem and in fact the attempts at its solution refer to the theory of state estimation of linear systems. If the model is linear and if it perfectly describes the dynamics of the variables involved (i.e., there is no process noise) and if there is no measurement noise on DO, then BOD values can be exactly determined if the system is completely observable (see Sec. 5-4). Since steady-state Streeter-Phelps models are completely observable even if a space discretization is performed, the first result is that the BOD values of a steady-state condition can be determined from samples of DO values of the same steady-state condition (see Koivo and Phillips, 1972). Moreover, the minimum number of samples needed to reconstruct the state vector is equal to the dimension of the state vector itself (see Sec. 5-4): thus, only two DO samples are necessary if a plug flow model is used, while four samples would be needed if dispersion were taken into account. More complex situations are obtained if reference is made to unsteady state conditions and to noise corrupted measurements, but still BOD can be reconstructed from continuous time and spatially discrete DO measurements as shown by Koivo and Koivo (1973), and from both time and spatially discrete DO measurements as shown by Tamura and Ueno (1975). All these results can possibly be generalized to the case of ecological models since the dissolved oxygen is in general sensitive to all state variables (loose interpretation of complete observability). Thus, a reasonable conjecture is, for example, the following: the variables describing the food chain of an n-th order ecological model in steady-state conditions can be determined from n spatial samples of DO.

A property similar to the one just investigated refers to the possibility of using only DO measurements in the calibration of the model as already discussed in Sec. 5-2 for the Bormida river. Results on this matter have been found by Koivo and Phillips (1971, 1972), Rinaldi and Soncini-Sessa (1974) and Rinaldi et al., (1976), but further research is certainly required for extensions and validation.

An identification problem substantially different from the problems considered until now is the identification of the pollution sources along the river. As shown by Whipple (1970) unrecorded loading is often a relevant part of the total loading so that schemes for its identification are of definite importance. Some contributions to this topic are available in the literature on linear distributed parameter systems. In particular, Ikeda et al., (1974) shows how this problem can be reduced to the solution of a *Fredholm integral equation* if a Streeter-Phelps model is used. Although the numerical results are very satisfactory, the applicability of Ikeda's procedures to real cases is quite limited since all the parameters of the model must be known in advance.

Finally, there are also studies concerned with the selection of the order of the system. Indeed, for some of the mathematical models described in Secs. 3-3, 4-3, namely, those in which the river is interpreted as a sequence of pools (and channels), the order is nothing but a parameter to be estimated. For some of these models (e.g., the Nash model) the selection of the order can be done in a relatively simple and intuitive way, while for other cases there is a definite need for a criterion for selection. Among the many criteria proposed it seems worth while to mention *Akaike's criterion* (Akaike, 1970) which takes into account the error covariance and the dimension of the data base. This criterion has been used by Tamura and Kawaguchi (1977) to estimate the order of the distributed-lag model by means of a recursive scheme.

REFERENCES

Section 5-1

Åström, K. J. and Eykhoff, P. (1971). System Identification: A Survey. *Automatica*, 7, 123–162.
Eykhoff, P. (1974). *System Identification*. John Wiley, London.
Isermann, R. (1974). *Process Identification* (in German). Springer, Berlin.
Mehra, R. K. and Tyler, J. S. (1973). Case Studies in Aircraft Parameter Identification, In *Identification and System Parameter Estimation, Part 1, Proc. 3rd IFAC Symposium, The Hague/Delft, The Netherlands. 12–15 June 1973.* North-Holland, Amsterdam.
Raiffa, H. (1968). *Decision Analysis: Introductory Lectures on Choices under Uncertainty.* Addison-Wesley, Reading, Mass.
Strobel, H. (1975). *Experimental Systems Analysis* (in German). Akademie-Verlag, Berlin.
Unbehauen, H., Göhring, B., and Bauer, B. (1974). *Parameter Estimation Techniques for Systems Identification* (in German). R. Oldenbourg, München, W. Germany.

Section 5-2

Marchetti, R. and Provini, A. (1969). Dissolved Oxygen Profile in Streams under Different Pollution Loads (in Italian). *Aria Acqua,* 7, 1–12.
Metcalf and Eddy, Inc. (1972). *Wastewater Engineering.* McGraw-Hill, New York.
Rinaldi, S., Romano, P., and Soncini-Sessa, R. (1976). Parameter Estimation of a Streeter-Phelps Type Water Pollution Model. In *Identification and Systems Parameter Estimation, Proc. 4th IFAC Symposium on Identification and System Parameter Estimation, Tbilisi, USSR, Sept. 1976.* Pergamon Press. London (to be published in *J. Env. Eng. Div., Proc. ASCE*).

Section 5-3

Bansal, M. K. (1973). Atmospheric Reaeration in Natural Streams. *Water Research*, 7, 769–782.
Bellman, R. E., Kagiwada, H., and Kalaba, R. E. (1966). Inverse Problems in Ecology. *J. Theoret. Biol.*, 11, 164–167.
Bellman, R. E. and Kalaba, R. E. (1965). *Quasilinearization and Nonlinear Boundary-Value Problems.* American Elsevier, New York.
Eykhoff, P. (1974). *System Identification.* John Wiley, London.
Lee, S. E. (1968). *Quasilinearization and Invariant Imbedding.* Academic Press, New York.
Löffler-Ertel, I. and Reichert, J. (1975). *Data Collection on the Water Balance of the Federal Republic of Germany* (in German). Institut für Systemtechnik und Innovationsforschung der Frauenhofer-Gesellschaft, Karlsruhe, W. Germany.
Mataušek, M. R. and Milovanović, M. D. (1973). Identification by Pseudosensitivity Functions and Quasilinearization. In *Identification and System Parameter Estimation, Part II* (P. Eykhoff, ed.), *Proc. 3rd IFAC Symposium, The Hague, 12–15 June 1973.* North-Holland, Amsterdam.
Negulescu, M. and Rojanski, V. (1969). Recent Research to Determine Reaeration Coefficient. *Water Research*, 3, 189–202.
Roberts, P. V. and Kreijci, V. (1975). *Cost-Benefit Analysis for Water Protection Planning* (in German). Eidg. Anstalt für Wasserversorgung, Abwasserreinigung und Gewässerschutz, Zürich, Switzerland.
Stehfest, H. (1973). Mathematical Modelling of Self-Purification of Rivers (in German; English translation available as report IIASA PP-77-11, International Institute for Applied Systems Analysis, Laxenburg, Austria). *Report KFK 1654 UF,* Kernforschungszentrum Karlsruhe, Karlsruhe, W. Germany.
Stehfest, H. (1978). On the Relationship Between Flow Rate and Organic Pollution Load of the Rhine River (in German). *GWF-Wasser/Abwasser,* 119, 302–305.

Section 5-4

Kalman, R. E. (1960). A New Approach to Linear Filtering and Prediction Problems. *Trans. ASME, Series D, J. of Basic Engineering*, **32**, 34–45.

Kwakernaak, H. and Sivan, R. (1972). *Linear Optimal Control Systems*. John Wiley, New York.

Luenberger, D. G. (1969). *Optimization by Vector Space Method*. John Wiley, New York.

Luenberger, D. G. (1971). An Introduction to Observers. *IEEE Trans. Automatic Control*, **AC-16**, 596–602.

Singh, M. G. (1973). *Some Applications of Hierarchical Control for Dynamical Systems*. Ph.D. Thesis, Cambridge University, Cambridge, Great Britain.

Singh, M. G. (1975). Multilevel State Estimation. *Int. J. Systems Sci.*, **6**, 533–555.

Tamura, H. (1976). Distributed-Lag Model of River Quality and Estimation of the Lag Coefficients (in Japanese). *Systems and Control* (J. Japan. Assoc. Automatic Control Engineers), **20**, 253–261.

Tamura, H. and Ueno, N. (1975a). Suboptimal Recursive Filter for the Distributed-Lag Model (SMART) (in Japanese). *Trans. Soc. Instr. Control Engineers of Japan*, **11**, 296–302.

Tamura, H. and Ueno, N. (1975b). Sequential Filter for the Distributed-Lag Model with Serially Connected Structure and its Application to State Estimation of Water Quality in a River (in Japanese). *Trans. Soc. Instr. Control Engineers of Japan*, **11**, 377–383.

Section 5-5

Beck, M. B. (1975). The Identification of Algal Population Dynamics in a Fresh Water Stream. In *Modeling and Simulation of Water Resources Systems* (G. C. Vansteenkiste, ed.). North-Holland, Amsterdam.

Beck, M. B. and Young, P. C. (1975). A Dynamic Model for DO–BOD Relationships in a Non-tidal Stream. *Water Research*, **9**, 769–776.

Beck, M. B. and Young, P. C. (1976). Systematic Identification of DO–BOD Model Structure. *J. Env. Eng. Div.*, *Proc. ASCE*, **102**, 909–927.

Fawcett, A., Taylor, N., Whitehead, P. G., and Young, P. C. (1974). Great Ouse Associated Committee, Bedford Ouse Study, First Annual Progress Report. Great Ouse River Authority, Huntington, England.

Jazwinski, A. H. (1970). *Stochastic Processes and Filtering Theory*. Academic Press, New York.

Lettenmaier, D. P. and Burges, S. J. (1976). Use of State Estimation Techniques in Water Resources System Modeling. *Water Resour. Bull.*, **12**, 83–99.

Section 5-6

Akaike, H. (1970). Statistical Predictor Identification. *Annals of the Institute of Statistical Mathematics*, **22**, 203–217.

Beck, M. B. (1974). Maximum Likelihood Identification Applied to DO–BOD–Algae Models for a Freshwater Stream. Report 7431 Lund Institute of Technology, Lund, Sweden.

Beck, M. B. (1975). The Identification of Algal Population Dynamics in a Freshwater Stream. In *Computer Simulation of Water Resources System* (G. C. Vansteenkiste ed.), North-Holland, Amsterdam.

Beck, M. B. (1976). The Identification of Biological Models in à Water Quality System. In *Identification and Systems Parameter Estimation*, *Proc. 4-th IFAC Symposium, Tbilisi, USSR, Sept. 1976*. Pergamon Press, London.

Ikeda, S., Miyamoto, S., and Sawaragi, Y. (1974). Identification Method in Environmental Systems and its Application to Water Pollution. *Int. J. of Systems Sci.*, **5**, 707–723.

Koivo, A. J. and Phillips, G. R. (1971). Identification of Mathematical Models for DO and BOD Concentrations in Polluted Streams from Noise Corrupted Measurements. *Water Resour. Res.*, **7**, 853–862.

Koivo, A. J. and Phillips, G. R. (1972). On Determination of BOD and Parameters in Polluted Stream Models from DO Measurements Only. *Water Resour. Res.*, **8**, 478–486.

Koivo, H. N. and Koivo, A. J. (1973). Optimal Estimation of Polluted Stream Variables. In *Identification and Systems Parameter Estimation, Proc. IFAC Symposium, The Hague Delft, The Netherlands, 12–15 June 1973*. North-Holland, Amsterdam.

Rinaldi, S., Romano, P., and Soncini-Sessa, R. (1976). Parameter Estimation of a Streeter-Phelps Type Water Pollution Model. In *Identification and Systems Parameter Estimation, Proc. 4-th IFAC Symposium, Tbilisi, USSR, Sept. 1976*. Pergamon Press, London (to be published in *J. Env. Eng. Div., Proc. ASCE*).

Rinaldi, S. and Soncini-Sessa, R. (1974). Sensitivity and Parameter Estimation Problems in River Pollution Modelling (in Italian). In *Environmental Systems Engineering* (S. Rinaldi, ed.). CLUP, Milano, Italy.

Shastry, J. S., Fan, L. T., and Erickson, L. E. (1973). Non-Linear Parameter Estimation in Water Quality Modeling. *J. Env. Eng. Div., Proc. ASCE*, **95**, 315–331.

Tamura, H. and Kawaguchi, T. (1977). On-Line Recursive System Identification by Multidimensional AR Models (in Japanese). *Trans. Soc. Instr. Control Engineers of Japan*, **13**, 14–20.

Tamura, H. and Ueno, N. (1975). Sequential Filter for the Distributed-Lag Model with Serially Connected Structure and its Application to State Estimation of Water Quality in a River (in Japanese). *Trans. Soc. Instr. Control Engineers of Japan*, **11**, 377–383.

Whipple, W. (1970). BOD Mass Balance and Water Quality Standards. *Water Resour. Res.*, **6**, 827–837.

Young, P. C. and Whitehead, P. G. (1975). A Recursive Approach to Time Series Analysis for Multivariable Systems. In *Computer Simulation of Water Resources Systems* (G. C. Vansteenkiste, ed.). North-Holland, Amsterdam.

CHAPTER
SIX

GENERAL REMARKS ON CONTROL

6-1 CONTROL PROBLEMS

Control problems arise every time the need arises to force a given system to behave in a desired way by suitably selecting its input function. That part of the input variable which must be selected is called *control* and denoted by **u** while the other is often called *disturbance* (or noise) and denoted by **v**. In distributed parameter systems the control variable **u** is, in general, a function of time and space, i.e., $\mathbf{u} = \mathbf{u}(t, l)$, so that the control problem consists in selecting the function $\mathbf{u}(\cdot, \cdot)$ in such a way that the behavior of the system is satisfactory. More precisely, control problems can be said to be characterized by the following four elements.

1. *The object to be controlled.* This is, in general, a dynamic system with two inputs, **u** and **v**, and an output **y** as shown in Fig. 6-1-1. The control **u** could be, for example, the flow released into the river from a reservoir in order to

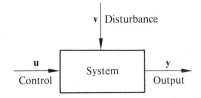

Figure 6-1-1 The system to be controlled.

augment the flow of the river and dilute pollutants, the disturbance **v** could be the natural flow rate of the river, the temperature of the water, and the rate of discharge of BOD at some given point, while the output **y** could be the DO concentration along the stretch under consideration. It is important to note that the decomposition of the input into control and disturbance does not, in general, depend upon the system but only on the particular alternative considered for the control action. For example, in the case mentioned above the control **u** could be taken as the rate of discharge of BOD if a wastewater treatment plant were to be selected, while it would be necessary to consider the flow released from the reservoir as a disturbance if the main purpose of the reservoir were flood control. Thus the decomposition of the input of the system into control and disturbance will be done on the basis of the facilities which can be used for water pollution control in each particular case.

2. *The desired behavior of the system.* This element of the control problem is often quite difficult to establish because it may be that the particular desired behavior cannot be obtained in practice. For example, we may wish the DO to be at the saturation level at any point of the river, but this would be an impossible target. Nevertheless, even if the desired behavior of the system is not attainable it must be specified in order that the control problem is well defined.

3. *The constraints.* As mentioned in Sec. 1-1, input **u**, state **x** and output **y** of a system are in general constrained to belong to particular sets **U**, **X**, and **Y**. In the general definition (see Sec. 1-1) these constraints are essential components of the system itself; thus they should be thought of as a part of the first element of the control problem (see point 1 above). Nevertheless, other constraints must in general, be considered when solving a control problem. In river pollution control it is possible to have, for example, limitations on the damage to the environment induced by water pollution. Constraints of this kind are very difficult to express in quantitative terms since the environmental damage is a complex function of the input and state of the system. Therefore, a well established and safe way of circumventing this problem is to substitute constraints of this kind by suitable constraints on some of the variables of the system. Examples are upper bounds on the rate of discharge of BOD, or on its concentration in the wastewater (*effluent standards*), constraints on BOD and DO at some particular point of the river (*stream standards*), or combinations of both. Other constraints take into account the limitations due to the particular technology used to control the system and possible budget limitations.

4. *The criterion.* This is, in general, a suitable criterion which allows discrimination among many solutions satisfying the constraints when no one of them corresponds to the desired behavior of the system. *Stability* of the system (namely ability of the system to absorb perturbations of the initial state) and *insensitivity* of the output with respect to the disturbances have been for years the basic criteria in different areas of application of control theory (see, for example, Truxal, 1955; Chestnut and Mayer, 1963; Gibson, 1963; Hsu and Meyer, 1968; and Cruz, 1972).

More recently, *optimality* of the solution has been used as a more rational criterion for selecting the control function. This approach needs the definition of a *performance index* which must be minimized or maximized with respect to all input functions satisfying the constraints. The *performance index* is often a quantitative measure of the deviation of the behavior of the system from the *desired behavior* and is therefore a functional of input, state, and output functions. In these cases, the problem is called an *optimal control problem* and quite sophisticated mathematical tools such as calculus of variations, dynamic programming, maximum principle, and Hamilton-Jacoby theory are required for its solution (see, for example, Pontryagin et al., 1962; Athans and Falb, 1966; Lee and Markus, 1967; Anderson and Moore, 1971; and Bryson and Ho, 1975). Unfortunately, optimal control problems can be solved analytically only in very special cases (linear system and quadratic performance index) and the general mathematical tools mentioned above are often of little help for the actual solution of the problem. This is particularly true in water pollution problems because of the many constraints and of the strong nonlinearities involved. Therefore, the approach which has been most widely used in optimal control of river pollution is that of discretizing the variables in space and time in order to transform the functional problem into a finite dimensional optimization problem. This approach, which could be reasonably called "brute force approach," has already been proved to work on problems of relatively large dimensions and its validity will possibly be confirmed by further investigations.

The optimization problem obtained by discretizing the optimal control problem is in general of the form

$$\min_{\mathbf{z}} J(\mathbf{z}) \qquad (6\text{-}1\text{-}1\text{a})$$

subject to the constraint

$$\mathbf{z} \in \mathbf{Z} \qquad (6\text{-}1\text{-}1\text{b})$$

where $\mathbf{z}$ is a finite dimensional vector. The *feasibility set* $\mathbf{Z}$ is usually defined by a finite number of equality and inequality constraints on the vector $\mathbf{z}$, and reflects the system behavior and the constraints on the problem (see point 3). Problems of this form belong to the area of *mathematical programming* which is briefly surveyed in the next section.

Given the four elements constituting a general control problem, different subproblems can be obtained if some further simplifying assumptions are made.

A natural way of describing the disturbance would be to assume that $\mathbf{v}$ is a suitable stochastic process (see Chapter 5). This assumption implies the need to deal with a *stochastic control problem*. Although these problems have been extensively analyzed in the last decade (see, for example, Feldbaum, 1965; Sworder, 1966; and Aoki, 1967) their solution still requires a prohibitive amount of work. This is the reason why *deterministic control problems*, where the noise is assumed to be a priori known (nominal value) or at least measurable, have been considered more often in the literature.

Of course, the selection of the nominal value of the disturbance can be done in many different ways; nevertheless, the "mean value of **v**" and the "worst possible value of **v**" are the nominal values most widely used in all applications of control theory. This is also the case in river pollution, where reference is made to the mean value of the river flow rate, or, more often, to the critical low flow–high temperature conditions which are a priori assumed as the worst possible conditions, although this can be false as has been shown via sensitivity analysis in Secs. 4-1 and 5-3.

Another particular problem is the one known as *steady state control* as opposed to the general control problem, called *unsteady state control* problem. In steady state control problems all variables are assumed to be constant in time so that the problem consists in finding a satisfactory (or optimal) equilibrium. If lumped parameter systems of the kind

$$\dot{\mathbf{x}}(t) = \mathbf{f}(\mathbf{x}(t), \mathbf{u}(t)) \qquad \mathbf{x}(t) \in X \qquad \mathbf{u}(t) \in U$$

$$\mathbf{y}(t) = \boldsymbol{\eta}(\mathbf{x}(t)) \qquad \mathbf{y}(t) \in Y$$

are considered, the optimal steady state control is formulated as

$$\min J(\mathbf{u}, \mathbf{x}, \mathbf{y})$$

subject to the constraints

$$\mathbf{f}(\mathbf{x}, \mathbf{u}) = \mathbf{0}$$

$$\mathbf{y} - \boldsymbol{\eta}(\mathbf{x}) = \mathbf{0}$$

$$\mathbf{u} \in U \qquad \mathbf{x} \in X \qquad \mathbf{y} \in Y$$

where J is the performance index. This problem is obviously a mathematical programming problem since it is of the form (6-1-1) with $\mathbf{z} = [\mathbf{x}^T \mathbf{u}^T \mathbf{y}^T]^T$. Thus, optimal steady state control problems of lumped parameter systems do not need any discretization in order to be transformed into mathematical programming problems. Unfortunately, river pollution models are distributed parameter models so that steady state control problems are still of the functional type. Space discretization will in general be required in order to transform such problems into more manageable problems.

In the following, a distinction will be made between steady state control problems which are investigated in Chapter 8 and unsteady state control problems, to which Chapter 9 is devoted. While the first problems can be solved by straightforward application of the methodologies of mathematical programming, the latter actually require a deeper knowledge of the main ideas of control theory. Some of these ideas and the alternatives in the implementation of the control functions (open-loop, feedback, feedforward control) are therefore described in Sec. 9-1.

The problems dealt with in Chapters 8 and 9 are characterized by a unique decision maker and a unique objective function, while river quality decision problems are often characterized by more sophisticated structures. This is, in particular, true for river basin management problems, some of which are therefore briefly considered in the last chapter of the book.

6-2 MATHEMATICAL PROGRAMMING

In this section the essential ideas and results of that branch of mathematics called *mathematical programming* are briefly described. The basic problem consists in finding the minimum of a function (called the *objective function*) subject to equality and inequality constraints, i.e.,

$$\min_{\mathbf{z}} [J(\mathbf{z})] \qquad (6\text{-}2\text{-}1a)$$

subject to

$$g_i(\mathbf{z}) = 0, \quad i = 1, 2, \ldots, m \qquad (6\text{-}2\text{-}1b)$$

$$h_j(\mathbf{z}) \le 0, \quad j = 1, 2, \ldots, l \qquad (6\text{-}2\text{-}1c)$$

where the n-dimensional vector

$$\mathbf{z} = [z_1 z_2 \ldots z_n]^T$$

is called the *decision vector*. Obviously, maximization problems can be converted into equivalent minimization problems, since $\min [J(\mathbf{z})] = -\max [-J(\mathbf{z})]$. Depending upon the form of J, g_i, and h_j particular problems are obtained like linear programming, quadratic programming, geometric programming, convex programming, nonlinear programming, and so forth.

Linear Programming

When all functions J, g_i, and h_j are linear, the problem becomes a *linear programming* problem. These problems can always be written in the form

$$\min_{\mathbf{z}} [J(\mathbf{z})] = \min_{\mathbf{z}} [r_1 z_1 + r_2 z_2 + \cdots + r_n z_n] \qquad (6\text{-}2\text{-}2a)$$

subject to

$$\begin{aligned} p_{11} z_1 + p_{12} z_2 + \cdots + p_{1n} z_n &= q_1 \\ p_{21} z_1 + p_{22} z_2 + \cdots + p_{2n} z_n &= q_2 \\ &\vdots \\ p_{m1} z_1 + p_{m2} z_2 + \cdots + p_{mn} z_n &= q_m \end{aligned} \qquad (6\text{-}2\text{-}2b)$$

$$z_1 \ge 0 \quad z_2 \ge 0 \quad \ldots \quad z_n \ge 0 \qquad (6\text{-}2\text{-}2c)$$

or in vector-matrix form

$$\min_{\mathbf{z}} [J(\mathbf{z})] = \min_{\mathbf{z}} [\mathbf{r}^T \mathbf{z}] \qquad (6\text{-}2\text{-}3a)$$

subject to

$$Pz = q \quad (6\text{-}2\text{-}3b)$$

$$z \geq 0 \quad (6\text{-}2\text{-}3c)$$

Linear programming problems of the form (6-2-2) (or (6-2-3)) are said to be in *standard form* (all the constraints are equality constraints and all variables z_j are assumed to be nonnegative). If the original problem is not in this form, it can easily be transformed into the standard form by using *slack variables* for inequality constraints, and by using two nonnegative variables instead of one unconstrained variable.

The set

$$\mathbf{K} = \{\mathbf{z}: \mathbf{Pz} = \mathbf{q}, \mathbf{z} \geq \mathbf{0}\}$$

is the set of all feasible solutions and is therefore called *feasibility set*. In a linear programming problem the feasibility set $\mathbf{K}$ is always a convex set. In fact if $\mathbf{z}^{(1)}$ and $\mathbf{z}^{(2)}$ are two feasible solutions then

$$\mathbf{Pz}^{(1)} = \mathbf{q} \qquad \mathbf{Pz}^{(2)} = \mathbf{q}$$

$$\mathbf{z}^{(1)} \geq \mathbf{0} \qquad \mathbf{z}^{(2)} \geq \mathbf{0}$$

Then, if $\mathbf{z}$ is a convex combination of $\mathbf{z}^{(1)}$ and $\mathbf{z}^{(2)}$, i.e.,

$$\mathbf{z} = \alpha \mathbf{z}^{(1)} + (1 - \alpha)\mathbf{z}^{(2)} \quad \text{with} \quad 0 \leq \alpha \leq 1$$

we have $\mathbf{z} \geq \mathbf{0}$ and

$$\mathbf{Pz} = \mathbf{P}[\alpha \mathbf{z}^{(1)} + (1-\alpha)\mathbf{z}^{(2)}] = \alpha \mathbf{Pz}^{(1)} + (1-\alpha)\mathbf{Pz}^{(2)} = \alpha \mathbf{q} + (1-\alpha)\mathbf{q} = \mathbf{q}$$

Therefore, $\mathbf{z}$ is also a feasible solution, which is equivalent to saying that $\mathbf{K}$ is convex. Furthermore, since $\mathbf{K}$ is determined by the intersection of a finite number of linear constraints, $\mathbf{K}$ can either be an empty set, a *convex polyhedron* or a convex region which is unbounded in some direction. In general, in well posed problems, $\mathbf{K}$ is a convex polyhedron characterized by a finite number of extreme points. In Fig. 6-2-1a the projection of a set $\mathbf{K}$ on the subspace of two decision variables z_1 and z_2 is shown as an example.

Since the objective function $J(\mathbf{z})$ is linear its contour lines $J(\mathbf{z}) = \text{const.}$ are represented by hyperplanes in the decision space, as shown in Figs. 6-2-1b and 6-2-1c. Thus, the objective function $J(\mathbf{z})$ assumes its minimum at an extreme point of the convex polyhedron $\mathbf{K}$. If it assumes its minimum at more than one extreme point (see Fig. 6-2-1c), then it takes the same value for every convex combination of the optimal extreme points.

In order to obtain the optimal solution of a linear programming problem, it is only necessary to look at the extreme points of the convex polyhedron $\mathbf{K}$. Consistently, the best known solution method, called the *simplex method*, is a scheme for generation of extreme points. By using this method the optimal

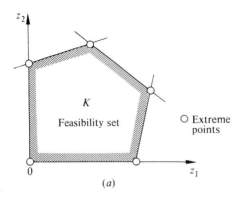

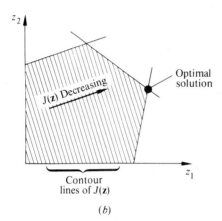

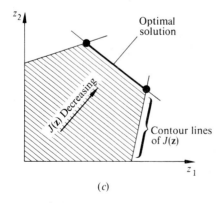

Figure 6-2-1 Geometric interpretation of linear programming:
(a) convex polyhedron and its extreme points
(b) unique optimal solution
(c) optimal solution at two extreme points.

solution is obtained in a finite number of iterative steps starting from any extreme point of **K**. The simplex method is composed of the following two phases:

Phase 1. Find an initial extreme point (starting point).
Phase 2. Proceed from the starting point to a new extreme point characterized by a smaller value of the objective function $J(\mathbf{z})$ and then iterate until the optimal solution is obtained.

The simplex method and its variations have been implemented on all general purpose computers. The most powerful programs can solve problems with thousands of constraints and variables.

Associated with the linear program (6-2-3), called *primal problem*, is a *dual problem* which is as follows:

$$\max_{\lambda} [\phi(\lambda)] = \max_{\lambda} [\lambda^T \mathbf{q}] \qquad (6\text{-}2\text{-}4a)$$

subject to

$$\lambda^T \mathbf{P} \leq \mathbf{r}^T \qquad (6\text{-}2\text{-}4b)$$

where λ is called the *dual variable*. Equations (6-2-3) and (6-2-4) are called *primal-dual pair* with primal in standard form. There is a one-to-one correspondence between the i-th dual variable λ_i and the i-th primal constraints in Eq. (6-2-2b)

$$p_{i1}z_1 + p_{i2}z_2 + \cdots + p_{in}z_n = q_i$$

The following is a *symmetric* primal-dual pair:

Primal	Dual
$\min_{\mathbf{z}} [J(\mathbf{z})] = \min_{\mathbf{z}} [\mathbf{r}^T \mathbf{z}]$	$\max_{\lambda} [\phi(\lambda)] = \max_{\lambda} [\lambda^T \mathbf{q}]$
subject to	subject to
$\mathbf{Pz} \geq \mathbf{q}$	$\lambda^T \mathbf{P} \leq \mathbf{r}^T$
$\mathbf{z} \geq \mathbf{0}$	$\lambda \geq \mathbf{0}$

The primal-dual relationships can be intuitively understood, even for more general mathematical programming problems, by looking at the geometric interpretation shown in Fig. 6-2-2 (Luenberger, 1969): the minimum distance of a reference point P from a convex set **K** is equal to the maximum distance from the point P to a hyperplane separating the point P and the convex set. Thus, the primal problem of minimization over vectors ($\mathbf{z}$) can be converted into the dual problem of maximization over hyperplanes (λ).

If $\mathbf{z}$ and λ are feasible primal and dual solutions and $\mathbf{z}^0$ and λ^0 are the optimal primal and dual solutions, then

$$\phi(\lambda) = \lambda^T \mathbf{q} \leq \lambda^{0T} \mathbf{q} = \mathbf{r}^T \mathbf{z}^0 \leq \mathbf{r}^T \mathbf{z} = J(\mathbf{z}) \qquad (6\text{-}2\text{-}5)$$

This relationship can also be seen in Fig. 6-2-2, where $J(\mathbf{z})$ corresponds to the distance of the reference point P from a point $\mathbf{z}$ in the convex set and $\phi(\lambda)$ corresponds to the distance of the reference point P from the hyperplane λ.

212 GENERAL REMARKS ON CONTROL

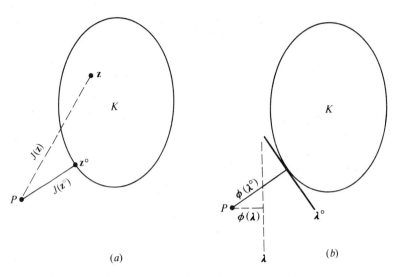

Figure 6-2-2 Geometric interpretation of primal–dual relationship.

Given any linear program, we can choose whether to solve the primal or the dual problem. In general, it is preferable to be oriented towards the one which has less constraints since it is easier to solve. The dual variables λ_i can be interpreted as *prices*, and are of great value in *sensitivity analysis*. In fact, from Eq. (6-2-5) we have

$$J(\mathbf{z}^0) = \sum_{i=1}^{m} \lambda_i^0 q_i \qquad (6\text{-}2\text{-}6)$$

from which it follows that if the i-th constraint is perturbed from q_i to $(q_i + \Delta q_i)$ the value of the objective function will be perturbed by $\lambda_i^0 \Delta q_i$ so that, going to the limit for $\Delta q_i \to 0$, we obtain

$$\frac{\partial J(\mathbf{z}^0)}{\partial q_i} = \lambda_i^0 \qquad i = 1, 2, \ldots, m$$

Hence, λ_i^0 can be interpreted as the "price" one has to pay (in terms of the objective function) if the i-th constraint is relaxed.

Quadratic Programming

Quadratic programming problems can be stated as follows:

$$\min_{\mathbf{z}} [J(\mathbf{z})] = \min_{\mathbf{z}} [\mathbf{r}^T \mathbf{z} + \tfrac{1}{2} \mathbf{z}^T \mathbf{S} \mathbf{z}] = \min_{\mathbf{z}} \left[\sum_{j=1}^{n} r_j z_j + \frac{1}{2} \sum_{i=1}^{n} \sum_{j=1}^{n} s_{ij} z_i z_j \right] \qquad (6\text{-}2\text{-}7a)$$

subject to

$$\mathbf{P}\mathbf{z} = \mathbf{q} \qquad (6\text{-}2\text{-}7b)$$

$$\mathbf{z} \geq \mathbf{0} \quad (6\text{-}2\text{-}7c)$$

where the $n \times n$ matrix $\mathbf{S}$ is assumed to be symmetric and positive semidefinite. This problem can be converted into a particular linear programming problem by using the well-known *Kuhn-Tucker conditions* for optimality, and efficient algorithms like the methods of Wolfe (1959) and Beale (1959) are available for solving it. The computer programs for quadratic programming problems are not as popular as those for linear programs, although the size of the problems which can be solved by using these programs is quite relevant (a few hundreds of constraints).

Geometric Programming

Geometric programming treats the problem of minimizing *posynomial* functions. The primal problem is as follows:

$$\min_{\mathbf{z}} [J(\mathbf{z})] = \min_{\mathbf{z}} [g_0(\mathbf{z})] \quad (6\text{-}2\text{-}8a)$$

subject to

$$g_k(\mathbf{z}) \leq 1 \quad k = 1, 2, \ldots, p \quad (6\text{-}2\text{-}8b)$$

$$z_i \geq 0 \quad i = 1, 2, \ldots, n \quad (6\text{-}2\text{-}8c)$$

where each function g_k, $k = 0, 1, \ldots, p$, is given by

$$g_k(\mathbf{z}) = \sum_{i \in \Omega[k]} c_i z_1^{a_{i1}} z_2^{a_{i2}} \ldots z_n^{a_{in}} \quad (6\text{-}2\text{-}9)$$

with

$$\Omega[k] = \{n_k, n_k + 1, n_k + 2, \ldots, m_k\}$$

where

$$m_0, m_1, \ldots, m_p(= m)$$

are given integers of increasing magnitude and

$$n_0 = 1, n_1 = m_0 + 1, n_2 = m_1 + 1, \ldots, n_p = m_{p-1} + 1$$

In Eq. (6-2-9) the exponents a_{ij} are arbitrary real numbers, while the coefficients c_i are assumed to be positive. With these restrictions the functions $g_k(\mathbf{z})$ are called *posynomials*, while the $m \times n$ matrix $[a_{ij}]$ is called the exponent matrix.

The dual formulation of a geometric programming problem is the following:

$$\max_{\lambda} [\phi(\lambda)] = \max_{\lambda} \left[\prod_{i=1}^{m} \left(\frac{c_i}{\lambda_i} \right)^{\lambda_i} \prod_{k=1}^{p} \mu_k(\lambda)^{\mu_k(\lambda)} \right] \quad (6\text{-}2\text{-}10a)$$

subject to

$$\sum_{i=1}^{m} a_{ij} \lambda_i = 0 \quad j = 1, 2, \ldots, n \quad (6\text{-}2\text{-}10b)$$

$$\sum_{i \in \Omega[0]} \lambda_i = 1 \quad \lambda_i \geq 0 \quad (6\text{-}2\text{-}10c)$$

where
$$\mu_k(\lambda) = \sum_{i \in \Omega[k]} \lambda_i \qquad k = 1, 2, \ldots, p$$

Each nontrivial primal problem can be formulated in such a way that its exponent matrix has rank n (the number of primal variables) with n being strictly less than m. The *degree of difficulty* d of a geometric programming problem is defined as

$$d = m - n - 1$$

and indicates the number of independent variables over which the dual function $\phi(\lambda)$ is to be maximized. If $d = 0$, the dual problem has a unique feasible solution, which is easily determinable, because the dual constraints are linear. If $d > 0$ more effort is required (see Duffin et al., 1967).

Dynamic Programming

The problems to which *dynamic programming* applies are the so-called multistage decision problems. These problems are characterized by a recursive structure of the constraints (each one defining one stage) and by the fact that the objective function can be written as the sum of many components, each one being associated with a stage of the problem. A simple and understandable presentation of dynamic programming is obtained by decomposing the vector $\mathbf{z}$ into two groups of variables. Since these variables will in the following correspond to the state and control variables of the system we will name them $\mathbf{x}$ and $\mathbf{u}$. More precisely, the multistage decision problems to which we apply dynamic programming are

$$\min_{\mathbf{x},\mathbf{u}} [J(\mathbf{x}, \mathbf{u})] = \min_{\mathbf{x},\mathbf{u}} \left[\sum_{k=0}^{K} g_k(\mathbf{x}(k), \mathbf{u}(k)) \right] \qquad (6\text{-}2\text{-}11\text{a})$$

subject to

$$\mathbf{x}(k+1) = \mathbf{f}_k(\mathbf{x}(k), \mathbf{u}(k)) \qquad (6\text{-}2\text{-}11\text{b})$$

$$\mathbf{x}(k) \in \mathbf{X}_k \qquad \mathbf{u}(k) \in \mathbf{U}_k \qquad k = 0, 1, \ldots, K \qquad (6\text{-}2\text{-}11\text{c})$$

Since state and control variables are not independent (see Eq. (6-2-11b)) the minimization problem (6-2-11a) is often written in the form

$$\min_{\mathbf{u}} [J(\mathbf{x}, \mathbf{u})]$$

Assume now that a particular subproblem of problem (6-2-11) has been solved, namely the one corresponding to the constraints $k, k+1, \ldots, K$ and suppose that this subproblem was solved for all possible values of $\mathbf{x}(k)$. Thus, the function

$$H_k \mathbf{x}(k) = \min_{\mathbf{u}(k)\ldots} \left[\sum_{j=k}^{K} g_j(\mathbf{x}(j), \mathbf{u}(j)) \right] \qquad (6\text{-}2\text{-}12)$$

is known.

A basic iterative relation for H_k called *functional equation* can easily be derived as follows

$$\begin{aligned}
H_k(\mathbf{x}(k)) &= \min_{\mathbf{u}(k)\ldots} \left[g_k(\mathbf{x}(k), \mathbf{u}(k)) + \sum_{j=k+1}^{K} g_j(\mathbf{x}(j), \mathbf{u}(j)) \right] \\
&= \min_{\mathbf{u}(k)} \min_{\mathbf{u}(k+1)\ldots} \left[g_k(\mathbf{x}(k), \mathbf{u}(k)) + \sum_{j=k+1}^{K} g_j(\mathbf{x}(j), \mathbf{u}(j)) \right] \\
&= \min_{\mathbf{u}(k)} \left[g_k(\mathbf{x}(k), \mathbf{u}(k)) + \min_{\mathbf{u}(k+1)\ldots} \sum_{j=k+1}^{K} g_j(\mathbf{x}(j), \mathbf{u}(j)) \right] \\
&= \min_{\mathbf{u}(k)} \left[g_k(\mathbf{x}(k), \mathbf{u}(k)) + H_{k+1}(\mathbf{f}_k(\mathbf{x}(k), \mathbf{u}(k))) \right] \qquad (6\text{-}2\text{-}13)
\end{aligned}$$

Equation (6-2-13) describes an iterative relation for determining H_k from the knowledge of H_{k+1}. The optimal decision $\mathbf{u}^0(k)$ is determined for each $\mathbf{x}(k)$ by minimizing the quantity in brackets in Eq. (6-2-13). The functional equation (6-2-13), known as Bellman's *principle of optimality*, says that the optimal decision at each stage is that one which minimizes the sum of the component of the objective function associated with the stage and the minimum of the objective function associated with the remaining stages.

To start the computation, it is necessary to determine $H_K(\mathbf{x}(K))$ for all $\mathbf{x}(K) \in \mathbf{X}_K$. Usually, no control is applied at $k = K$, and hence the cost function at K depends only on the final state $\mathbf{x}(K)$, i.e.,

$$H_K(\mathbf{x}(K)) = g_K(\mathbf{x}(K))$$

The advantage of dynamic programming is that it can handle nonlinear systems, as well as state and control constraints. However, the disadvantage is that it has enormous computational and storage requirements even for fairly low dimensional problems. Ways of reducing the computational requirements are described in Larson (1968).

Nonlinear Programming

In nonlinear programming problems it is important to distinguish between global and local minima. We say that a function $J(\mathbf{z})$ has a *global minimum* at $\mathbf{z}^0$ if

$$J(\mathbf{z}^0) \leq J(\mathbf{z}) \qquad \text{for all } \mathbf{z}$$

while we say that $J(\mathbf{z})$ has a *local minimum* at $\mathbf{z}^0$ if there exists a positive ε so that

$$J(\mathbf{z}^0) \leq J(\mathbf{z}) \qquad \text{for all } \mathbf{z} \text{ such that } \|\mathbf{z} - \mathbf{z}^0\| \leq \varepsilon$$

Unfortunately, with the majority of the available methods, one can generally only be sure of obtaining a local minimum.

If $J(\mathbf{z})$ is a convex function, and if the feasibility set $\mathbf{K}$ is convex, the nonlinear programming problem

$$\min_{\mathbf{z}} [J(\mathbf{z})]$$

subject to

$$\mathbf{z} \in \mathbf{K}$$

is called a *convex programming problem*. Problems of this type have only one minimum, hence the global minimum will be obtained for sure with any convergent algorithm.

Unconstrained nonlinear programming Problems of this kind are simply given by

$$\min_{\mathbf{z}} [J(\mathbf{z})] \tag{6-2-14}$$

As is well known, if $J(\mathbf{z})$ is continuous and has continuous derivatives, a first order necessary condition for a local minimum in $\mathbf{z}^0$ is

$$\left[\frac{\partial J}{\partial \mathbf{z}} \right]_{\mathbf{z}^0} = \left[\frac{\partial J}{\partial z_1} \quad \frac{\partial J}{\partial z_2} \quad \cdots \quad \frac{\partial J}{\partial z_n} \right]_{\mathbf{z}^0} = \mathbf{0} \tag{6-2-15}$$

while a sufficient condition for a local minimum in a point $\mathbf{z}^0$ satisfying the first order condition given by Eq. (6-2-15) is that the Hessian matrix

$$\left[\frac{\partial^2 J(\mathbf{z})}{\partial \mathbf{z}^2} \right]_{\mathbf{z}^0} = \begin{bmatrix} \frac{\partial^2 J}{\partial z_1^2} & \frac{\partial^2 J}{\partial z_1 \partial z_2} & \cdots & \frac{\partial^2 J}{\partial z_1 \partial z_n} \\ \vdots & \vdots & & \vdots \\ \frac{\partial^2 J}{\partial z_n \partial z_1} & \frac{\partial^2 J}{\partial z_n \partial z_2} & \cdots & \frac{\partial^2 J}{\partial z_n^2} \end{bmatrix}_{\mathbf{z}^0}$$

is positive definite.

A straightforward approach for solving problem (6-2-14) would be to find all solutions of Eq. (6-2-15), which is a set of n nonlinear equations with n unknowns, and then determine which are the corresponding Hessian matrices that are positive definite. However, the solution of a system of n equations in n unknowns is, in general, a problem which is as difficult as the original problem (6-2-14) so that no advantage is obtained by using the necessary and sufficient conditions for optimality. This is why direct search methods like *steepest descent* method, *conjugate gradient* method, *variable metric* method and others are usually preferred. All these methods are iterative and are characterized by a recursive equation of the form

$$\mathbf{z}^{i+1} = \mathbf{z}^i + \alpha_i \mathbf{s}^i \qquad i = 1, 2, \ldots \tag{6-2-16}$$

where $\mathbf{z}^i$ and $\mathbf{z}^{i+1}$ are the solutions of the i-th and $(i+1)$-th iterations, $\mathbf{s}^i$ is the *search direction* (n-dimensional vector) and α_i is the *step length*. The way of determining the search direction $\mathbf{s}^i$ varies from one method to another. For example, in the *steepest descent* method the search direction is normal to the contour line of the objective function

$$\mathbf{s}^i = -\left[\frac{\partial J(\mathbf{z})}{\partial \mathbf{z}}\right]^T_{\mathbf{z}^i}$$

while in *Newton's method*

$$\alpha_i \mathbf{s}^i = -\left[\frac{\partial^2 J(\mathbf{z})}{\partial \mathbf{z}^2}\right]^{-1}_{\mathbf{z}^i}\left[\frac{\partial J(\mathbf{z})}{\partial \mathbf{z}}\right]^T_{\mathbf{z}^i}$$

which corresponds to approximating $J(\mathbf{z})$ with a quadratic function and then selecting the perturbation $\alpha_i \mathbf{s}^i$ in such a way that the new point $\mathbf{z}^{i+1}$ is the minimum of the approximating function. The steepest descent method is also called first-order gradient method since it uses only first order derivatives, while Newton's method is called second-order gradient method. Second order gradient methods are more complex to apply but their speed of convergence is generally greater. For this reason, some methods like *variable metric method* (Fletcher and Powell, 1963) and *conjugate gradient method* (Fletcher and Reeves, 1964) have been developed to obtain the speed of convergence of Newton's method without using the second derivatives of the objective function. In all direct search methods, which do not use second derivatives, the step length α_i must be carefully selected at each iteration in order to obtain a high speed of convergence. For this purpose the step length minimizing the objective function in the search direction $\mathbf{s}^i$ is usually determined, which means that the following one-dimensional optimization problem must be solved:

$$\min_{\alpha_i} J(\mathbf{z}^i + \alpha_i \mathbf{s}^i)$$

where $\mathbf{z}^i$ and $\mathbf{s}^i$ are given. Very efficient methods like *cubic interpolation, quadratic interpolation, Fibonacci search* and others are available for solving this kind of minimization problem.

Constrained nonlinear programming Among the *constrained nonlinear programming* problems, the easiest one to analyze is the problem with *equality constraints* only, since this problem can be transformed into an unconstrained one by means of the *classical Lagrange multiplier method*. The problem to be solved is

$$\min_{\mathbf{z}} [J(\mathbf{z})] \qquad (6\text{-}2\text{-}17a)$$

subject to

$$g_i(\mathbf{z}) = 0 \qquad i = 1, 2, \ldots, m < n \qquad (6\text{-}2\text{-}17b)$$

The Lagrange multiplier method converts the constrained problem (6-2-17) into the equivalent unconstrained problem

$$\min_{\mathbf{z},\lambda} [L(\mathbf{z}, \lambda)] = \min_{\mathbf{z},\lambda} \left[J(\mathbf{z}) + \sum_{i=1}^{m} \lambda_i g_i(\mathbf{z}) \right] \qquad (6\text{-}2\text{-}18)$$

where the function $L(\mathbf{z}, \lambda)$ is the *Lagrangian* and the λ_i are the *Lagrange multipliers*. The first order necessary condition for the local minimum is then given by

$$\frac{\partial L(\mathbf{z}, \lambda)}{\partial \mathbf{z}} = \frac{\partial J(\mathbf{z})}{\partial \mathbf{z}} + \sum_{i=1}^{m} \lambda_i \frac{\partial g_i(\mathbf{z})}{\partial \mathbf{z}} = \mathbf{0} \qquad (6\text{-}2\text{-}19a)$$

$$\frac{\partial L(\mathbf{z}, \lambda)}{\partial \lambda} = [g_1(\mathbf{z}) \quad g_2(\mathbf{z}) \quad \cdots \quad g_m(\mathbf{z})]^T = \mathbf{0} \qquad (6\text{-}2\text{-}19b)$$

The geometric interpretation of the necessary conditions (6-2-19) is shown in Fig. 6-2-3: Equation (6-2-19b) is exactly the equality constraint (6-2-17b), while Eq. (6-2-19a) says that at the optimal solution the gradient of the objective function must be a linear combination of the gradients of all equality constraints. Thus, the techniques described above for the unconstrained case can be applied to Eq. (6-2-18), although specific and efficient methods exist for the solution of problem (6-2-17) (in particular for the case of linear constraints g_i). When there are inequality constraints, the problem may still be transformed into an unconstrained one, which, however is more difficult from an analytical point of view. In this case the *Kuhn-Tucker conditions*, a generalization of the first order Lagrange necessary conditions (Eq. (6-2-19)), can be used under suitable regularity assumptions for the constraints (Kuhn-Tucker qualification).

Computational methods for constrained problems vary according to the form of the constraints. For the problem of minimizing nonlinear objective functions subject to linear constraints, one efficient approach is the *convex simplex method* (Zangwill, 1967) which repeatedly applies the simplex method to the linearized objective function. Another approach is the *method of feasible direction* which is also relatively efficient. The idea is to pick up a starting point satisfying the

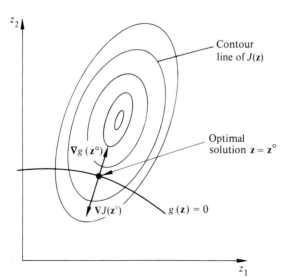

Figure 6-2-3 Geometric interpretation of the classical Lagrange multiplier method.

constraints and to find a search direction, called feasible direction, such that a small move in that direction does not violate any constraints and improves the objective function.

For the problem of minimizing nonlinear objective functions subject to nonlinear constraints, one of the most efficient approaches is the *generalized reduced gradient method* (Abadie and Carpentier, 1969). The idea of this method is identical to that of the convex simplex method (the linear constraints being replaced by a linearization of the nonlinear constraints). Another approach is the so-called *penalty function method* which transforms a constrained problem into an unconstrained one by adding to the objective function a suitably weighted penalty term corresponding to the constraints. Then any gradient technique for unconstrained problems can be directly used. The limitation of this method is that the resulting unconstrained problem will be ill conditioned when the penalty becomes very high.

All the computational techniques described above generate only local optimal solutions for nonconvex problems. Thus, more sophisticated approaches based on generalized Lagrangian functions (see Hestenes, 1969; and Nakayama et al., 1975) are necessary to be sure of obtaining global optimal solutions.

Large-Scale Mathematical Programming

Mathematical programming problems formulated for large problems (many variables and constraints) have almost always a very special structure. *Large-scale mathematical programming* takes into account the special structure of the problem as much as possible, and decomposes it into a set of smaller problems. Dantzig and Wolfe (1961) *decomposition technique* for linear programming initiated the area of large-scale mathematical programming and extensive work has followed in both linear and nonlinear programming.

In river quality control problems we have to cope with *multistage decision* type *problems* where each stage of the problem, characterized by one decision variable, can be associated either with a reach of the river (steady state control) or with a time interval (unsteady state control). For multistage linear programming problems, the *nested decomposition technique* recently developed by Glassey (1973) can be effectively applied, while for multistage nonlinear programming problems efficient algorithms are not yet available.

Further Reading

In this section the key ideas of mathematical programming were described only very briefly. Books suggested for further reading are Gass (1975) for linear programming; Luenberger (1973) and the first chapter of Lasdon (1970) for linear and nonlinear programming; Himmelblau (1972) for computational aspects of nonlinear programming; Lasdon (1970), Wismer (1971), Geoffrion (1972), and Himmelblau (1973) for large-scale mathematical programming; and Varaiya (1972) for the relationships between optimal control and mathematical programming.

REFERENCES

Section 6-1

Anderson, B. D. O. and Moore, J. B. (1971). *Optimal Control Systems*. Prentice-Hall, Englewood Cliffs, N.J.
Aoki, M. (1967). *Optimization of Stochastic Systems*. Academic Press, New York.
Athans, M. and Falb, P. L. (1966). *Optimal Control: An Introduction to the Theory and its Applications*. McGraw-Hill, New York.
Bryson, A. E. and Ho, Y. C. (1975). *Applied Optimal Control*. John Wiley, New York.
Chestnut, H. and Mayer, R. W. (1963). *Servomechanisms and Regulating System Design*. John Wiley, New York.
Cruz, J. B. (1972). *Feedback Systems*. McGraw-Hill, New York.
Feldbaum, A. A. (1965). *Optimal Control Systems*. Academic Press, New York.
Gibson, L. E. (1963). *Non-Linear Automatic Control*. McGraw-Hill, New York.
Hsu, J. C. and Meyer, A. V. (1968). *Modern Control Principles and Applications*. McGraw-Hill, New York.
Lee, E. B. and Markus, L. (1967). *Foundations of Optimal Control Theory*. John Wiley, New York.
Pontryagin, L. S., Boltyanskii, V. G., Gamkrelidze, R. V., and Mishchenko, E. F. (1962). *The Mathematical Theory of Optimal Processes*. John Wiley, New York.
Sworder, D. (1966). *Optimal Adaptive Control Systems*. Academic Press, New York.
Truxal, J. G. (1955). *Automatic Feedback Control Synthesis*. McGraw-Hill, New York.

Section 6-2

Abadie, J. and Carpentier, J. (1969). Generalization of the Wolfe Reduced Gradient Method to the Case of Nonlinear Constraints. In *Optimization* (R. Fletcher, ed.). Academic Press, London.
Beale, E. M. L. (1959). On Quadratic Programming. *Naval Research Logistics Quarterly*, **6**, 227–243.
Dantzig, G. B. and Wolfe, P. (1961). The Decomposition Algorithm for Linear Programming. *Econometrica*, **29**, 767–778.
Duffin, R. J., Peterson, E. L., and Zener, C. (1967). *Geometric Programming: Theory and Application*. John Wiley, New York.
Fletcher, R. and Powell, M. J. D. (1963). A Rapidly Convergent Descent Method for Minimization. *Computer J.*, **6**, 163–168.
Fletcher, R. and Reeves, C. M. (1964). Function Minimization by Conjugate Gradients. *Computer J.*, **7**, 149–153.
Gass, S. I. (1975). *Linear Programming: Methods and Applications*. McGraw-Hill, New York.
Geoffrion, A. M., ed. (1972). *Perspectives on Optimization*. Addison-Wesley, Reading, Mass.
Glassey, C. R. (1973). Nested Decomposition and Multistage Linear Programs. *Management Sci.*, **20**, 282–292.
Hestenes, M. R. (1969). Multiplier and Gradient Methods. *J. Optimization Theory and Appl.*, **4**, 303–320.
Himmelblau, D. M. (1972). *Applied Nonlinear Programming*. McGraw-Hill, New York.
Himmelblau, D. M., ed. (1973). *Decomposition of Large-Scale Problems*. North-Holland, Amsterdam.
Larson, R. E. (1968). *State Increment Dynamic Programming*. American Elsevier, New York.
Lasdon, L. S. (1970). *Optimization Theory for Large Systems*. Macmillan, New York.
Luenberger, D. G. (1969). *Optimization by Vector Space Method*. John Wiley, New York.
Luenberger, D. G. (1973). *Introduction to Linear and Nonlinear Programming*. Addison-Wesley, Reading, Mass.
Nakayama, H., Sayama, H., and Sawaragi, Y. (1975). A Generalized Lagrangian Function and Multiplier Method. *J. Optimization Theory and Appl.*, **7**, 211–227.
Varaiya, P. P. (1972). *Notes on Optimization*. Van Nostrand-Reinhold, New York.
Wismer, D. A., ed. (1971). *Optimization Methods for Large-Scale Systems*. McGraw-Hill, New York.
Wolfe, P. (1959). The Simplex Method for Quadratic Programming. *Econometrica*, **27**, 382–398.
Zangwill, W. I. (1967). The Convex Simplex Method. *Management Sci.*, **14**, 221–238.

CHAPTER
SEVEN
A SHORT SURVEY OF WATER POLLUTION CONTROL FACILITIES

7-1 WASTEWATER TREATMENT

The various waste treatment methods can be broadly classified into physical, chemical, and biological. Methods of treatment in which the application of physical forces predominate are known as *unit operations*, while, when the removal of contaminants is brought about by chemical or biological agents, the methods are known as *unit processes*. The most common unit operations and processes are listed in Table 7-1-1. It would be beyond the scope of this book to give a detailed definition and description of all the operations and processes listed in the table, and thus only a rough outline is presented. For further information the

Table 7-1-1 Some common operations and processes for treating industrial and municipal wastes

Unit operation	Unit chemical processes	Unit biological processes
Screening	Precipitation	Aerobic processes
Mixing	Neutralization	Anaerobic processes
Sedimentation	Adsorption	Aerobic–anaerobic processes
Flocculation	Disinfection	
Flotation	Chemical oxidation	
Filtration	Chemical reduction	
Drying	Incineration	
Distillation	Ion exchange	
Centrifuging	Electrodialysis	
Freezing		
Reverse osmosis		

reader is referred to the handbooks of wastewater engineering (e.g., Metcalf and Eddy, 1972). In general, the same processes are going on in wastewater treatment plants as during self-purification of rivers, though in a concentrated and controlled form. Hence, most physical, chemical, and biological phenomena mentioned in the following have already been discussed in Chapters 2 and 3.

Unit Operations

Most unit operations are used to remove suspended matter, both degradable and nondegradable. Suspended solids can be removed by sedimentation (if necessary with the aid of flocculants like polyelectrolytes), by filtration through various kinds of screens and filters (e.g., bar racks and grit chambers), or by centrifuging. For nondegradable suspended solids a very high degree of removal can be obtained, but there is still a residual in the form of a concentrated brine or a semisolid sludge, the ultimate disposal of which is often a problem. When the waste contains biodegradable organic matter the residual is a wet sludge which is difficult to handle, and is usually digested in heated anaerobic tanks before final disposal (or dried and burned). When sedimentation tanks are operated adequately the BOD load of a typical municipal wastewater can be reduced by 30 to 35 percent. The most efficient unit operations are distillation, freezing, and reverse osmosis. They remove practically all contaminants to a very high degree (95–99 percent), but, owing to the high operation costs, their use is severely limited.

Chemical Unit Processes

Chemical precipitation was a well established method of sewage treatment as early as 1870, but, with the development of biological treatment the use of chemicals was gradually abandoned. Only chlorination maintained a wide application for disinfection of the final effluent. Currently, there is a renewed interest in the use of chemical treatment to obtain a better quality effluent by eliminating those substances which are not, or are little, affected by conventional biological treatment. Of particular importance in this context are inorganic nutrient substances, such as nitrogen and phosphorus compounds, which may direct the receiving body of water towards eutrophic forms. Phosphorus can be removed via precipitation by adding coagulants such as alum, lime, or polyelectrolytes, either to the sedimentation tanks or to the biological stage of treatment. Organic nitrogen can be eliminated by activated carbon adsorption or ion exchange.

Most types of chemical processes involve low capital costs and high operating costs, therefore they are particularly useful for specialized treatment problems and for exceptional application during periods of adverse conditions.

Biological Unit Processes

As shown in Table 7-1-1, the biological processes are classified as *aerobic*, *anaerobic*, and *aerobic–anaerobic processes,* according to the oxygen dependence of the microorganisms responsible for waste degradation. In essence, they are nothing but a controlled and accelerated version of the oxidation processes which would occur naturally in the river (aerobic processes) and in the sediments (aerobic–anaerobic and anaerobic processes).

Aerobic processes The most common aerobic treatment units are the *activated sludge plant*, the *trickling filter*, and the *aerobic stabilization ponds*. The activated sludge process is used almost exclusively in large cities. The trickling filters are often used for smaller towns and for biodegradable industrial wastes. The aerobic stabilization ponds are used where large land areas are available. These systems differ somewhat in the detention times, in the oxygen requirements, and in the utilization of the biological slimes, but essentially the same biochemistry occurs within each one of them. The two most important biochemical phenomena are the conversion of the pollutants into biomass of a microbial community, which is called the synthesis stage, and the decrease of the biomass through endogenous respiration and death, which is called the respiration stage. Even with the synthesis stage alone, wastewater purification may be achieved, since the pollutants are converted into a settleable form. In this case not only biochemical synthesis, but also other processes, such as adsorption to cell walls and flocculation, play an important role.

The synthesis stage is predominant in the processes using *activated sludge*, which consists of a gelatinous matrix in which filamentous bacteria are imbedded and on which protozoa crawl and feed. In this process (see Fig. 7-1-1), only part of the activated sludge which has settled in the sedimentation tank is sent to the anaerobic digester, while the remaining part is recycled (*sludge recycle*) and mixed with the influent water to form a mixed liquor which is first aerated for 4 to 6 hours in an aeration tank. Then the liquor flows into the sedimentation tank whose overflow leaves the process as effluent. The sludge recycle flow rate depends upon the desired treatment efficiency and upon considerations related to growth kinetics. Other important design variables are the length of the aeration period and the amount of air introduced. A rapid turnover of waste and sludge saves space and power, but requires well-trained personnel and a continuous monitoring of the plant performance. Finally, the efficiency of BOD removal is improved by increasing the quantity of air pumped into the aeration tank. In a well designed activated sludge process the total removal of organic material is between 80 and 90 percent. Even higher efficiencies can be obtained, but the costs rapidly increase as the removal of BOD is pushed above 95 percent.

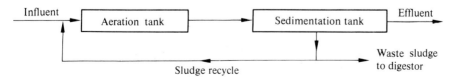

Figure 7-1-1 Structure of the conventional activated-sludge process.

The second major aerobic biological treatment option is the *trickling filter*, which is a bed of highly permeable media, generally constituted by a layer of broken stones. The wastewater is sprinkled intermittently over the filter and allowed to trickle through the bed. The bottom of the filter bed is provided with an underdrainage system to remove the filter effluent, supply ventilation, and maintain aerobic conditions in the filter. The organic material present in the wastewater is degraded by a population of microorganisms (*biological slime*) which grows like a film on the surface of the filter medium. In the outer portion of the biological slime layer the organic matter is aerobically degraded, while new cell biomass is synthesized. As the microorganism population grows and the slime layer increases in thickness, the oxygen is consumed before it can penetrate the full depth of the layer. Thus, anaerobic conditions are established on the media surface. When the thickness of the slime layer increases even further, the organic matter also cannot reach the microorganisms near the media surface, since it is already completely metabolized. As a consequence, the microorganisms on the media surface enter an endogenous phase of respiration and lose their ability to adhere to the media surface. The liquid then washes the layer away (*sloughing off*) and a new slime layer starts to grow. Generally, the filter effluent is passed through a settling tank to retain the settleable solids sloughed off from the filter. Recirculation may be applied, but it is not as important as in the activated sludge process. The majority of the active microorganisms are in fact clinging to the filter media surface and are not contained in the effluent as is the case in the activated sludge process. The primary purpose of recirculation (when applied) is to dilute influent waste and to recover the effluent in contact with the biological slime for further treatment.

There are many alternative flow patterns for trickling filtration plants. Some of the most common ones are presented in Fig. 7-1-2. A universally applicable structure does not exist and one may easily find plants which use different flow patterns under different operating conditions (load, temperature, type of waste,

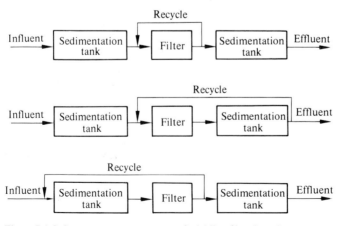

Figure 7-1-2 Some common structures of trickling filtration plants.

etc.). The hydraulic loading rate (daily wastewater flow per unit area of filter surface), which accounts for shear velocities, and the organic loading rate (daily mass flow of BOD per unit volume of filter medium), which accounts for the rate of metabolism in the slime layer, are the two main design parameters of a trickling filter. Based on the hydraulic and organic loading rates, filters are usually classified as low rate (low loading) and high rate (high loading) filters. A low rate filter is relatively simple to operate. The biological community in the filter contains a large proportion of nitrifying bacteria, so that the effluent has a high concentration of nitrites and nitrates. Usually, recirculation is not provided for. The drawbacks are odors (especially in the warm season) and filter flies. The BOD removal efficiency is 80 to 85 percent. A high rate filter is characterized by a lower efficiency (65 to 85 percent) and by the presence of a recirculation flow. The control of this flow is of fundamental importance, since the absence of odors and flies and the efficiency of the process strongly depend upon the selection of the right recirculation flow. Thus, from an operational point of view a high rate trickling filter is more complex than a low rate one and a certain skill is necessary to obtain good results. Finally, to summarize the advantages and disadvantages of trickling filters, advantages include the relatively high nitrifying effect, the relatively low operating and maintenance costs, and the ability to operate under extreme weather conditions. Disadvantages include the odor and fly nuisances (low rate filters), the large surface area required, and the relatively high construction costs.

The third type of aerobic process facility is the *aerobic stabilization* (or oxidation) *pond*, which is a relatively shallow body of water contained in an earthen basin. The process undergone by waste in a stabilization pond is approximately equivalent to an activated sludge process without recycle, the difference being that the sludge withdrawn is small enough in volume to be discharged with the final effluent, and much more of the solid matter is in a mineralized form. The oxygen is supplied by natural surface aeration and by algal photosynthesis and sometimes by artificial aeration devices. A cyclic symbiotic relationship exists between aerobic bacteria and algae, since the first utilize the oxygen released by the algae through the process of photosynthesis, while the latter use the nutrients and carbon dioxide produced in the degradation stage by the bacteria. To achieve the best results with aerobic ponds, it is necessary to periodically mix their contents to prevent stratification and sedimentation of suspended solids. The efficiency of BOD removal in aerobic ponds is high, reaching up to 95 percent. However, it is worth while noting that, even if the BOD load has been removed, the pond effluent may contain considerable amounts of algae. If it is not removed, this algal mass can ultimately exert a biological oxygen demand which is comparable to the original one. Moreover, stabilization ponds differ from other types of treatment in that their effluents contain virtually all the plant nutrients originally in the waste discharge. Thus, the effluents from this type of treatment process fertilize the receiving waters more than effluents from other treatment processes. However, if the algal mass is harvested, the major portion of the plant nutrients can be extracted from the wastewater. Finally, these ponds are becoming more and more popular with small rural communities since, when land is easily available, their low construction and operating costs offer significant advantages over other treatment methods.

Aerobic–anaerobic processes are essentially performed by *facultative ponds*. These are ponds, in which the water is not periodically mixed and suspended solids, are allowed to settle down in such a way that an anaerobic sludge layer is present on the bottom. The facultative pond, therefore, has an aerobic upper layer, where algae and aerobic bacteria operate, and an anaerobic bottom layer, where anaerobic and facultative bacteria live. The maintenance of the aerobic zone serves to minimize odor problems, since many odorous decomposition products generated in the bottom layer are utilized by the aerobic organisms of the upper layer. The efficiency of BOD removal is, as for aerobic stabilization ponds, around 95 percent, it has the advantage of doing without the equipment, power, and labour necessary to mix the water periodically.

Anaerobic processes Anaerobic waste treatment involves the decomposition of organic matter in the absence of free oxygen. The major application of this type of process is in the stabilization (digestion) of concentrated sewage sludge coming from the sedimentation tanks. The digestion of sewage sludge is brought about in two stages by two different groups of bacteria. In the first stage there is an intensive production of organic acids, while in the second stage the organic acids are converted into methane gas and carbon dioxide. The bacteria responsible for this conversion are strict anaerobes, which all have very slow growth rates. As a consequence the digestion process requires a very long time. The detention time of the waste sludge in a completely mixed reactor (*digestor*) is between 10 and 30 days, depending on how the system is operated. When the treatment system is sufficiently large, the methane gas produced is burned to heat the digesting sludge and used to run gas engines, which mix the liquid. Both operations have the effect of accelerating the digestion process. The solid matter resulting from the process is well stabilized and suitable for disposal, after drying or dewatering, in dumps or on land as a solid conditioner or humus like material.

Modeling and Design of Treatment Plants

Traditionally, unit operations and processes are aggregated into so-called *primary* and *secondary treatment*. In the primary treatment (see Fig. 7-1-3) the floating and

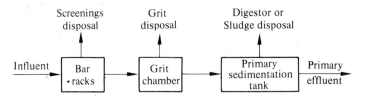

Figure 7-1-3 Typical structure of a primary treatment plant.

settleable solids contained in wastewater are removed by means of physical operations, such as screening and sedimentation. In the secondary treatment, biological and chemical processes are used to remove most of the organic waste (in Fig. 7-1-4 two examples of secondary treatment are shown). Recently the term *"tertiary treatment"* or *"advanced treatment"* has been applied to the operations and processes used to remove contaminants which are unaffected or have been insufficiently removed by primary and secondary treatment (e.g., inorganic nitrogen and phosphorous compounds).

Quite often, only one treatment structure is considered during the design phase, since legislation often directly prescribes the type of treatment process for each type of wastewater. Even if the law is not so precise, and only quantitatively fixes the characteristics **u** of the effluents discharged into the receiving body of water (*effluent standards*), only one structure is usually considered by the designer of the treatment plant. In fact, since the characteristics **U** of the raw wastewater influent (e.g., flow rate and composition) are usually given, the "optimal" structure of the treatment plant is practically known from preceding analogous experiences. Thus, the design consists in determining the least costly plant among all plants of a certain structure which will transform the inlet characteristics **U** into the outlet characteristic **u**. The cost of the "optimal" plant obviously depends upon the input and output characteristics and will, therefore, be indicated by $\mathscr{C}(\mathbf{U}, \mathbf{u})$ in the following discussion.

When the aim of installing a treatment plant is to satisfy a given *stream standard* the function $\mathscr{C}(\mathbf{U}, \cdot)$ must be determined since the characteristics **u** of the

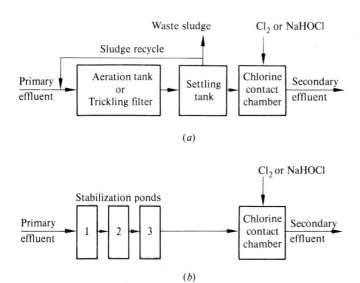

Figure 7-1-4 Typical structures of secondary treatment plants:
(a) activated sludge process or trickling filter
(b) stabilization ponds.

output of the plant are not fixed a priori (rather, they must be obtained by minimizing an overall performance index). In other cases (see, for example, Sec. 10-4), it is also important to determine how the cost of the best treatment plant is influenced by the characteristics U of the raw influent.

In any case, the basic elementary problem is the determination of the cost $\mathscr{C}(\mathbf{U}, \mathbf{u})$ of the least costly treatment plant characterized by a given structure. However, an analysis of the design procedures used up to now indicates (see, for instance, Parkin and Dague, 1972) that for a long time the most common practice has been to fix a priori the input and output characteristics of each component (or unit) of the plant and then to determine the least costly units with those characteristics. Obviously, this way of proceeding does not give rise to the least costly plant since the trade-off between the different units is not solved in an optimal way. Only in the last fifteen years has the problem of the optimal design of treatment plants been analysed from a more rational point of view (see, for instance, Lynn et al., 1962; Thomas and Burden, 1963; Evenson et al., 1969; Shihi and DeFilippi, 1970; Ecker and McNamara, 1971; Parkin and Dague, 1972).

In order to apply systems analysis to the design of treatment plants it is, in principle, necessary to model all the different unit processes and operations. This is not an easy task since the majority of the unit processes (in particular all the biological processes) and some of the unit operations are dynamic systems, i.e., their output does not only depend on the present input but also on the past history of the unit. For this reason a lot of research activities has been directed towards this subject in the last decade, and a certain degree of success has been obtained (see Keinath and Wanielista, 1975; Olsson, 1978). However, almost all these dynamical models have not yet been validated on full scale operation pilot plants (Olsson, 1978), and this is the reason why these modeling techniques are still far from being applied in the design of wastewater treatment plants. Nevertheless empirical performance functions for unit operations and processes, warranted by long practical experience, are nowadays available. These performance functions are, in general, of the form

$$\sigma = \mathbf{F}(\mathbf{s}, \mathbf{d})$$

and give the relationships existing in steady state conditions between the input characteristics **s** of the unit (e.g., flow rate and concentration of contaminants), the "design variables" **d**, and the output characteristics σ.

Determination of the Cost Functions

The design problem of a treatment plant can now be stated as a mathematical programming problem. As already mentioned, the criterion by means of which different alternatives are compared is the total annual cost of the treatment plant, i.e., the sum of capital, maintenance, and operation costs. The constraints under which the least costly treatment plant has to be determined are the characteristics U and **u** of influent and effluent, the kind and number N of units constituting the plant, and the structure of the plant. The i-th unit is characterized by its performance function $\sigma_i = \mathbf{F}_i(\mathbf{s}_i, \mathbf{d}_i)$ and by its total annual cost $\mathscr{C}_i(\mathbf{s}_i, \mathbf{d}_i)$. The structure of the plant is specified by expressing the input characteristics $\mathbf{s}_i$ of each unit as a function of $\mathbf{U}$ and of the output characteristics σ_i of all N units (including the i-th, owing to the possibility of effluent recycle), i.e.,

$$\mathbf{s}_i = \mathbf{G}_i(\mathbf{U}, \sigma_1, \ldots, \sigma_N)$$

Thus, the optimal treatment plant is determined by solving the following mathematical programming problem

$$\mathscr{C}(\mathbf{U}, \mathbf{u}) = \min \sum_{i=1}^{N} \mathscr{C}_i(\mathbf{s}_i, \mathbf{d}_i) \tag{7-1-1a}$$

subject to

$$\sigma_i = \mathbf{F}_i(\mathbf{s}_i, \mathbf{d}_i) \qquad i = 1, \ldots, N \tag{7-1-1b}$$

$$\mathbf{s}_i = \mathbf{G}_i(\mathbf{U}, \sigma_1, \ldots, \sigma_N) \qquad i = 1, \ldots, N \tag{7-1-1c}$$

$$\mathbf{d}_i \in \mathbf{D}_i \qquad i = 1, \ldots, N \tag{7-1-1d}$$

$$\mathbf{u} = \mathbf{H}(\sigma_1, \ldots, \sigma_N) \tag{7-1-1e}$$

Equation (7-1-1d) expresses the condition that, owing to technical factors, land availability, and aesthetic restraints, the design variables $\mathbf{d}_i$ must belong to a prescribed set $\mathbf{D}_i$, while Eq. (7-1-1e) says that the output characteristics of the plant are functions of the outputs of the units.

It is important to note that if the structure of the least costly plant with given input and output characteristics is not known in advance, the "optimal" structure can be determined by solving a mathematical programming problem of the preceding kind. In fact, if it is necessary to choose among a few particular structures, it is always possible to define a general structure which contains all the preceding ones as particular, degenerate cases. The optimal solution of the problem will indicate which units must not be present in the optimal plant, thus giving the best structure from among those considered.

Lynn et al., (1962) were the first to solve a mathematical programming problem of type (7-1-1) for the optimal design of a treatment plant. The formulation they give is linear (i.e., the functions $\mathscr{C}_i$, $\mathbf{F}_i$, $\mathbf{G}_i$, and $\mathbf{H}_i$ are linear, and the set $\mathbf{D}_i$ can be expressed through linear inequalities) so that the Simplex method can be used to determine the optimal solution. The use of linear programming is somewhat restrictive, since economies of scale in the cost functions $\mathscr{C}_i$ of the individual units cannot easily be represented. In order to overcome this difficulty, Evenson et al., (1969), and later on Shihi and DeFilippi (1970), showed, by means of some examples, how problem (7-1-1) can, in general, be solved by means of dynamic programming. A different approach was proposed by Ecker and McNamara (1971), who solved a particular case of problem (7-1-1) by means of geometric programming.

It must be emphasized that the solution of the mathematical programming problem (7-1-1) gives the cost of the optimal treatment plant for a particular input and output pair (**U**, **u**). Therefore, if the function $\mathscr{C}(\cdot,\cdot)$ must be determined, the mathematical programming problem (7-1-1) must be solved for a suitable number of grid points in the (**U**, **u**) space. For this reason, the cost function $\mathscr{C}(\cdot,\cdot)$ has, to date, only been determined in the very simple case in which **U** and **u** are scalar variables.

It is very difficult to derive some properties of the cost function by analyzing the mathematical structure of Eq. (7-1-1). Nevertheless, the numerical results obtained in many particular cases and the examples reported in the literature (see, for instance, Kneese and Bower, 1968) enable some general conclusions to be drawn. In particular, when U and u are the pollutant mass flow rates of the influent and effluent respectively (note that U and u are now scalars), in a suitable range of the (U, u) plane (roughly defined by $u/U \leq 0.6$) the cost function $\mathscr{C}(U, u)$ of the treatment plant has the shape shown in Fig. 7-1-5, so that the following properties hold.

(a) $\mathscr{C}(U, u)$ is convex and decreasing with respect to u for any U.
(b) $\mathscr{C}(U, u)$ is convex and increasing with respect to U for any u.
(c) $\mathscr{C}(U, \alpha U)$ is concave and increasing with respect to U.

These properties are somehow intuitive. For example, the third one means that at a fixed efficiency ($u/U = \alpha$ = constant implies that the efficiency $r = (U - u)/U = 1 - \alpha$ is also constant) the cost of treating one unit of BOD decreases with increasing plant size (increasing U), i.e., the treatment plant cost exhibits economies of scale. Thus, the cost function $\mathscr{C}(U, u)$ is not convex in the space (U, u) (see Fig. 7-1-5), although it is convex with respect to U and u, separately. These properties of the cost function $\mathscr{C}$ will be used in the following (see Sec. 8-2 and 10-4).

Often (see, for instance, Lynn et al., 1962; and ReVelle et al., 1968), when the pollutant mass flow rate U of the plant influent is fixed ($U = \bar{U}$), the cost function $\mathscr{C}(\bar{U}, u)$ is approximated with a linear relationship of the kind

$$\mathscr{C}(U, u) = \alpha + \beta u$$

This approximation turns out to be particularly acceptable for treatment efficiencies ranging between 40 and 90 percent ($0.1\bar{U} \leq u \leq 0.6\bar{U}$).

7-2 TEMPERATURE CONTROL

The importance of temperature as a parameter influencing water quality has been emphasized many times throughout the preceding sections. In almost all rivers of the industrialized countries, water temperature deviates considerably from its natural level because of human actvities, some of which have already been mentioned in Sec. 2-2. Control of those activities represents another means of controlling river quality.

The main anthropogenic temperature effect on rivers is due to the *waste heat*

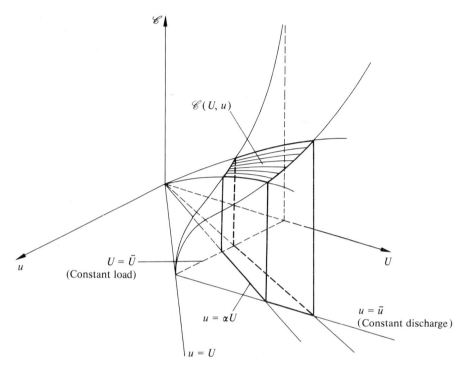

Figure 7-1-5 The cost function $\mathscr{C}(U, u)$ of a treatment plant.

production of thermal electric power plants. The easiest and cheapest way to get rid of the waste heat, inevitably produced in those plants, is to have it carried away by rivers. But there are alternatives, the most important one being discharge into the air (see page 232). The importance of the effect of waste heat from electric power plants on river temperature may be evaluated through the following rough calculation, which is based on values for Central Europe. The total annual electricity production per capita is 4×10^3 kWh/a, the waste heat production is about twice as high. This corresponds to a waste heat production density of about 0.6 kWh/a m². On the other hand, the mean annual precipitation is 500 mm, of which about one half is surface runoff. Comparing these two figures one finds that the temperature of the surface runoff could be artificially increased by 2 °C through the waste heat from electricity production if there were no natural cooling. The local effects of electrical production on river temperature may be considerably higher than this global figure. Considering, on the other hand, the impact of temperature changes on river quality (see Secs. 2-3, 3-5) one can conclude that river quality may be greatly influenced by controlling the waste heat disposal of electric power plants.

The amount of waste heat discharged into a river can be varied either by *throttling* the power plant or by changing the type of cooling system. If a power plant is run below its nominal capacity the *efficiency* is diminished, which means

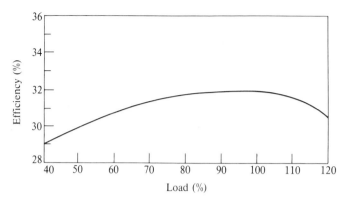

Figure 7-2-1 Dependence of efficiency of a thermal electric power plant on load (after Beck and Göttling, 1973).

that the electricity costs increase. Figure 7-2-1 gives an example of how the efficiency depends on load (Beck and Gottling, 1973). If a power plant is throttled because of its impact on a river, either more electricity generating capacity is needed somewhere else or the electricity demand cannot be met. In either case an additional cost results.

The cooling options for a power plant are shown schematically in Fig. 7-2-2 (see, for example, Guerney and Cotter, 1966; Zathureczky, 1972). The cooling water runs through the *condenser*, where the expanded steam, coming from the turbine, is condensed. In this way, the cooling water is heated, and the temperature increase is usually around 10 °C. The heated cooling water may be discharged directly into the river (*once-through cooling*; the water flow is indicated by the letter *a* in Fig. 7-2-2) or it may run through a *cooling tower*. The water leaving the cooling tower may be discharged into the river (*open loop*; option *b* in Fig. 7-2-2) or fed back to the condenser (*closed loop*; option *c*). The cooling water may trickle through the cooling tower, in which case heat is carried away mainly as latent heat (*evaporative cooling*). Or it may be piped through cooling coils in the cooling tower which implies that heat is exchanged only through convection and radiation (*dry cooling*). This type of cooling requires much bigger cooling towers than evaporative cooling, i.e., the investment costs are higher, and it is still quite rare. It is useful only in closed-loop arrangements, because the cooling pipes in the cooling towers have to be quite narrow and would frequently become choked up if ordinary river water were used. The draft, which is necessary within the cooling tower to carry away the waste heat, can be achieved either through the lifting force of the heated air (*natural-draft cooling tower*) or through fans (*mechanical-draft cooling tower*). Natural-draft cooling towers have to be much more voluminous than mechanical-draft ones of the same capacity. Hence the investment costs are smaller for the mechanical-draft cooling towers; but its operational costs are higher because of the energy used to drive the fans. Which one of these two options is ultimately less costly depends on the climate and on the actual costs of labor, electricity, etc.

It is already obvious from Fig. 7-2-2 that the heat discharge into the river can be relatively easily controlled by switching from one cooling option to another or even by mixing different options appropriately. Plants using a combination of cooling options are becoming quite numerous, e.g., the power plants under construction or planned on the Rhine river will all be equipped with both once-through cooling and evaporative cooling with or without cooling water feedback. The high investment costs due to the realization of more than one cooling option are tolerated because of the possibility of using options with lower operational costs under favorable conditions (see page 234).

Even with closed-loop cooling a certain amount of waste heat has usually to be discharged into the river. Because of evaporative losses the salinity of the cooling water increases, which is undesirable mainly because of corrosion. Therefore, a certain amount of cooling water has to be discharged continuously (*blow-down water*). The amount q_x of blow-down water can easily be derived from the balance equations of the cooling system (see Fig. 7-2-2):

$$q_x = \frac{q_0 - q_e}{1 - \dfrac{q_0}{q_e}\left(\dfrac{s_m}{s_i} - 1\right)}$$

where q_0 = water flow through the condenser
q_e = water flow evaporated in the cooling tower
s_i = salinity of the make-up water
s_m = maximum permissible salinity in the condenser

For power plants on small rivers with high salinity the possibilities for controlling river temperature may be very limited because of the necessity to release the blow-down water, in particular under low flow conditions.

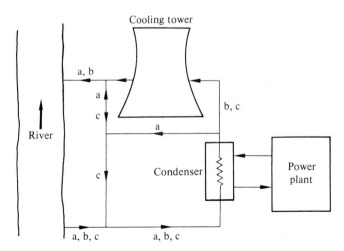

Figure 7-2-2 Cooling options for a thermal electric power plant:
(*a*) once-through cooling
(*b*) cooling tower in open-loop arrangement
(*c*) cooling tower in closed-loop arrangement

The decrease in temperature of the cooling water when passing through the cooling tower is a complex function of cooling water temperature and weather conditions. During most of the time the final temperature is higher than the river temperature. This means the *condenser temperature* is usually higher with closed-loop cooling (option *c*) than with cooling options *a* and *b* (see Fig. 7-2-2). A typical value for this increase is 15 °C for evaporative cooling; with dry cooling it is even higher. The increased condenser temperature implies a lower efficiency of the plant, i.e., higher operational costs. Figure 7-2-3 gives an example of how the efficiency may change if the condenser temperature changes (Beck and Göttling, 1973). The operational costs of electricity production are also different for the various cooling options because of the pumping and maintenance effort required, which is obviously higher if cooling towers are used.

Considering all cost relationships between cooling options one can state that the total costs of electricity production increase in the following order:

once-through cooling;
open-loop evaporative cooling;
closed-loop evaporative cooling;
dry cooling.

(No general statements can be made about the costs connected with combinations of these options unless the frequency distribution of the various modes of operation is specified.) The amount of waste heat discharged into the river decreases in the same order, which, from a river quality point of view, is usually considered an increasing benefit. Another benefit is attached to open-loop evaporative cooling: the water leaving the cooling tower is saturated with oxygen and therefore the oxygen conditions in the receiving river may be improved. Oxygen saturation of the released water can easily be achieved with once-through cooling, too, but the benefit is smaller because the saturation concentration in this case is lower (due to the higher temperature, see Sec. 3-5).

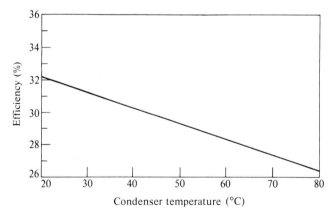

Figure 7-2-3 Dependence of efficiency of a thermal electric power plant on condenser temperature (after Beck and Göttling, 1973).

7-3 ARTIFICIAL AERATION

In the development of water quality management programs it is usually assumed that the BOD load coming from "recorded" effluents constitutes the major fraction of the total BOD load entering the river system. Thus, it seems to be sufficient to select the right degree of removal on the recorded sources in order to achieve any desired level of water quality in the river. Unfortunately, it has been shown (see, for instance, Whipple, 1970) that even in well administered areas the recorded effluents hardly represent more than one half of the total load. When this is the case, improvement of the wastewater treatment efficiencies may be insufficient. Moreover, it may be the case that high treatment levels are required only to prevent the occurrence of too low DO levels during short periods of adverse wastewater assimilation characteristics, while less costly treatment plants would be sufficient to obtain the desired water quality during the rest of the year. In both cases *artificial instream aeration* turns out to be more effective than advanced wastewater treatment (see, for instance, Cleary, 1966; Susag et al., 1966; Whipple and Yu, 1971; Ortolano, 1972). Obviously, advanced treatment affects pollutants other than the ones related to oxygen depletion (e.g., phosphates can be removed), but if these effects are not very important the artificial aeration can be considered a good control alternative.

Two different types of aeration units are available, the diffusion (or bubble) aerators and the mechanical (or surface) ones. Both types of equipment are also used in activated sludge process treatment plants and in aerated stabilization ponds.

Diffusion aerators consist of a continuous pipeline through which air is pumped by a blower with spargers that allow the air to enter the liquid in the form of a train of bubbles. The oxygen transfer takes place as the bubble is formed, as it rises to the surface, and as it breaks the surface. It has been shown that efficiency varies with depth, with spacing, and particularly with the fineness of bubbles (Bewtra and Nicholas, 1964).

The most commonly used *mechanical surface aerators* consist of an impeller located at the water surface, driven by an electric or diesel engine. The impeller draws the water and casts it out radially (see Fig. 7-3-1a). The oxygen transfer takes place owing to the negative pressure generated by the rotation, and to the increased interfacial area between air and water in the highly turbulent zone around the aerator. The circulation pattern created by such an aerator in a small river is shown in Fig. 7-3-1b. For more details on aeration equipment, the interested reader can refer to any handbook on wastewater treatment (e.g., Eckenfelder and O'Connor, 1961; or Metcalf and Eddy, 1972). The *standard efficiency* η_s (mg O_2/kWh) of a mechanical unit is defined as the amount of oxygen transferred per unit power employed and per hour under "standard conditions," namely temperature = 20 °C, pressure = 1 atmosphere, DO concentration in the water to be aerated = 0 mg/l, DO saturation level = 9.2 mg/l, and water quality of a typical tap water. Once the standard efficiency η_s is known, the equipment

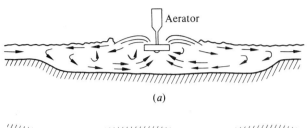

(a)

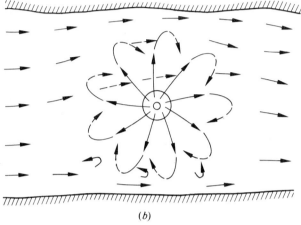

(b)

Figure 7-3-1 Effects of mechanical aerator on a stream:
(a) the impeller and the water cast
(b) the circulation pattern.

differential efficiency under field conditions is defined by

$$\frac{d(Qu)}{d\pi} = \frac{1}{K}[c_s - (c + u)] \tag{7-3-1}$$

where

$$\frac{1}{K} = \eta_s \frac{\mu \lambda^{(T-20)}}{9.2}\left(\frac{p}{760}\right)$$

c = average DO level immediately upstream from the aerator turbulent area in (mg/l)
c_s = DO saturation level at temperature T (°C) in (mg/l)
u = DO increment in (mg/l), i.e., the difference between the downstream $(c + u)$ and upstream (c) DO concentrations
Q = flow rate in (l/h) (thus Qu is the oxygen transfer rate)
π = equipment power in (kW)
p = air pressure in (mm Hg)
μ, λ = empirical correction factors used to account for the dependence of the oxygen transfer rate on water quality and temperature, respectively (for an operative definition see Ortolano, 1972).

Then the total power necessary to increase the oxygen concentration from c to $c + u$ can be obtained by integrating Eq. (7-3-1), which yields

$$\pi = \pi(u, c, Q) = KQ \ln \left[\frac{(c_s - c)}{(c_s - (c + u))} \right] \quad (7\text{-}3\text{-}2)$$

It has been noted (Whipple et al., 1970) that the conversion factor K decreases with stream velocity. In fact, the circulation pattern generated by the equipment tends to bring the aerated water back into the impeller, while an increase in velocity helps to counteract this tendency and extends the turbulent plume downstream. Moreover, the efficiency of two aeration units can be greatly reduced if the distance between them is too small (Price et al., 1973). This effect must be carefully taken into account if the optimal aerator positions are to be selected.

Let w_c be the cost purchase and installation of a unit and w_o the yearly operational cost. Both w_c and w_o are functions of the power π. More precisely, $w_c(\pi)$ is a concave function (presence of economies of scale) which can be well approximated by the linear relationship $w_c = \alpha + \beta\pi$ for high values of π, while w_o is directly proportional to π, i.e., $w_o = \gamma\pi$. Then the yearly cost $\mathscr{C}$ of a mechanical aerator of power π is given by

$$\mathscr{C} = \mathscr{C}(\pi) \simeq \frac{\alpha}{Y_a} + \left(\frac{\beta}{Y_a} + \gamma \right) \pi \quad (7\text{-}3\text{-}3)$$

where Y_a is the life time of the aerator and discounting has been neglected. From Eqs. (7-3-2) and (7-3-3) it follows that the cost $\mathscr{C}$ is given by

$$\mathscr{C} = \mathscr{C}(u, c, Q) = AQ \ln \left[\frac{(c_s - c)}{(c_s - (c + u))} \right] + B \quad (7\text{-}3\text{-}4)$$

where A and B are suitable constants. It is worth while noting that $\mathscr{C}$ depends linearly upon Q and tends to infinity when the induced increment u approaches the deficit $(c_s - c)$, i.e., Eq. (7-3-4) reflects the difficulty of attaining supersaturation. Moreover, the same amount u exhibits increasing transfer costs for increasing DO levels.

The aeration cost $\mathscr{C}$ is said to exhibit the property of economies of scale if the cost of an aerator, improving the DO level from c to $(c + u)$, is lower than the sum of two aerators in series giving rise to the same effect (i.e., two aerators of which the first improves the DO level from c to $(c + u')$ and the second from $(c + u')$ to $(c + u)$). For example, the cost function given by Eq. (7-3-4) exhibits the property of economies of scale since

$$\mathscr{C}(u, c, Q) < \mathscr{C}(u', c, Q) + \mathscr{C}(u - u', c + u', Q)$$

However, if the second term on the righthand side of Eq. (7-3-4) can be neglected with respect to the first one (i.e., if $B = 0$ in Eq. (7-3-4)), it turns out that the cost $\mathscr{C}$ does not exhibit the property of economies of scale, since

$$\mathscr{C}(u, c, Q) = \mathscr{C}(u', c, Q) + \mathscr{C}(u - u', c + u', Q)$$

In order to select the "optimal" power and location of aeration devices along a stretch of river it is necessary to describe quantitatively their effects on the DO levels of the receiving water. A reasonable approximation to the real effects is to assume that instream aeration constitutes only a source of DO, i.e., to assume that artificial aeration does not influence the value of the natural reaeration coefficient and the BOD decay rate. However, the impact of instream aeration on water quality indicators other than DO is still an open problem. Moreover, it is usually assumed that a mechanical aerator is equivalent to a point source of dissolved oxygen and can therefore be represented by an impulse function (δ) in the DO equation

$$\frac{\partial c}{\partial t} + v \frac{\partial c}{\partial l} - D \frac{\partial^2 c}{\partial l^2} = -k_1 b + k_2(c_s - c) + u(t)\delta(l - \bar{l})$$

where $\bar{l}$ is the aerator location and $u(t)$ the time dependent increment of DO concentration induced at point $\bar{l}$. On the other hand, diffusion aerators are equivalent to a distributed DO source and are consistently represented by a continuous input function u depending upon time and space, i.e.,

$$\frac{\partial c}{\partial t} + v \frac{\partial c}{\partial l} - D \frac{\partial^2 c}{\partial l^2} = -k_1 b + k_2(c_s - c) + u(t, l)$$

where $u(t, l)$ is the rate of artificial instream aeration in point l at time t.

7-4 OTHER SOLUTIONS

The quality of receiving waters can be improved either by reducing waste discharges or by increasing and making better use of their assimilative capacity. Wastewater treatment plants (Sec. 7-1) and aerators (Sec. 7-3) are classical examples of these alternative approaches. Other control methods are briefly presented in this section. Some of them will also be referenced in Secs. 8-3 and 9-3.

Methods for Reducing Waste Discharges

The reduction of waste discharges can be achieved either by modifying the waste generating processes or by reducing the waste after generation. Technically, the ways of obtaining the latter solution are *material recovery, by-product production, waste treatment*, and *effluent reuse* for land disposal or groundwater recharge.

For industrial waste the methods based on process modification are more rational and often less costly than waste treatment. Therefore, these methods must always be considered as possible alternatives, in spite of the well established (or market induced) tendency towards treatment.

For organic wastes an attractive alternative to advanced treatment is *land disposal*. Nutrient substances (primarily nitrogen and phosphorous compounds) present in the soil are synthetized into living biomass by plants, and plants are used as food by animals and men. Therefore, when in the old agricultural system excrements of animals and men were deposited on the ground, the activity of the decomposers returned the nutrient substances to the land. Thus, the ecological pathway was perfectly closed. But when the population is concentrated in urban areas and animals are bred in pens, the cheapest and most widely adopted solution is to dispose of their wastes into the waters. In this case not only is the soil ecosystem exploited but the overloaded water ecosystem is increasingly turned towards eutrophic forms. Since eutrophication is caused both by raw and treated waste, from an ecological point of view the only rationale is to dispose of wastewaters on land and not in water. However, the many chemical contaminants in present-day wastewater, as well as health considerations, dictate against the direct disposal of raw waste. Thus, in practice, land disposal is only considered as an alternative to advanced treatment. Another drawback of land disposal is the increase of water consumption due to the evapotranspiration of the irrigated crops.

To evaluate the usefulness of the above mentioned alternatives to treatment, it is necessary to know their costs and benefits, in particular the different effects they induce on river water quality. This can be relatively easily achieved for process modifications, while the same information is more difficult to obtain for land disposal (see, for instance, Haith 1973) and groundwater recharge.

Methods for Increasing Assimilative Capacity

The assimilative capacity of a given river can be improved by artificial reaeration (see Sec. 7-3) as well as by modification of the time pattern of river flow in order to bridge over periods of low flow (the low flow (LQ) is generally defined as that value of the daily flow which is exceeded with a 0.95 probability). This solution, known as *low-flow augmentation*, rests on the assumption of the existence of a positive correlation between water quality and water quantity. Hence, during periods of natural low-flow conditions a common practice is to increase the flow, by controlled releases from reservoirs, up to a level at which assimilation can take place without violating a given quality standard.

Unfortunately, low-flow augmentation does not always lead to better water quality. Its effectiveness strongly depends on the type of waste, on the conditions of the receiving water, and on the location of the pollution sources and of the reservoir. In the case of organic wastes the effectiveness of low-flow augmentation as evaluated by means of a Streeter-Phelps model depends on the values of the reaeration and deoxygenation parameters as well as on the values of their derivatives with respect to flow rate (Rinaldi and Soncini-Sessa, 1978). In the case of suspended solids which settle out, an increase of the flow would improve the conditions near the source and worsen them further downstream, since the augmented velocity and the reduced flow time negatively affect sedimentation. On the other hand, the concentration of nondegradable dissolved substances (e.g., chloride), is lowered everywhere. Finally, it must be noticed that low-flow augmentation does not lead to a reduction of the concentration in the biota of those compounds which are absorbed in the food web (e.g. mercury, DDT).

The effects of the reservoir storage itself on water quality must be carefully taken into account. On the one hand, biodegradable wastes tend to be reduced, suspended materials are settled and, in summertime, the release is cooler than normal streamflows; on the other hand, many projects failed (see Kneese and Bower, 1968) because the impact of the impoundment on the dissolved oxygen concentration of the release was not carefully assessed in advance. In fact, in the deepest part of the reservoir, the DO concentration can be very low due to the combined impact of BOD demand and thermal stratification. This drawback can be overcome either by mixing equipment, the purpose of which is to destratify the impoundment, or by installing a multiple outlets system, so that water can always be released from the surface. In both cases costs are higher than for an equivalent reservoir built for other purposes.

Like the majority of the water uses, low-flow augmentation is concerned with what is termed a *firm yield*. The firm yield of a reservoir is that daily release value which can be guaranteed with a 0.95 probability. If the reservoir were to be used for quality control only, the costs of obtaining a low-flow condition of $\underline{Q}$ m³/day is exactly the cost of constructing and operating a reservoir which ensures a firm yield of $\underline{Q}$ m³/day. Bramhall and Mills (1966) have shown that in this case low-flow augmentation is more expensive than advanced treatment. In the case of a multi-purpose reservoir, the cost of low-flow augmentation is a fraction of the total cost of the reservoir since the cost must be properly shared among all the benefits induced. Then low-flow augmentation may be a reasonable alternative to treatment.

The cost of low-flow augmentation can obviously be obtained when the relationship between storage capacity and cost, and between firm yield and storage capacity are known. Little need be said about the first relationship since its determination lies in the classical realm of hydraulic engineering. Suffice to say that the construction and operation costs depend to a large extent on the topographic and physiographic characters of the area in which the dam has to be built. Depending on the dam site characteristics, the marginal cost can be either increasing or decreasing (Hall and Dracup, 1970). The relationship between firm yield and storage capacity is given by the solutions of a mathematical programming problem parametric in the low-flow value $\underline{Q}$. More precisely, let V be the storage capacity, $a(i)$ the inflow, $s(i)$ the storage, and $r(i)$ the release of the i-th day of the year. Then, the firm yield–storage capacity relationship is the solution $V^0(\underline{Q})$ of the following (stochastic) mathematical programming problem (parametric in $\underline{Q}$)

$$V^0(\underline{Q}) = \min V \qquad (7\text{-}4\text{-}1\text{a})$$

$$0 \le s(i) \le V \qquad \text{for all } i \qquad (7\text{-}4\text{-}1\text{b})$$

$$s(i+1) = s(i) + a(i) - r(i) \qquad \text{for all } i \qquad (7\text{-}4\text{-}1\text{c})$$

$$P\big(r(i) \ge \underline{Q}\big) \ge 0.95 \qquad \text{for all } i \qquad (7\text{-}4\text{-}1\text{d})$$

where the constraint (7-4-1c) is the continuity equation for the reservoir and $P(\alpha)$ denotes the probability that the logic statement α is verified. Alternatively, we may assume the reservoir to be operated in closed-loop form (Blower et al., 1962; Young, 1967), i.e., the release $r(i)$ is a function (*operating rule*)

$$r(i) = g(s(i), a(i), V, \theta)$$

of the storage $s(i)$ and inflow $a(i)$, as well as of the storage capacity V and of a vector θ of parameters to be determined. Often the linear operating rule

$$r(i) = s(i) + a(i) - \theta_i \qquad (7\text{-}4\text{-}1e)$$

is used (Revelle et al., 1969; Revelle and Kirby, 1970). In any case the solution of the mathematical programming problem (7-4-1) depends upon the characteristics of the stochastic process $a(i)$ and upon the initial value of the storage. Different proposals can be found in the literature for the solution of the problem. As far as the simplest and oldest approach is concerned (Rippl, 1883), reference can be made to the works by Dorfman (1962) and Loucks (1976), while the papers by Revelle et al., (1969) and Revelle and Kirby (1970) provide a good introduction to the most recent approach.

Methods for Making Better Use of Assimilative Capacity

A better use of the self-purification capacity of the river can be made using a rational policy of *effluents redistribution*. The redistribution can be either in time by means of *regulated discharge* (see Sec. 9-3) or in space by piping the effluents to well selected discharge locations (*by-pass piping*, see Sec. 8-3).

Regulated discharge involves the temporary storage of waste effluent for discharge at times when more water is available for dilution (dual alternative of low-flow augmentation). Some degradation of the biodegradable waste may take place during the temporary storage, in particular when the storage lasts some weeks. The efficiency of this control facility strongly depends upon the data collection system which provides continuous information on the future quantity and quality of both the receiving water and the waste (see Sec. 9-3).

Usually each individual polluter discharges its effluent into the nearest section of the river. With the use of regional treatment facilities, wastewater partially treated at the origin, would be piped to a central treatment plant. After several possible stages of treatment the effluent would be piped to one or more outflow points along the river. With this setup the discharge points are no longer constrained to be in the neighborhood of the original sources. Since many water quality variables, like dissolved oxygen, are very sensitive to the discharge pattern, the location of the discharge points can be designed in order to minimize the impact on river quality. Graves et al., (1969) showed, by analyzing a case study, that by-pass piping can give rise to very efficient solutions. In fact, they showed that the simple spatial redistribution of the effluents on the Delaware estuary can induce substantial improvements in heavily polluted sections at low cost and with only slight deterioration of the less polluted sections.

The cost of wastewater transmission reflects the costs of land (or right-of-way), pipelines, pumping stations, and the costs of power, operation, and maintenance. The principal factors influencing these costs are capacity and transmission distance. Linaweaver and Clark (1964) have shown that the unit cost of transmission (dollars per unit capacity per mile) is practically independent of distance, while it significantly decreases with capacity, i.e., economies of scale are present with respect to capacity but not with respect to distance.

REFERENCES

Section 7-1

Ecker, J. G. and McNamara, J. R. (1971). Geometric Programming and the Preliminary Design of Industrial Waste Treatment Plants. *Water Resour. Res.*, **7**, 18–23.

Evenson, D. E., Orlob, G. T., and Monser, J. R. (1969). Preliminary Selection of Waste Treatment Systems. *J. WPCF*, **41**, 1845–1858.

Keinath, T. M. and Wanielista, M. P. (1975). *Mathematical Modelling for Water Pollution Control Processes*. Ann Arbor Science, Ann Arbor, Mich.

Kneese, A. V. and Bower, B. T. (1968). *Managing Water Quality: Economics, Technology, Institutions*. John Hopkins Press, Baltimore, Md.

Lynn, W. R., Logan, J. A., and Charnes, A. (1962). Systems Analysis for Planning Wastewater Treatment Plants. *J. WPCF*, **34**, 565.

Metcalf and Eddy, Inc. (1972). *Wastewater Engineering*. McGraw-Hill, New York.

Olsson, G. (1978). Estimation and Identification Problems in Wastewater Treatment. In *Recent Developments in Real-Time Forecasting on Control of Water Resources Systems, Proc. IIASA Workshop, 18–20 Oct. 1976*. John Wiley, New York.

Parkin, J. G. and Dague, R. R. (1972). Optimal Design of Wastewater Treatment Systems by Enumeration. *J. San. Eng. Div., Proc. ASCE*, **98**, 833–851.

ReVelle, C. S., Loucks, D. P., and Lynn, W. R. (1968). Linear Programming Applied to Water Quality Management. *Water Resour. Res.*, **4**, 1–9.

Shihi, C. S. and DeFilippi, J. A. (1970). System Optimization of Waste Treatment Plant Process Design. *J. San. Eng. Div., Proc. ASCE*, **96**, 409–421.

Thomas, Jr., H. A. and Burden, R. P. (1963). *Operations Research in Water Quality Management*. Harvard Water Resources Group, Cambridge, Mass.

Section 7-2

Beck, P. and Göttling, D. (1973). *Energy and Waste Heat* (in German). Beiträge zur Umweltgestaltung, Heft B8, Erich Schmidt Verlag, Berlin.

Guerney, J. D. and Cotter, I. A. (1966). *Cooling Towers*. Maclaren, London.

Zathureczky, A. (1972). *Construction and Operation of Cooling Towers* (in German). W. Ernst, Berlin.

Section 7-3

Bewtra, J. K. and Nicholas, W. R. (1964). Oxygenation from Diffuser Air in Aeration Tanks. *J. WPCF*, **36**, 1195–1224.

Cleary, E. J. (1966). The Re-Aeration of Rivers. *Industrial Water Engineering*, **3**, 16–21.

Eckenfelder, W. W. Jr. and O'Connor, D. J. (1961). *Biological Waste Treatment*. Pergamon Press, New York.

Metcalf and Eddy, Inc. (1972). *Wastewater Engineering*. McGraw-Hill, New York.

Ortolano, L. (1972). Artificial Aeration as a Substitute for Wastewater Treatment. In *Models for Managing Regional Water Quality* (R. Dorfman et al., eds.). Harvard University Press, Cambridge, Mass.

Price, K. S., Conway, R. A., and Cheely, A. H. (1973). Surface Aerator Interactions. *J. Env. Eng. Div., Proc. ASCE*, **99**, 283–299.

Susag, R. H., Polta, R. C., and Schroepfer, G. J. (1966). Mechanical Surface Aeration of Receiving Waters. *J. WPCF*, **38**, 53–68.

Whipple, W. Jr. (1970). BOD Mass Balance Analysis and Water Quality Standards. *Water Resour. Res.*, **6**, 827–837.

Whipple, W. Jr., Coughlan, F. P. Jr., and Yu, S. L. (1970). Instream Aerators for Polluted Rivers. *J. San. Eng. Div., Proc. ASCE*, **96**, 1153–1165.

Whipple, W. Jr. and Yu, S. L. (1971). Alternative Oxygenation Possibilities for Large Polluted Rivers. *Water Resour. Res.*, **7**, 566–579.

Section 7-4

Blower, B. T., Hufschmidt, M. M., and Reedy, W. W. (1962). Operating Procedures: Their Role in the Design of Water Resource Systems by Simulation Analyses. In *Design of Water Resources Systems* (A. M. Mass et al., eds.). Harvard University Press, Cambridge, Mass.

Bramhall, D. F. and Mills, E. S. (1966). Alternative Methods of Improving Stream Quality: An Economic Policy Analysis. *Water Resour. Res.*, **2**, 355–363.

Dorfman, R. (1962). Basic Economic and Technologic Concepts: A General Statement. In *Design of Water Resource Systems* (A. M. Mass et al., eds.). Harvard University Press, Cambridge, Mass.

Graves, G. W., Hatfield, G. B., and Whinston, A. B. (1969). Water Pollution Control Using By-Pass Piping. *Water Resour. Res.*, **5**, 13–47.

Haith, D. A. (1973). Optimal Control of Nitrogen Losses from Land Disposal Areas. *J. Env. Eng. Div., Proc. ASCE*, **99**, 923–937.

Hall, H. and Dracup, J. (1970). *Water Resources Systems Engineering*. McGraw-Hill, New York.

Kneese, A. V. and Bower, B. T. (1968). *Managing Water Quality: Economics, Technology, Institutions*. John Hopkins Press, Baltimore, Md.

Linaweaver, F. P. and Clark, C. S. (1964). Costs of Water Transmission. *J. Am. Water Works Assn.*, **56**, 1549–1560.

Loucks, D. P. (1976). Surface-Water Quantity Management Models. In *Systems Approach to Water Management* (A. K. Biswas, ed.). McGraw-Hill, New York.

Revelle, C., Joeres, E., and Kirby, W. (1969). The Linear Decision Rule in Reservoir Management and Design 1: Development of the Stochastic Model. *Water Resour. Res.*, **5**, 767–777.

Revelle, C. and Kirby, W. (1970). The Linear Decision Rule in Reservoir Management and Design 2: Performance Optimization. *Water Resour. Res.*, **4**, 1033–1044.

Rinaldi, S. and Soncini-Sessa, R. (1978). Sensitivity Analysis of Generalized Streeter Phelps Models. *Advances in Water Resources*, **1**, 141–146.

Rippl, W. (1883). The Capacity of Storage Reservoirs for Water Supply. *Proc. I.C.E.*, 71.

Young, G. (1967). Finding Reservoir Operating Rules. *J. Hydr. Div., Proc. ASCE*, **93**, 297–321.

CHAPTER
EIGHT
STEADY STATE CONTROL

8-1 GENERAL REMARKS

As already mentioned in Sec. 6-1 steady state control problems refer to the case in which it is necessary to select the best steady state behavior among many possible alternatives.

These alternatives are characterized by various decisions on control actions, such as the selection of the degree of treatment of a set of wastewater treatment plants, or the selection of the amount of flow to be released from a reservoir. Of course, all relevant effects related to these decisions must be investigated and taken into account while solving the problem. Thus, *profits* of polluters must be considered if the sizes and the locations of some industries are under discussion, since the profit of each one of these industries will be, in general, related to the amount of pollutant discharged into the river. Second, the *costs* of the wastewater treatment plants or, more generally, the costs of the control facilities must be taken into account. Finally, the third main effect which must be kept in mind is the *environmental damage* which is usually associated with any kind of action taken in a river basin. Unfortunately, the preceding attributes are often difficult to express in quantitative terms. The only exception is constituted by the cost of the control facilities, and this is actually the reason why profits of polluters and environmental damages are not explicitly mentioned in the majority of the technical reports which have been written on this matter. Since these terms, and in particular environmental damages, cannot be neglected without transforming the problem into a degenerate exercise, suitable upper bounds, more often called *standards*, are often imposed on those variables which are recognized to be directly responsible for the environmental damages. Thus, it is possible to have *effluent standards* if the rate of discharge of a pollutant or its effluent concentration is constrained to be smaller than or equal to a prescribed value, or *stream standards* if some suitable water pollution index, like dissolved oxygen deficit, is required to be smaller than or equal to a standard value (see Sec. 2-3).

Inequality constraints can also be imposed on the values of each control action in order to bound the cost of each control facility. For example, if we are in the phase of determining the size of a set of wastewater treatment plants we can a priori say that we are not interested in plants with more than 95 percent efficiency in order to keep the cost of each plant reasonably low. However, it could easily be the case that by using a plant with very high efficiency somewhere in the river basin more can be saved on other plants. Thus, if there are budget limitations on building the plants it would be better to express this constraint in a global form by just saying that the total cost of the plants must be smaller than or equal to a certain value (see Sec. 8-2). Of course, having solved a problem of this kind the total cost of the plants must then be shared among the polluters and this can easily be done if the polluters form a kind of cooperative team or if a supervisory control agency, like a regional authority, can force them to behave properly by means of a suitable legal mechanism. Problems of this kind, characterized by the presence of many decision makers, are better classified as management problems and will be dealt with in Chapter 10.

A great variety of steady state control problems could now be derived since there are many different ways of defining a particular problem, namely the kind of control facilities, the river quality model, the constraints, and the performance. Moreover, the same basic problem can often be formulated in different ways. For example, the problem of minimizing the total cost of a set of plants with a constraint S_1 on the environmental damage could be considered or, as well, the problem in which it is necessary to select the plants in such a way that the environmental damage is minimized while satisfying a budget constraint, S_2, i.e.,

Problem 1	*Problem 2*
min [cost of plants]	min [environmental damage]
subject to	subject to
[environmental damage] $\leq S_1$	[cost of plants] $\leq S_2$

The optimal solutions of these two problems will satisfy the equality relationship in the constraint equation. Recalling that an optimization problem with equality constraints can be transformed into an unconstrained problem by simply adding the constraint to the original performance (see Sec. 6-2), the following two reformulated problems are obtained.

Problem 1'
  min $\{[\text{cost of plants}] + \lambda_1([\text{environmental damage}] - S_1)\}$

Problem 2'
  min $\{[\text{environmental damage}] + \lambda_2([\text{cost of plants}] - S_2)\}$

where the weighting factors (*Lagrangian multipliers*) λ_1 and λ_2 are to be determined. The optimal solution of Problem 1' also gives the optimal weight λ_1^0 and if the standard S_1 is varied the weight λ_1^0 will also vary. Thus, the standards S_1 and S_2 can always be adjusted in such a way that $\lambda_1^0 = 1/\lambda_2^0$, or in other words in such a way that the four problems have the same solution. This means that if Problems 1' and 2' are considered as original problems with the λ's fixed, the selection of the relative weighting factors λ_1 and λ_2 between cost of the plants and environmental damage is, in principle, equivalent to the selection of the standards S_1 and S_2. Since it is often unclear what the numerical values of the constraints and of the weighting factors could be a priori, their selection is, in general, not an easy problem to solve. Of course, it would be possible to gain some insight into the problem by solving the same steady state problem for many different values of the weighting factors, but still the problem of finally selecting its proper value would remain.

Since this is one of the typical tasks of a central authority the last chapter of the book will return to this problem where management problems will be briefly discussed (see, in particular, Sec. 10-2). Thus, in conclusion, the only serious limitation of this chapter is that of making reference to an idealized cooperative situation in which all economic terms can be quantitatively compared.

8-2 LINEAR PROGRAMMING

This section shows how relatively general steady state river quality control problems, which are defined in terms of BOD-DO models, can be described as linear programming problems. More precisely, the two following kinds of problems are dealt with.

1. Determine the BOD mass flow rate of the effluents of a set of treatment plants such that a given *stream standard* is met at *minimum treatment cost*.
2. Determine the BOD mass flow rate of the effluents of a set of treatment plants such that the maximum value of a given pollution indicator is *minimized* under a *budget constraint* on the total cost of the treatment plants.

First, it is shown that these two types of problems are *convex optimization problems* if the river is described by a linear model and if the cost functions of the treatment plants are convex with respect to the BOD mass flow rate of the effluents. Then, to obtain linear programming problems it is necessary to discretize the model in space, and to use a linear approximation of the cost functions of the treatment plants.

Recall that any linear steady state model is of the form

$$\frac{d\mathbf{x}(l)}{dl} = \mathbf{F}(l)\mathbf{x}(l) + \mathbf{v}(l) + \mathbf{u}(l) \qquad (8\text{-}2\text{-}1)$$

where $\mathbf{x}(l)$ is the state vector, which for the rest of this section is assumed to be

given by
$$\mathbf{x}(l) = [b(l) \quad d(l)]^T$$
$\mathbf{v}(l)$ represents all uncontrollable inputs, such as distributed BOD loads and photosynthetic oxygen production, and $\mathbf{u}(l)$ is the control vector associated with the effluents of N treatment plants. If the oxygen content of the effluent of the treatment plants is assumed to be negligible we can write

$$\mathbf{u}(l) = \left[\sum_{i=1}^{N} \delta(l - l_i) \frac{u_i}{Q(l_i)} \quad 0 \right]^T$$

where δ is the impulse function and u_i is the mass flow rate of BOD discharged by the i-th treatment plant in point l_i.

The solution of Eq. (8-2-1) is (see Eqs. (1-2-2) and (1-2-4))

$$\mathbf{x}(l) = \boldsymbol{\Phi}(0, l)\mathbf{x}_0 + \boldsymbol{\Theta}_u(0, l)\mathbf{u}(\cdot) + \boldsymbol{\Theta}_v(0, l)\mathbf{v}(\cdot) \quad (8\text{-}2\text{-}2)$$

where $\mathbf{x}_0 = [b_0 \quad d_0]^T$ is the upstream boundary condition of BOD and DO deficit. Equation (8-2-2) can be written in the more compact form

$$\mathbf{x}(l) = \mathbf{a}(l) + \mathbf{B}(l)\mathbf{u}$$

where

$$\mathbf{u} = [u_1 \cdots u_N]^T$$

while the vector $\mathbf{a}(l)$ and the matrix $\mathbf{B}(l)$ can easily be derived from Eq. (8-2-2). Then, the optimal control problem of meeting a quality standard

$$\bar{\mathbf{x}} = [\bar{b} \quad \bar{d}]^T$$

along a river stretch $\mathscr{L} = \{l : 0 \leq l \leq L\}$ while minimizing the total treatment cost may be stated as

$$\min_{\mathbf{u}} [J(\mathbf{u})] = \min_{\{u_i\}} \left[\sum_{i=1}^{N} \mathscr{C}_i(u_i) \right] \quad (8\text{-}2\text{-}3a)$$

subject to

quality standard constraints $\quad \mathbf{x}(l) \leq \bar{\mathbf{x}} \quad \forall l \in \mathscr{L} \quad (8\text{-}2\text{-}3b)$

BOD discharge constraints $\quad \underline{\mathbf{u}} \leq \mathbf{u} \leq \bar{\mathbf{u}} \quad (8\text{-}2\text{-}3c)$

where $\mathscr{C}_i(u_i)$ is the cost of a treatment plant with output u_i (the load U_i entering the treatment plant is assumed to be given), and the vectors $\underline{\mathbf{u}}$ and $\bar{\mathbf{u}}$ given by $\underline{\mathbf{u}} = [\underline{u}_1 \cdots \underline{u}_N]^T$ and $\bar{\mathbf{u}} = [\bar{u}_1 \cdots \bar{u}_N]^T$ represent the lower and upper bounds imposed on the discharges of the treatment plants either by law or for technological reasons.

As shown in Sec. 7-1, the treatment costs $\mathscr{C}_i(U_i, u_i)$, which is here indicated by $\mathscr{C}_i(u_i)$ since U_i is constant, is generally neither a convex nor a concave function of u_i, but for particular values of $\underline{u}_i$ and $\bar{u}_i$, it may be assumed to be a convex function or even a linear one (see Fig. 8-2-1). When the cost functions $\mathscr{C}_i(u_i)$ of all the treatment plants are convex, the optimization problem (8-2-3) turns out to be

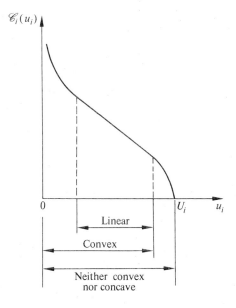

Figure 8-2-1 Treatment cost curve.

well posed, having one and only one solution. To prove this statement, it may be noted that when the cost functions $\mathscr{C}_i(u_i)$ are convex in $[\underline{u}_i, \bar{u}_i]$ the set

$$\mathscr{U}_{\bar{J}} = \{\mathbf{u} : \underline{\mathbf{u}} \leq \mathbf{u} \leq \bar{\mathbf{u}}, J(\mathbf{u}) \leq \bar{J}\}$$

is convex and such that

$$\mathscr{U}_{\bar{J}_1} \subseteq \mathscr{U}_{\bar{J}_2} \quad \text{if } \bar{J}_1 \leq \bar{J}_2$$

as shown in Fig. 8-2-2. Moreover, the *set of feasible BOD discharge* $\mathscr{U}_{\bar{\mathbf{x}}}$, defined as

$$\mathscr{U}_{\bar{\mathbf{x}}} = \{\mathbf{u} : \underline{\mathbf{u}} \leq \mathbf{u} \leq \bar{\mathbf{u}}, \quad \mathbf{x}(l) = \mathbf{a}(l) + \mathbf{B}(l)\mathbf{u} \leq \bar{\mathbf{x}} \quad \forall\, l \in \mathscr{L}\} \quad (8\text{-}2\text{-}4)$$

is also a convex set. In fact, consider two different feasible discharges $\mathbf{u}'$ and $\mathbf{u}''$ ($\mathbf{u}' \in \mathscr{U}_{\bar{\mathbf{x}}}, \mathbf{u}'' \in \mathscr{U}_{\bar{\mathbf{x}}}$) and let

$$\mathbf{u} = \lambda \mathbf{u}' + (1 - \lambda)\mathbf{u}'' \qquad (0 \leq \lambda \leq 1)$$

Then $\underline{\mathbf{u}} \leq \mathbf{u} \leq \bar{\mathbf{u}}$ and

$$\mathbf{a}(l) + \mathbf{B}(l)\mathbf{u} = \mathbf{a}(l) + \mathbf{B}(l)[\lambda \mathbf{u}' + (1 - \lambda)\mathbf{u}'']$$
$$= \lambda[\mathbf{a}(l) + \mathbf{B}(l)\mathbf{u}'] + (1 - \lambda)[\mathbf{a}(l) + \mathbf{B}(l)\mathbf{u}'']$$
$$\leq \lambda \bar{\mathbf{x}} + (1 - \lambda)\bar{\mathbf{x}} = \bar{\mathbf{x}} \quad \forall\, l \in \mathscr{L}$$

which means that $\mathbf{u} \in \mathscr{U}_{\bar{\mathbf{x}}}$ (i.e., $\mathscr{U}_{\bar{\mathbf{x}}}$ is a convex set). Thus problem (8-2-3) turns out to be a convex optimization problem as is shown in Fig. 8-2-2 and, as it is well known, such a problem has one and only one solution (point A of Fig. 8-2-2).

Similarly, the problem of meeting a budget constraint while one is maximizing the river quality (ReVelle et al., 1969) may be formally stated as

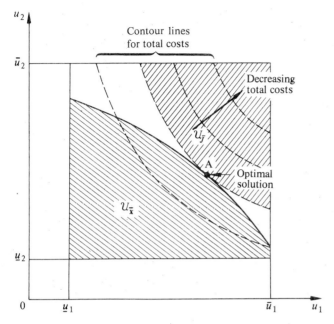

Figure 8-2-2 Geometric interpretation of the convex optimization problem (8-2-3).

$$\min_{\mathbf{u}} [J(\mathbf{u})] = \min_{\mathbf{u}} \left[\max_{l \in \mathscr{L}} \left[\max \{\alpha b(l), d(l)\} \right] \right] \qquad (8\text{-}2\text{-}5a)$$

subject to

BOD discharge constraints $\qquad \underline{\mathbf{u}} \leq \mathbf{u} \leq \bar{\mathbf{u}} \qquad (8\text{-}2\text{-}5b)$

budget constraint $\qquad \sum_{i=1}^{N} \mathscr{C}_i(u_i) \leq \bar{C} \qquad (8\text{-}2\text{-}5c)$

The positive weight α appearing in Eq. (8-2-5a) is a measure of the importance assigned to BOD with respect to DO so that the term

$$\max_{l \in \mathscr{L}} \left[\max \{\alpha b(l), d(l)\} \right]$$

is a pollution indicator. Again it can be proved that the optimization problem (8-2-5) has one and only one solution under the hypothesis that the cost functions $\mathscr{C}_i(u_i)$ are convex in the interval $[\underline{u}_i, \bar{u}_i]$. In fact, the set of feasible BOD discharges

$$\mathscr{U}_{\bar{C}} = \left\{ \mathbf{u} : \underline{\mathbf{u}} \leq \mathbf{u} \leq \bar{\mathbf{u}}, \sum_{i=1}^{N} \mathscr{C}_i(u_i) \leq \bar{C} \right\}$$

is a convex set, and the set

$$\mathscr{U}_{\bar{J}} = \{\mathbf{u} : \underline{\mathbf{u}} \leq \mathbf{u} \leq \bar{\mathbf{u}}, J(\mathbf{u}) \leq \bar{J}\}$$

250 STEADY STATE CONTROL

is convex and such that

$$\mathcal{U}_{\bar{J}_1} \subseteq \mathcal{U}_{\bar{J}_2} \quad \text{if } \bar{J}_1 \leq \bar{J}_2$$

as it is easy to check. Figure 8-2-3 shows the geometric interpretation for the convex optimization problem (8-2-5). By comparing Fig. 8-2-2 with Fig. 8-2-3, a *dual relationship* can be seen between the optimization problems (8-2-3) and (8-2-5) (see also Sec. 8-1).

The optimization problem (8-2-3) is not in standard form since it is not characterized by a finite number of constraints (see Eq. (8-2-3b)). In order to transform the problem into a standard mathematical programming problem the constraints given by Eq. (8-2-3b) can be discretized over space, as proposed by Revelle et al., (1968), thus imposing that the standards are satisfied only in a finite number of points, i.e.,

$$\mathbf{x}(l_{ij}) \leq \bar{\mathbf{x}} \quad i = 1, \ldots, N \quad j = 1, \ldots, n_i \quad (8\text{-}2\text{-}6)$$

where $l_{ij}, j = 1, \ldots, n_i$ are n_i points located between the i-th and the $(i + 1)$-th treatment effluent. This operation corresponds to approximating the convex set $\mathcal{U}_{\bar{x}}$ given by Eq. (8-2-4) by means of a convex polyhedron. To further simplify the problem it is necessary to linearize the cost functions $\mathcal{C}_i$ and write

$$\mathcal{C}_i(u_i) = \beta_i + \gamma_i u_i$$

where γ_i is the marginal cost of the treatment plant. As already mentioned in Sec. 7-1 this approximation turns out to be particularly reasonable when the

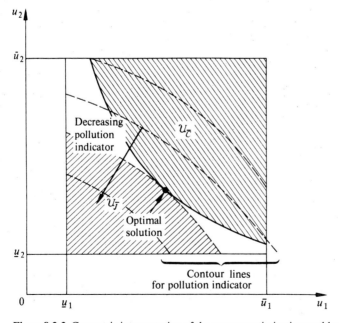

Figure 8-2-3 Geometric interpretation of the convex optimization problem (8-2-5).

interval $[\underline{u}_i, \bar{u}_i]$ corresponds to treatment efficiencies between 40 and 90 percent. Thus, the objective function (8-2-3a) can be given the simple linear form

$$J(\mathbf{u}) = \sum_{i=1}^{N} \gamma_i u_i \qquad (8\text{-}2\text{-}7)$$

and Eq. (8-2-3) becomes a linear program. The problem obtained in this way and the analogous one for Eq. (8-2-5) are, unfortunately, large linear programming problems since a high number of points l_{ij} must be selected if the solutions of the original problem and of the reformulated one are not to be too different. Therefore, it is important to find ways of simplifying the problem and two main approaches can be mentioned in this respect. The first approach consists of devising a suitable method for solving the linear programming problem which takes advantage of the special structure of the problem itself. This can be done if the constraints given by Eq. (8-2-6) are substituted by the constraints obtained by discretizing Eq. (8-2-1) in space. In this way the constraints are in recursive form and the corresponding problem is a *multistage linear programming problem* which can be solved in a very effective way by means of some suitable decomposition technique (Glassey, 1973). The second approach consists of reducing the number of constraints given by Eq. (8-2-6) by eliminating points l_{ij} in which the standards are, for some reason, certainly satisfied. Of course, these two approaches are completely independent, so that once the number of constraints has been reduced it is still possible to look for a suitable decomposition method, thus obtaining a very effective algorithm.

Two different techniques will now be briefly described for eliminating redundant constraints in the case in which standards are imposed only on DO deficit.

The first technique requires that the critical point l_c^i downstream of the i-th discharge in the absence of the downstream effluents $i + 1, \ldots, N$ (see Fig. 8-2-4) can be a priori given a lower bound $\underline{l}_c^i$ and an upper bound $\bar{l}_c^i$. In this case, in fact, the only check points to consider are the points of the intervals $[\underline{l}_c^i, \bar{l}_c^i]$ contained in the i-th reach (i.e., the reach extending from the i-th to the $(i + 1)$-th effluent and the final point of all reaches). Moreover, if $\mathbf{v}(l)$ in Eq. (8-2-1) is continuous the derivative of the deficit at any discharge point has a positive discontinuity so that the final points of the reaches can also be considered to be redundant with the exception of the last one (see Fig. 8-2-4). The computation of the intervals $[\underline{l}_c^i, \bar{l}_c^i]$ is particularly simple in the case in which the matrix $\mathbf{F}$ and the vector $\mathbf{v}$ in Eq. (8-2-1) are constant (see Secs. 4-1, 4-2). For example, the Streeter-Phelps models would give

$$\underline{l}_c^i = l_i + l_c\left(\frac{u_i}{Q}, \bar{d}\right)$$

$$\bar{l}_c^i = l_i + L_c$$

where the function $l_c(\cdot, \cdot)$ and its maximum L_c are given by Eqs. (4-1-5) and (4-1-8) while $\underline{u}_i/Q$ and $\bar{d}$ are, respectively, the lowest BOD and the highest DO deficit concentrations at the upstream boundary of the i-th reach.

252 STEADY STATE CONTROL

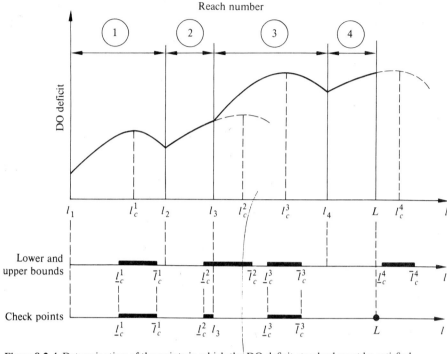

Figure 8-2-4 Determination of the points in which the DO deficit standard must be satisfied.

A second and more efficient way of reducing the number of constraints has been proposed by Arbabi and Elzinga (1975) and is now presented in the simple case in which both the matrix $\mathbf{F}(l)$ and the non controllable sources $\mathbf{v}(l)$ in Eq. (8-2-1) are constant (for more complex situations the interested reader can refer to the original paper by Arbabi and Elzinga although some statements in the paper are questionable). Under the preceding assumptions, the constraints (8-2-3b) are equivalent to (recall that standards are imposed only on DO deficit)

$$d_i \leq \bar{d} \qquad d_c^i(b_i, d_i) \leq \bar{d} \qquad i = 1, \ldots, N \qquad (8\text{-}2\text{-}8)$$

where $b_i = b(l_i)$, $d_i = d(l_i)$ are the BOD and DO concentrations at the upstream boundary of the i-th reach after mixing with the i-th effluent, while d_c^i is the critical deficit associated with the i-th BOD discharge, i.e., the critical deficit that would occur in the absence of other downstream effluents. As shown in Secs. 4-1, 4-2 the critical deficit d_c^i is a nonlinear function of the initial concentration (b_i, d_i), so that Arbabi et al., (1974) proposed approximating the function $d_c^i(b_i, d_i)$ with a linear function $\hat{d}_c^i$ in order to obtain a linear programming problem. If model (8-2-1) is the Streeter-Phelps model, the critical DO deficit is given by Eq. (4-1-6), which is repeated here for ease of reference

$$d_c^i = \frac{b_i}{f}\left\{f\left[1 - (f-1)\frac{d_i}{b_i}\right]\right\}^{1/(1-f)}$$

where $f = k_2/k_1$. Since d_c^i has continuous second partial derivatives and the Hessian matrix

$$\mathbf{H} = \begin{bmatrix} \dfrac{\partial^2 d_c^i}{\partial b_i^2} & \dfrac{\partial^2 d_c^i}{\partial b_i \, \partial d_i} \\ \dfrac{\partial^2 d_c^i}{\partial d_i \, \partial b_i} & \dfrac{\partial^2 d_c^i}{\partial d_i^2} \end{bmatrix} = \dfrac{f^2}{b_i} \left\{ f \left[1 + (1-f) \dfrac{d_i}{b_i} \right] \right\}^{\left(\frac{2f-1}{1-f}\right)} \begin{bmatrix} \left(\dfrac{d_i}{b_i}\right)^2 & -\dfrac{d_i}{b_i} \\ -\dfrac{d_i}{b_i} & 1 \end{bmatrix}$$

is positive semidefinite, the critical deficit d_c^i is a convex function of the initial BOD and DO deficit concentrations b_i and d_i. Moreover, when $d_i/b_i = \text{const.}$, d_c^i is a linear function of b_i (see Fig. 8-2-5). If precise adherence to a water quality standard is required, d_c^i must be approximated by a linear function that overestimates the true value, i.e.,

$$\hat{d}_c^i \geq d_c^i$$

for all appropriate values of b_i and d_i. One such approximation (see Fig. 8-2-5) is represented by the plane passing through the origin and sharing the two extreme rays given by $d_i = 0$ and $d_i/b_i = 1/f$ with the surface d_c^i (recall that a critical point exists if and only if $d_i/b_i \leq 1/f$, see Eq. (4-1-7)). The corresponding approximation is given by

$$\hat{d}_c^i = \delta_1 b_i + \delta_2 d_i$$

where the two parameters δ_1 and δ_2 do not depend upon the reach and are given by

$$\delta_1 = f^{f/(1-f)} \qquad \delta_2 = 1 - f^{1/(1-f)}$$

The maximum fractional deviation

$$\varepsilon = \max_{b_i, d_i} \left| \dfrac{\hat{d}_c^i - d_c^i}{d_c^i} \right|$$

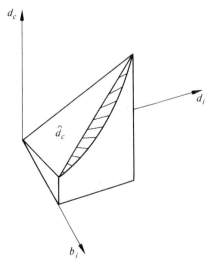

Figure 8-2-5 The surface $d_c^i(b_i, d_i)$ and the linear approximation $\hat{d}_c^i$.

occurs for $f = 1$ and is equal to 13.12 percent. An alternative linear approximation, which generates a lower fractional deviation, is obtained by minimizing the maximum possible fractional error

$$\min_{\delta_0,\delta_1,\delta_2} \left[\max_{b_i,d_i} \left| \frac{\hat{d}_c^i - d_c^i}{d_c^i} \right| \right]$$

with

$$\hat{d}_c^i = \delta_0 + \delta_1 b_i + \delta_2 d_i$$

For the Streeter-Phelps model, min–max fitting $(\delta_0, \delta_1, \delta_2)$ values can be obtained analytically (see Arbabi et al., 1974) and the maximum fractional error is about 6 percent. For more complex linear models, the parameters $(\delta_0, \delta_1, \delta_2)$ giving the min–max fit can be obtained numerically by using linear programming. In conclusion, the constraints given by Eq. (8-2-8) can be substituted by the following linear constraints $(i = 1, \ldots, N)$

$$d_i \leq \bar{d} \qquad \hat{d}_c^i = \delta_0^i + \delta_1^i b_i + \delta_2^i d_i \leq \bar{d}$$

The $\sum_{i=1}^{N} n_i$ constraints (8-2-6) have therefore been transformed into only $2N$ constraints, thus enormously reducing the time needed to solve the problem.

An important advantage of solving water quality problems as linear programming problems is that the dual variables (prices) associated with all constraints can be easily obtained. As already mentioned in Sec. 6-2, the knowledge of these variables allows the variation of the objective function induced by small variations of the constraints to be predicted. For example, in the case considered in this section, it is easy to determine the variations of the total cost of the plants due to a perturbation on the DO deficit standard $\bar{d}$ or, vice versa. If the budget constraint problem is considered, the variation of the maximum DO deficit induced by a small variation of the available budget can be determined and sometimes this extra information about the problem turns out to be even more important than the solution of the problem itself.

8-3 NONLINEAR PROGRAMMING

There exists a wide variety of applications of nonlinear programming techniques to river quality control problems, both in problem formulation and in solution methods. Compared with the linear programming formulation given in the preceding section, the nonlinear programming approach can handle more general river quality control problems. As mentioned in Sec. 6-2, the solution methods for nonlinear programming problems are, however, much less efficient. Therefore, for real multivariable problems the main task is to overcome the dimensionality difficulties.

In the problem formulation phase the main constituents (objective function, river quality model, and inequality constraints) of the problem can be roughly classified and exemplified as follows.

$$\text{Objective function} \begin{cases} \text{Linear} \ldots\ldots\ldots\ldots\ldots\ldots \text{Linear treatment costs} \\ \text{Nonlinear} \begin{cases} \text{Convex} \ldots\ldots\ldots \text{Convex treatment costs} \\ \text{Nonconvex} \ldots \text{General treatment costs} \end{cases} \end{cases}$$

$$\text{River quality model} \begin{cases} \text{Linear} \ldots\ldots\ldots\ldots\ldots \text{Streeter-Phelps model} \\ \text{Nonlinear} \ldots\ldots\ldots\ldots \begin{cases} \text{Modified Streeter-Phelps model} \\ \text{Ecological model} \end{cases} \end{cases}$$

$$\text{Inequality constraints} \begin{cases} \text{Linear} \ldots\ldots\ldots\ldots\ldots \begin{cases} \text{BOD, DO standards} \\ \text{Budget constraint for linear treatment costs} \end{cases} \\ \text{Nonlinear} \begin{cases} \text{Convex} \ldots\ldots\ldots \text{Budget constraint for convex treatment costs} \\ \text{Nonconvex} \ldots \text{Budget constraint for general treatment costs} \end{cases} \end{cases}$$

As already described in Sec. 6-2, appropriate solution methods must be chosen according to the type of problem formulated.

Convex Programming Problems

As explained in Sec. 6-2, if the objective function and the feasibility set are convex a *convex programming problem* is obtained. Fan et al., (1971) have dealt with a problem of this kind, in which the objective function was the total cost of wastewater treatment and drinking water production minus the (roughly estimated) recreational and aesthetic benefits. The river quality model used is a discretized linear dispersion model (Dobbins' model, see Sec. 4-2), and the inequality constraints are BOD and DO standards. The objective function is convex over the feasibility set. The method used for the solution of this problem is a search technique, namely the *sequential unconstrained minimization technique* (Fiacco and McCormick, 1964) in conjunction with an heuristic technique which brings the search variables back into the feasibility region. Since a dispersion model is used, the approach is especially suitable for estuaries or rivers with low flow velocity.

Hays and Gloyna (1972) have formulated a convex programming problem for the Houston Ship Channel, where the river quality model is a discretized Streeter-Phelps dispersion model for estuaries. They obtained the minimum treatment cost solution by *Rosen's* (1960) *gradient projection method*. Rosen's method is a powerful technique for optimizing convex objective functions under linear constraints.

Davidson and Bradshaw (1970) have applied a variational approach not described in Sec. 6-2 to the problem of artificial aeration. The river model is the

Streeter-Phelps plug flow model. The objective function is quadratic and has the form

$$J = \int_0^T [\alpha_1(\underline{c} - c)^2 + \alpha_2 u^2]\, d\tau \qquad (8\text{-}3\text{-}1)$$

where τ is flow time, α_1 and α_2 are positive weights and u is the artificially induced oxygen uptake of the river. Since the aeration effort for a certain oxygen uptake depends on c, J is hardly realistic as a cost function (see Secs. 7-3, 8-4, 9-2). Because no discretization was introduced the problem is an infinite dimensional convex programming problem, and a variational approach has to be used. Since no explicit inequality constraints are considered, the optimal solution can be obtained in closed form.

Sayama and Kameyama (1975) have applied a *generalized Lagrange multiplier method* (Nakayama et al., 1975) to the optimal artificial aeration problem, where the river quality model is a discretized Streeter-Phelps plug flow model. The objective function is a discrete version of Eq. (8-3-1), i.e., it is quadratic. Constraints are imposed on DO, on the capacity of each aerator, and on the sum of the capacities of the aerators. In principle, there is no need to use the relatively intricate generalized Lagrange multiplier method, which can be applied even to nonconvex problems. The problem formulated is essentially a quadratic programming problem, and Wolfe's (1959) or Beale's (1959) methods for quadratic programming would probably have been more effective in this case.

Nonconvex Programming Problems

A nonconvex programming problem results if at least one of the following two conditions is fulfilled:

(a) the objective function is nonconvex over the feasibility region
(b) the feasibility region is nonconvex

Condition (b) is fulfilled anyway if the quality model is nonlinear over the region defined by the inequality constraints.

Problems for which at least one of the conditions (a) and (b) is fulfilled are described, for instance, in Hwang et al., (1973) and Pingry and Whinston (1973). They are very general not only from the methodological point of view, but also as far as the number of control options is concerned. An even more general approach, which also takes into account air quality and solid wastes, is described by Spofford (1973).

To give an illustrative example of this type, the nonconvex steady state optimal control problem discussed by Pingry and Whinston (1973) is discussed in detail.

The most significant features of the problem are the following:

(1) A river temperature model is coupled with a quality model of Streeter-Phelps type, in which dispersion is neglected.

(2) The quality control options considered are
   by-pass piping for raw and treated wastewater
   flow augmentation
   treatment plants
   cooling towers.
(3) Standards are imposed on T, BOD, and DO.
(4) The objective is to find that combination of control facilities which achieves the standards at minimum cost.

First, the river is divided into N reaches. A new reach begins where one of the following occurs:

(a) effluent flow enters the river;
(b) incremental flow enters the river (tributary, ground water, etc.);
(c) the flow in the main channel ramifies;
(d) model parameters are altered.

Another consideration for the division into reaches is that the reach should not be so long as to allow the most critical part of the DO sag curve to be completely contained in it, because compliance with water quality standards is checked only at the end points of each reach.

The variables associated with reach i are the following (see Fig. 8-3-1):

q_{mi} = effluent flow from the the m-th BOD polluter into reach i, $m = 1, 2, \ldots, M$, $i = 1, 2, \ldots, N$
q_{mk} = effluent flow from the m-th BOD polluter into treatment plant k, $k = 1, 2, \ldots, K$
q_{ki} = effluent flow from treatment plant k into reach i
ε_k = percent removal of BOD at treatment plant k
η_i = percent removal of heat by cooling towers in reach i
Q_{i1} = incremental flow in reach i
Q_{i2} = sum of the effluent flows from BOD polluters and from treatment plants into reach i, $(Q_{i2} = \sum_m q_{mi} + \sum_k q_{ki})$
Q_{i3} = augmentation flow into reach i
Q_{i4} = effluent flow from thermal polluters in reach i

The quantities Q_{i1} and Q_{i4} are not subject to optimization.

The great variety of control variables (mainly due to the many by-pass possibilities) is subject to linear equality constraints expressing flow conservation. The flow conservation for treatment plants, for example, is

$$\sum_m q_{mk} = \sum_i q_{ki} \qquad k = 1, 2, \ldots, K$$

Moreover, upper and lower bounds are imposed on each control variable, i.e., a constraint of the type

$$\underline{\mathbf{u}} \leq \mathbf{u} \leq \bar{\mathbf{u}}$$

has to be met, where $\mathbf{u}$ is the control vector.

258 STEADY STATE CONTROL

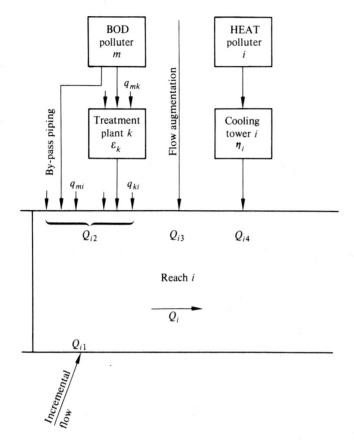

Figure 8-3-1 Sketch of the control facilities related to reach i.

In order to specify the water quality standards it is assumed that all discharges (both polluting and diluting) are located on the boundaries between reaches, and the standards have to be met at the beginning and at the end of each reach. Let $Q_i, b_i, c_i,$ and T_i be river flow rate, BOD, DO, and water temperature, respectively, at the beginning of reach i. If reach i is neither a tributary nor receives a tributary, nor is it right below a ramification point, the quality constraints for the upstream end of the i-th reach (just below the discharges) read

$$T_i = \left(\sum_{j=1}^{4} Q_{ij} T_{ij} + Q_{i-1} T'_{i-1} \right) \bigg/ Q_i \leq \bar{T}$$

$$b_i = \left(\sum_{j=1}^{4} Q_{ij} b_{ij} + Q_{i-1} b'_{i-1} \right) \bigg/ Q_i \leq \bar{b} \quad (8\text{-}3\text{-}2)$$

$$c_i = \left(\sum_{j=1}^{4} Q_{ij} c_{ij} + Q_{i-1} c'_{i-1} \right) \bigg/ Q_i \geq \underline{c}$$

with

$$Q_i = \sum_{j=1}^{4} Q_{ij} + Q_{i-1}$$

where $T'_{i-1}, b'_{i-1}, c'_{i-1}$ are temperature, BOD, and DO at the end of reach $i-1$, $\overline{T}, \overline{b}, \underline{c}$ are the standards for temperature, BOD, and DO, and $T_{ij}, b_{ij},$ and c_{ij} are self-explanatory (e.g., b_{i1} is the BOD concentration of the incremental flow in reach i). It is assumed that T_{ij} with $j \neq 4$, b_{ij} with $j \neq 2$, and c_{ij} are all constant, i.e., BOD polluters are not heat polluters and vice versa. The modifications of Eq. (8-3-2) for the reaches excluded above are straightforward. The constraints for the downstream end point of the i-th reach are

$$T'_i \leq \overline{T}$$
$$b'_i \leq \overline{b}$$
$$c'_i \geq \underline{c}$$

The model used for obtaining T'_i, b'_i, and c'_i from T_i, b_i, and c_i is (in flow time)

$$\dot{T} = -k_3(T - T_E) \qquad (8\text{-}3\text{-}3a)$$
$$\dot{b} = -k_1 b \qquad (8\text{-}3\text{-}3b)$$
$$\dot{c} = -k_1 b + k_2(c_s - c) \qquad (8\text{-}3\text{-}3c)$$

where T_E is the equilibrium water temperature (see Sec. 3-4). Flow time τ is transformed into distance l along the river by means of the relationship $l = v\tau$ where the velocity v is assumed to be of the form

$$v = \alpha Q^\beta \qquad (8\text{-}3\text{-}4)$$

where α and β are parameters to be estimated for each reach. For the temperature dependence of k_1 and k_2 Eq. (3-5-9) is used, and the dependence of k_2 upon Q is assumed to be of the same type as that expressed by Eq. (8-3-4) (see Sec. 3-5). Equations (8-3-3a), (8-3-3b), and (8-3-3c) can be solved in cascade.

Finally, the objective function to be minimized is the total cost $\mathscr{C}$ of all kinds of treatments, i.e.,

$$\mathscr{C} = \mathscr{C}^B + \mathscr{C}^A + \mathscr{C}^T + \mathscr{C}^C$$

where $\mathscr{C}^B, \mathscr{C}^A, \mathscr{C}^T$ and $\mathscr{C}^C$ represent the cost for by-passing, flow augmentation, effluent treatment, and cooling tower, respectively. These costs are nonlinear functions of the control variables (see Chap. 7) and can be written as

$$\mathscr{C}^B = \sum_m \sum_i \mathscr{C}^B_{mi}(q_{mi}) + \sum_m \sum_k \mathscr{C}^B_{mk}(q_{mk}) + \sum_k \sum_i \mathscr{C}^B_{ki}(q_{ki})$$

$$\mathscr{C}^A = \sum_i \mathscr{C}^A_i(Q_{i3})$$

$$\mathscr{C}^T = \sum_k \mathscr{C}_k^T \left(\sum_m q_{mk}, \varepsilon_k \right)$$

$$\mathscr{C}^C = \sum_i \mathscr{C}_i^C (\eta_i)$$

The explicit form of the cost functions can be found in Pingry and Whinston (1973).

Having formulated the nonlinear programming problem a few remarks on the solution technique are in order (Graves et al., 1972). The general form of the optimization problem is

$$\min J(\mathbf{z}) \tag{8-3-5a}$$

subject to

$$g_i(\mathbf{z}) \leq 0 \qquad i = 1, 2, \ldots, m \tag{8-3-5b}$$

because the equality constraints given by the quality model can be removed by using the analytical solution of the model throughout. The technique used is a direct search method (see Sec. 6-2), which means that approximate values $\mathbf{z}^{j+1}$ to the optimal $\mathbf{z}$ value are calculated iteratively according to

$$\mathbf{z}^{j+1} = \mathbf{z}^j + \alpha_j \mathbf{s}^j \qquad j = 0, 1, \ldots$$

The search direction $\mathbf{s}^j$ is determined by solving a linear programming problem which is obtained by linearization of problem (8-3-5) around $\mathbf{z}^j$. The scalar α_j specifies the step length and is determined by the minimum of J along the line specified by $\mathbf{s}^j$ (see Sec. 6-2). Since the algorithm is composed of a sequence of linear programming tasks, it is computationally efficient. However, it can find only *local optimal solutions* of the nonconvex programming problem.

The nonlinear programming approach described above has been applied to a 277 km section of the West Fork White river, Indiana, U.S.A. The river was divided into 62 reaches, the reach lengths varied from 0.16 km to 9.9 km. Thirteen major BOD polluters and three waste heat dischargers had to be taken into account. The optimization results can be found in Pingry and Whinston (1973); they are not reported here since they do not allow any particularly interesting discussion.

8-4 DYNAMIC PROGRAMMING

If the problem of optimal steady state control of river quality is considered from the flow time point of view it can be seen to be a *multistage decision problem*: each time the outlet of a control facility is passed, a decision has to be taken concerning the expenditure for this facility. If, in addition, the performance index can be written as a sum of terms each of which depends on the decision and the river quality at one stage only (e.g., sum of treatment costs), the problem has exactly the structure of a dynamic programming problem (see Sec. 6-2).

Since the multistage decision structure appears so naturally, dynamic programming algorithms have been applied many times to water quality control problems (see, for example, Liebman and Lynn, 1966; Dysart and Hines, 1970; Shih, 1970; Converse, 1972; Stehfest, 1978; Rinaldi and Soncini-Sessa, 1978). The main advantages of this approach are its simplicity and flexibility: it can easily take into account any kind of constraints, it can deal with stochastic phenomena, and one need not worry about being caught by a suboptimum. On the other hand, this approach has serious drawbacks: if the number n of state variables becomes too high ($n > 5$) the requirements for computer storage and computing time become, in general, prohibitive.

In this section dynamic programming is applied to two optimal steady state control problems of water quality. The first of these problems is solved with two different water quality models (Stehfest, 1978), namely the ecological model described in Sec. 4-4 and the Streeter-Phelps model. The problem consists in minimizing the total wastewater treatment expenditure on a river such that certain quality standards are met. The second application concerns the minimum cost design of a system of mechanical aerators such that the DO level is higher than a prescribed standard (Fioramonti et al., 1974, Rinaldi and Soncini-Sessa, 1978).

The applications shown are very different as far as the dimensionality of the optimal control problem is concerned. This is illustrated by Fig. 8-4-1 which shows the general structure of the biochemical submodel (see Sec. 3-5 and page 268). The submodel is made up of two parts: the first one describes the biochemical degradation and synthesis processes, and the second one is just the oxygen balance equation. As usual, it is assumed that there is no feedback from the second part to the first, i.e., the output (vector) $\mathbf{z}(\cdot)$ of the first part is completely independent from any input in the oxygen balance equation. This means that the problem of allocating mechanical aerators is a one-dimensional problem, with $\mathbf{z}(\cdot)$ as an input function which can be calculated in advance. On the other hand, a change of the pollutant input usually affects all variables of the biochemical submodel. Hence, the problem of optimal allocation of wastewater treatment effort has the same dimension as the biochemical submodel itself. This means that the use of the ecological model discussed in Sec. 4-4 generates an optimization problem which is barely solvable through dynamic programming on present-day computers.

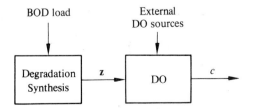

Figure 8-4-1 Usual structure of biochemical river quality models.

Optimal Allocation of Wastewater Treatment Effort

The two water quality models for the Rhine river that were discussed in Sec. 5-3 are now used for optimal allocation of wastewater treatment effort. The aim is to find the minimum cost distribution of treatment effort along the river in 1985 such that certain standards are met. As outlined in Sec. 5-3, the reason for solving the same problem twice with different river quality models is to discriminate between the models in view of the purpose for which they are built.

The algorithm The scheme for the dynamic programming solution of the problem is shown in Fig. 8-4-2. Only one state variable is depicted, on which a standard (in the form of a lower bound) is imposed, which varies in space. The river section considered is suitably divided into N reaches. As in Sec. 5-3, it is assumed that the composition of the wastewater is the same everywhere and that within each reach the discharge of (raw or treated) wastewater is uniformly distributed. The kind and extent of treatment within each reach is subject to the optimization; the treatment options considered are the same in all reaches (see p. 265).

As outlined in Sec. 6-2, the optimal pollution control is calculated in the following way: starting from the last reach, for all reaches $i, i = N, N - 1, \ldots, 1$, the functions $H_i(\mathbf{x}_i)$ are calculated, which give the minimum of treatment cost for reaches $i, i + 1, \ldots, N$ if one starts from state $\mathbf{x}_i$ at the upstream end of reach i. Each $H_i(\mathbf{x}_i)$ is calculated on the basis of $H_{i+1}(\mathbf{x}_{i+1})$ by solving the following

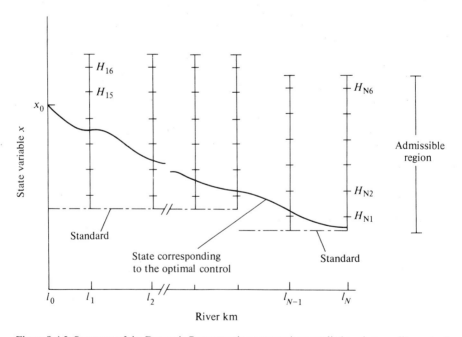

Figure 8-4-2 Structure of the Dynamic Programming approach as applied to river quality control.

minimization problem (cf. Eq. (6-2-13)):

$$H_i(\mathbf{x}_i) = \min_{\mathbf{u}_i} \left[\mathscr{C}_i(\mathbf{u}_i) + H_{i+1}(\mathbf{f}_i(\mathbf{x}_i, \mathbf{u}_i)) \right] \qquad (8\text{-}4\text{-}1)$$

where $\mathscr{C}_i(\mathbf{u}_i)$ gives the treatment cost in reach i and $\mathbf{f}_i(\mathbf{x}_i, \mathbf{u}_i)$ is the state resulting at the end of the reach i if the control action $\mathbf{u}_i$ is taken. Only those control actions $\mathbf{u}_i$ and state values $\mathbf{x}_i$ are considered for which $\mathbf{f}_i(\mathbf{x}_i, \mathbf{u}_i)$ satisfies the standards (they define the feasibility set). Note that the treatment cost $\mathscr{C}_i$ is independent of the state; this will not be the case for the artificial aeration problem discussed below. After having calculated in the backward direction all functions $H_i(\mathbf{x}_i)$, the optimal control strategy is constructed by starting from the prescribed initial state $\mathbf{x}_0$ at the upstream end of the river section considered, and calculating for each reach the control action which minimizes the sum of the treatment costs in this and all downstream reaches, using the functions H_i. During the backward calculation one could also record the optimal control actions belonging to each state. In this case the forward calculation would not require new minimizations. This scheme has certain computational drawbacks, however, so will not be applied here.

In general, the functions $H_i(\mathbf{x}_i)$ cannot be given in closed form, but can be evaluated only in certain grid points of the feasibility set (the value of $H_i(\mathbf{x}_i)$ in grid point $\mathbf{x}_{ik}$ has been denoted by H_{ik} in Fig. 8-4-2). This implies that for the calculation of $H_{i-1}(\mathbf{x}_{i-1})$ and for the final forward calculation it is necessary, in general, to interpolate between grid points. (In most practical cases the functions $H_i(\mathbf{x}_i)$ are continuous functions of $\mathbf{x}_i$, so that interpolation poses no particular problems. The optimal decision attached to each state during the backward calculation, however, is frequently not a continuous function of $\mathbf{x}_i$. This is one of the computational drawbacks mentioned above for the scheme in which the optimal decisions during backward calculation are recorded.)

A simplification of the computational scheme described consists in using the value of $H_i(\mathbf{x}_i)$ at the closest grid point instead of interpolating. Following this principle, and assuming that the control variables $\mathbf{u}_i$ are also discretized, the backward calculation becomes superfluous. Starting from the prescribed initial value, all possible policies (i.e., sets of control actions) may be directly followed in the forward direction, where for each grid point reached the accumulated costs and the last decision are stored. If two policies lead to one and the same grid point, the more costly policy is discarded, which is the application of the principle of optimality. Finally, the optimal policy is constructed in the backward direction starting from that state in the last reach which has the lowest accumulated cost attached to it. This simple scheme has sometimes been used for water quality control studies (e.g., Liebman and Lynn, 1966; Newsome, 1972), but no clear advantage can be seen in comparison with the scheme using interpolation, because the avoidance of interpolation has to be paid for with a finer grid, which requires more computations and storage. For example, for the ecological model ($n = 6$) the storage requirements were prohibitive.

A crucial point for the numerical calculation is the selection of the *admissible region* for each reach, i.e., the subset of the feasibility region covered by the grid.

The admissible region should be as large as possible in order to be sure not to exclude any possible control policy. On the other hand, the number of grid points is limited by the necessity to obtain a reasonable computing time and by the amount of storage available. Hence, if the admissible regions are large, the distance between the grid points is large, too, and consequently, interpolation may yield inaccurate results. For high dimensional problems it may be impossible to resolve this trade-off in a satisfactory way. Then, one has to resort to iterative techniques, which may lead, however, to suboptimal solutions. The optimal control problem can first be solved with sufficiently large admissible regions; because of great interpolation errors only a crude approximation to the optimal solution is obtained. Then, a better approximation is calculated by solving the problem again with smaller admissible regions which are centered around the previously obtained solution (*coarse–fine grid method*). This procedure may be repeated. Another way of approaching the solution of the dynamic programming problem iteratively starts with admissible regions which are sufficiently small for accurate interpolation. The solution is enforced to remain within these regions by imposing appropriate constraints. If those constraints turn out to be active a second run is made with the admissible regions centered around the first solution, etc. This procedure is continued until the solution is no longer deformed by those boundaries of the admissible regions which do not coincide with a water quality standard (*relaxation method*).

Application to the Rhine river For the application in which the ecological model is used, a combination of both refining techniques has been used, while the corresponding application based on the Streeter-Phelps model could be solved in one shot. Each variable of the ecological model was discretized into 6 values, and the section of the Rhine river considered was divided into 16 reaches. Hence, for the problem with the ecological model the values of $H_i(\mathbf{x}_i)$, $i = 1, \ldots, N$, in $16 \times 6^6 \simeq 7.5 \times 10^5$ grid points had to be evaluated, each value requiring the solution of a minimization problem of type (8-4-1). Minimization was carried out by discretizing the decision variables (i.e., the total capacity of the treatment plants in each reach) and comparing the costs given by the bracket expression in Eq. (8-4-1) for all admissible control actions. The state transition function $\mathbf{f}_i(\mathbf{x}_i, \mathbf{u}_i)$ for the ecological model was evaluated by solving system (5-3-7) between the boundaries of reach i by means of a Runge-Kutta integration scheme. For the Streeter-Phelps model the analytical solution of Eq. (5-3-6) could have been used. In order to be able to compare the two solutions consistently, however, the same integration procedure (with the same step width) was used. The function $\mathbf{f}_i(\mathbf{x}_i, \mathbf{u}_i)$ for the ecological model need not be evaluated in this way for each grid point because of the special structure of the model (see Fig. 8-4-1); since the equation for oxygen concentration c is linear in c, and because the other equations are independent of c, the evolution of c can be split up into a free motion and a forced motion (see Sec. 1-2). Consequently, the whole system (5-3-7) has to be integrated only once for all grid points which differ only with respect to the value of the state coordinate c. The resulting value of $\mathbf{f}_i(\mathbf{x}_i, \mathbf{u}_i)$ is then corrected for the various initial oxygen values c_i using the simple analytic solution of the homogeneous part of the oxygen equation. The same argument applies to the equation for the nondegradable pollutants, so that for each control option $\mathbf{u}_i$ the full system (5-3-7) has to be solved only 6^4 times, instead of 6^6 times as one may expect at first glance. For interpolation a simple linear scheme was used.

The values of the wastewater production in the reaches were derived from the values given in Sec. 5-3 by applying factors which account for the growth from 1970 to 1985; the growth factors do not differ from reach to reach, but they are different for the three categories of pollutants. The relative growth of the three waste components is shown in Fig. 8-4-3, which corresponds to the present trends in wastewater production. Two types of treatment were considered (see Sec. 7-1):

(a) mechanical and biological treatment
(b) precipitation with lime and adsorption with activated carbon

Of course, type (b) is considered only as an addition to (a). The costs differ from reach to reach because the plant sizes are assumed to be different. The underlying plant sizes were estimated mainly on the basis of industrial and population density within the reaches. These densities determine how the sewage transportation costs increase if the plant size increases and hence they determine the optimal plant size (see also Sec. 7-1). The exclusion of the possibility of modifying the waste generating processes gives rise to overestimates of the cost of abatement. Nevertheless, it is necessary to restrict the analysis to wastewater treatment alone since reliable data on costs of changing the production are not available.

Standards were imposed on both oxygen concentration and nondegradable

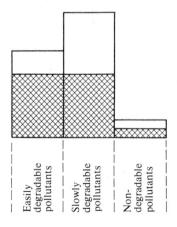

 Wastewater production 1970

 Wastewater production 1985

Figure 8-4-3 Assumed increase of wastewater production from 1970 to 1985 (COD production, in arbitrary units).

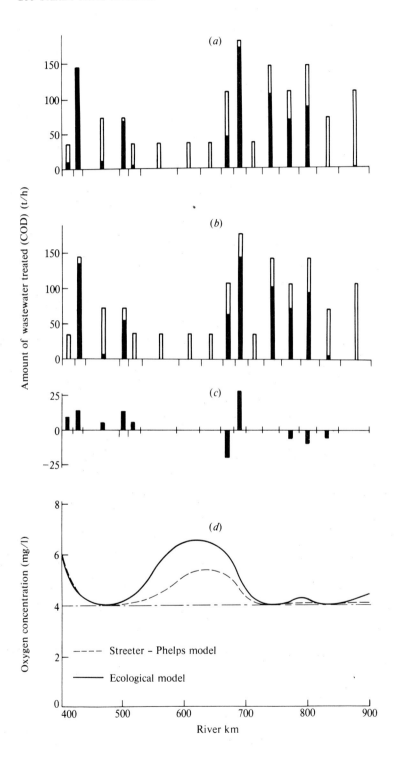

pollutant concentration. The first is to guarantee acceptable conditions for the aquatic life while the second is to limit the expenditure necessary for drinking water production. The river conditions under which the standards have to be met are $T = 20\,°C$ and $Q = 0.56MQ$, where MQ denotes the average flow rate; the latter value corresponds approximately to the flow rate which is exceeded with 80 percent probability at Köln. These conditions have been chosen because it is usually believed that the joint occurrence of high temperature and low flow is the most unfavorable event with regard to water quality.

In Fig. 8-4-4 the optimization results for the two models are shown under the constraints (standards) $w_3 \leq 20$ mg/l and $c \geq 4$ mg/l. For the sake of simplicity, an example has been chosen where the treatment effort on the tributaries was held fixed (it is not shown in Fig. 8-4-4), the standard for w_3 is barely not active, and the solution did not contain advanced treatment plants. If the two solutions are compared (cf. Fig. 8-4-4c), the most obvious feature is that with the ecological model more treatment effort is located in the upstream part of the river section. The total annual costs for the two solutions are

$$\mathscr{C}^1 = 1.08 \times 10^9 \text{ DM/a for the Streeter-Phelps model}$$

$$\mathscr{C}^2 = 1.16 \times 10^9 \text{ DM/a for the ecological model}$$

(DM = German Marks). Although the differences between the two solutions are not dramatic in this example, a decision maker may be faced with the problem of selecting one of the two optimal controls for realization. This generalized type of *model discrimination* (see Sec. 5-5) will be discussed in Sec. 10-1.

Optimal Allocation of Artificial Instream Aeration

In the preceding example the least costly system of wastewater treatment plants has been determined under the constraint that a DO standard of 4 mg/l has to be satisfied during a steady state regime characterized by a temperature of $20\,°C$ and a flow rate of $Q = 0.56MQ$, a flow rate which is exceeded with 80 percent probability at Köln. Suppose now, that a more severe DO standard is set: for example, 4 mg/l with a 95 percent probability of being met at $20\,°C$. In order to satisfy the standard the previously exposed dynamic programming algorithm may be applied once more and a new and more powerful system of treatment plants be determined. However, higher treatment levels would be required only to prevent

---

Figure 8-4-4 Optimal allocation of wastewater treatment effort along the Rhine river ($Q = 0.56MQ$, $w_3 \leq 20$ [mg/l], $c \geq 4$ [mg/l]).
The black part of the columns indicate the amount of wastewater treated, while the total columns give the amount of wastewater produced in each reach:
(a) treatment effort determined with the ecological model
(b) treatment effort determined with the Streeter-Phelps model
(c) difference between (a) and (b)
(d) oxygen concentration for optimally allocated treatment effort.

the occurrence of excessively low DO levels during rare periods of adverse wastewater assimilation characteristics, while the already determined, less costly treatment system is sufficient to obtain the desired water quality most of the time. In this case, as already pointed out in Sec. 7-3, artificial instream aeration turns out to be more effective than advanced wastewater treatment.

In this subsection, therefore, the problem of optimal allocation and design of a system of mechanical aerators is formulated as an optimal control problem and solved via dynamic programming. First, for a large class of plug flow river quality models, some simple structural properties of the solution of the problem are proved on the assumption of uniform flow rate. By means of these properties the allocation problem is simplified, and, when the aeration cost does not exhibit economies of scale, the problem is decomposed into a set of independent subproblems which are easier to solve. By means of the same properties an efficient algorithm is devised. Then, the condition of the uniformity of the flow is removed and it is shown how the above mentioned structural properties can be extended to the case of a whole river basin, with different tributaries and pollutional sources. Finally, the optimal allocation of aeration units on the Rhine river is briefly analyzed.

The optimal control problem may be formulated as follows: the number N of units to be used, their location l_i ($i = 1, \ldots, N$), and their power π_i, (which is a function of the DO increment u_i), have to be determined in such a way that a fixed DO standard is not violated at any point of a given river stretch $\mathscr{L}$ ($\mathscr{L} = \{l: 0 \leq l \leq L\}$) and that the total aeration cost is minimized. Once the steady state design conditions (flow rate $Q(l)$, temperature T, BOD load, etc.) have been fixed, a river quality model can be selected to describe the system. Such a model is generally of the form

$$\frac{d\mathbf{z}(l)}{dl} = \mathbf{f}(\mathbf{z}(l), \mathbf{v}(l)) \tag{8-4-2a}$$

$$\frac{dc(l)}{dl} = k_2(l)(c_s - c(l)) + g(\mathbf{z}(l), w(l)) + u(l) \tag{8-4-2b}$$

where $\mathbf{z}(l)$ is a suitable n-th order vector describing the various stages in the degradation or synthesis of the organic matter. For example, on the one extreme (Streeter-Phelps model) $\mathbf{z}(l)$ is simply the BOD, while on the other extreme (complex ecological model) $\mathbf{z}(l)$ stands for the concentrations of different types of pollutants and for the biomasses of various nodes of the food web. The variables $\mathbf{v}(l)$ and $w(l)$ take into account all external sources and sinks of the components of $\mathbf{z}$ and of dissolved oxygen, respectively, while

$$u(l) = \sum_{i=1}^{N} u_i \delta(l - l_i)$$

is the artificial instream aeration (δ is the impulse function and u_i is the difference between the DO concentrations downstream and upstream from the location l_i of the i-th aerator.) Finally, $\mathbf{f}$, g, k_2, $\mathbf{v}$, and w are continuous functions. In

Eq. (8-4-2) it is assumed that the aerators can be described as a set of point sources of DO and that the presence of the aerator does not influence either the natural aeration process or the process of self-purification (see Sec. 7-3).

The block diagram of the model is shown in Fig. 8-4-1. Since the function $\mathbf{z}(l)$ is completely independent of the input u it can be computed once and for all, so that the model employed in solving the optimization problem is always one-dimensional (Eq. (8-4-2b)), even if extremely sophisticated, high dimensional ecological models are selected to describe the river.

The objective function to be minimized is the sum of the costs of all aerators (see Sec. 7-3). It is worthwhile remembering that the cost $\mathscr{C}_i$ of any unit increases with u_i, $Q_i = Q(l_i)$, and $c_i = c_i(l_i^-)$, where $c_i(l_i^-)$ represents the DO concentration immediately upstream from the aerator. (In the following, for the sake of notational simplicity, l_i will often be used instead of l_i^-, the correct meaning always being clear from the context.)

The problem of determining the best aeration system can now be stated more formally as follows:

$$\min [J] = \min \left[\sum_{i=1}^{N} \mathscr{C}_i(u_i, c_i, Q_i) \right] \quad (8\text{-}4\text{-}3a)$$

with respect to

$$N \quad \{u_i\}_{i=1}^{N} \quad \{l_i\}_{i=1}^{N}$$

so that

$$\frac{dc}{dl} = k_2(l)(c_s - c) + g(\mathbf{z}(l), w(l)) + \sum_{i=1}^{N} u_i \delta(l - l_i) \quad (8\text{-}4\text{-}3b)$$

$$c(l) \geq \underline{c} \quad \forall l \in \mathscr{L} \quad (8\text{-}4\text{-}3c)$$

where the initial condition of Eq. (8-4-3b) is given, i.e.,

$$c(0) = \tilde{c}_0$$

In some cases the problem may be more complex, since additional constraints (e.g., an upper bound $\bar{N}$ on the number of units, see Fioramonti et al., 1974) or a stream standard varying over space might be imposed. It will be clear to the reader how the algorithm presented here can be modified to solve those more general problems.

Problem (8-4-3) is not in standard form, since it is not characterized by a finite number of constraints. In order to transform the problem into a standard mathematical programming problem it can simply be discretized over space, as proposed by many authors (Liebman and Lynn, 1966; Revelle et al., 1968; Chang and Yeh, 1973; Koivo and Phillips, 1975).

Some properties of the optimal solution of Problem (8-4-3) will now be proved for the particular case in which the flow rate $Q(l)$ is constant $(Q(l) = Q)$. On the basis of these properties, the problem will be simplified and, in the absence of economies of scale, decomposed into a set of subproblems. Remember (see Sec.

7-3) that the aeration cost $\mathscr{C}$ is said to exhibit the property of economies of scale if the cost of an aerator, improving the DO load from c to $(c + u)$, is lower than the sum of the costs of two aerators in series giving rise to the same effect.

In all points l between two aerators, Eq. (8-4-3b) is a linear differential equation with $u(l) = 0$. If $c(\cdot)$ and $c'(\cdot)$ are two solutions with initial value c_0 and $c_0 + \Delta c_0 (\Delta c_0 > 0)$ one obtains

$$\frac{d}{dl}(c'(l) - c(l)) = -k_2(l)(c'(l) - c(l)) < 0$$

which means that $c'(l) - c(l)$ is a decreasing function of l. On the basis of this fact it is possible to prove the following property:

Property A
The optimal solution must have all the aerator devices located in points l_i where $c_i = \underline{c}$.

In fact, if one aerator is not in such a position, it is possible to lower its cost, without violating the standard, by shifting it downstream as shown in Fig. 8-4-5 (recall that the cost function is increasing with u_i and c_i). This property can easily be understood by remembering that the natural reaeration process is more efficient for lower oxygen levels. Therefore, the least costly solution will be one where DO reaches its lowest possible value $\underline{c}$.

As a consequence, the optimal solution will be characterized by the absence of any aerators upstream from the point $\underline{l}_1$ where the natural oxygen profile (i.e., the solution c^1 of Eq. (8-4-2b), with $u = 0$ and $c^1(0) = \tilde{c}_0$) reaches the standard for the first time.

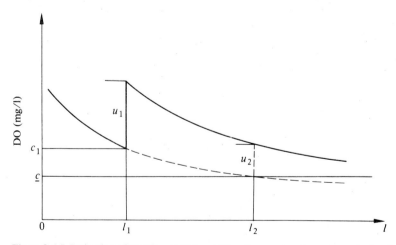

Figure 8-4-5 Reduction of aeration costs by shifting the aerator downstream to the first place where $c = \underline{c}$.

Determine now, if it exists, the point $\bar{l}_1$, such that

$$\bar{l}_1 = \min l \qquad (8\text{-}4\text{-}4a)$$

subject to

$$g(\mathbf{z}(l), w(l)) = -k_2(l)(c_s - \underline{c}) \qquad (8\text{-}4\text{-}4b)$$

$$\frac{d}{dl}g(\mathbf{z}(l), w(l)) < -(c_s - \underline{c})\frac{d}{dl}k_2(l) \qquad (8\text{-}4\text{-}4c)$$

$$l \in [\underline{l}_1, L] \qquad (8\text{-}4\text{-}4d)$$

Note that $\bar{l}_1$ is the first point downstream from $\underline{l}_1$ where, for a suitable value of $c(\underline{l}_1)$, the minimum of the DO sag curve is tangent to the standard $\underline{c}$. Then, let c^2 be the solution of Eq. (8-4-2b) for $l \geq \bar{l}_1$, with $u = 0$ and initial condition $c^2(\bar{l}_1) = \underline{c}$. Determine, if it exists, the point $\underline{l}_2$, such that

$$\underline{l}_2 = \min l$$

subject to

$$c^2(l) = \underline{c} \qquad l \in [\bar{l}_1, L]$$

(i.e., determine the first point downstream from $\bar{l}_1$ where $c^2(l)$ is equal to the standard). Finally, determine the point $\bar{l}_2$, such that

$$\bar{l}_2 = \min l$$

subject to constraints (8-4-4b), (8-4-4c) and

$$l \in [\underline{l}_2, L]$$

and continue in the same way until all the stretch $\mathscr{L}$ is worked out and a finite number p of segments

$$[\underline{l}_1, \bar{l}_1], [\underline{l}_2, \bar{l}_2], \ldots, [\underline{l}_p, \bar{l}_p]$$

is obtained (see Fig. 8-4-6 where $p = 2$). Then the following property holds.

Property B

The optimal solution of problem (8-4-3) is characterized by the presence of aerators only in the p segments $[\underline{l}_k, \bar{l}_k), k = 1, \ldots, p$ (notice that the segments are open on the right).

The proof of this property is very simple. In fact, let $c^*(l)$ be the optimal oxygen profile. Obviously, $c^*(\bar{l}_{k-1}) \geq \underline{c}$ and this implies

$$c^*(l) \geq c^k(l) \qquad \forall l \in (\bar{l}_{k-1}, \underline{l}_k) \qquad k = 1, \ldots, p$$

since $c^k(\bar{l}_{k-1}) = \underline{c}$. But

$$c^k(l) > \underline{c} \qquad \forall l \in (\bar{l}_{k-1}, \underline{l}_k) \qquad k = 1, \ldots, p$$

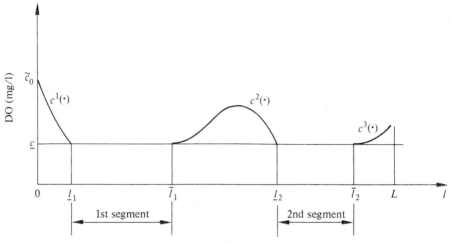

Figure 8-4-6 Determination of the segments $[\underline{l}_k, \bar{l}_k]$.

so that it can be concluded that

$$c^*(l) > \underline{c} \qquad \forall\, l \in (\bar{l}_{k-1}, \underline{l}_k) \qquad k = 1, \ldots, p$$

From property A it follows that no aerator will be present within the segment $(\bar{l}_{k-1}, \underline{l}_k)$. Moreover, no aerator will be placed in the point $\bar{l}_{k-1}$, since the same effect can be obtained at a lower cost without violating the standard by shifting the aerator to the point $\underline{l}_k$ (see Fig. 8-4-7 where $u' > u''$).

Finally the following property is worth mentioning since it allows a nice decomposition of the problem

Property C

If there are no economies of scale problem (8-4-3) can be decomposed into the following p independent subproblems ($k = 1, \ldots, p$)

$$\min\,[J_k] = \min\left[\sum_{i=1}^{N_k} \mathscr{C}_i(u_i, c_i, Q)\right] \tag{8-4-5a}$$

with respect to

$$N_k \qquad \{u_i\}_{i=1}^{N_k} \qquad \{l_i\}_{i=1}^{N_k}$$

so that

$$c(l) \geq \underline{c} \qquad \forall\, l \in [\underline{l}_k, \bar{l}_k) \tag{8-4-5b}$$

where $c(l)$ is the solution of Eq. (8-4-2b) with initial condition $c(\underline{l}_k) = \underline{c}$.

To prove this property, it is sufficient to show that, in the absence of economies of scale, $c^*(\underline{l}_k) = \underline{c}$, since then the result follows immediately from property B. Thus, consider the optimal solution c^* and, in particular, the first aerator placed upstream from the point $\bar{l}_{k-1}$, and suppose, absurdly, that its DO increment $(u' + u'')$ is such that (see Fig. 8-4-8a)

$$c^*(l_k) > \underline{c} \tag{8-4-6}$$

Since there are no economies of scale, the cost of that aerator which improves the DO level from $\underline{c}$ to $\underline{c} + u' + u''$ is equal to the cost of two aerators in series improving the DO level from $\underline{c}$ to $\underline{c} + u'$ and from $\underline{c} + u'$ to $\underline{c} + u' + u''$. (Notice that u' is such that $c(\bar{l}_{k-1}) = \underline{c}$.) Since $u'' > u'''$ (see Fig. 8-4-8a), the cost of instream aeration can be reduced, without violating the standard, by shifting the second aerator to point $\underline{l}_k$ and this contradicts Eq. (8-4-6).

The algorithm For the sake of simplicity, the algorithm is presented with reference to the case in which the aeration cost does not exhibit the property of economies of scale, so that Problem (8-4-3) can be decomposed into subproblems. The extension to the other case is very simple and will be briefly outlined in the following.

In order to apply dynamic programming to each one of the subproblems in which Problem (8-4-3) has been decomposed, it is necessary (Ortolano, 1972;

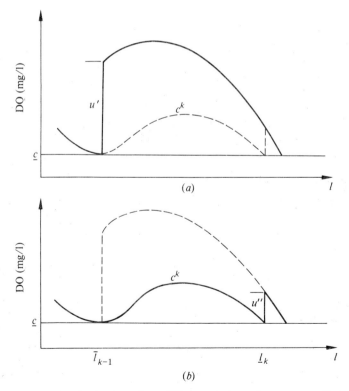

Figure 8-4-7 Comparison of two different solutions: the oxygen profile in (b) is obtained at a lower cost than the one in (a), since $\mathbf{u}'' < \mathbf{u}'$.

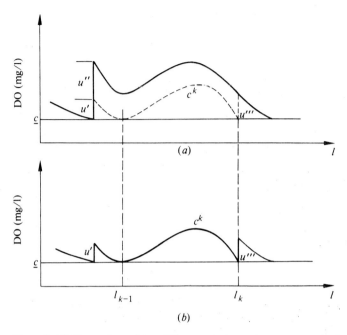

Figure 8-4-8 Comparison of two different solutions: the oxygen profile in (b) is less costly than the one in (a) if there are no economies of scale, since $u''' < u''$.

Chang and Yeh, 1973; Koivo and Phillips, 1975) to discretize constraint (8-4-5b) and restrict the decision process to consideration of only a finite number of positions in which an aerator can be placed. For this purpose, it is worth noting that an aerator, working downstream from another one, is affected by a remarkable loss of efficiency if the two aerators are too close (Price et al., 1973). Hence, the distance between two aerators must be greater than or equal to a critical distance Δ, so that the interval $[\underline{l}_k, \overline{l}_k]$ can be subdivided into M sub-intervals, where M is given by the integer part of $(\overline{l}_k - \underline{l}_k)/\Delta$. The end-points of these subintervals will be indexed by j in the following. Then, it is assumed that an aerator can be placed at any one of these points and, consistently, the compliance of the DO level with the standard is checked only in these points.

Therefore, the continuous model (8-4-2) is replaced by a discrete model. Denoting the DO concentration at the downstream end of the j-th interval by c_j and the difference between the DO levels upstream and downstream from the j-th point by U_j ($U_j = 0$ means that no aerator is located in point j), such a model consists of the following difference equation

$$c_{j+1} = \phi_j(c_j + U_j) + \theta_j \tag{8-4-7}$$

where ϕ_j and θ_j are coefficients derived from the solution of Eq. (8-4-2b). For the sake of notational simplicity, Eq. (8-4-7) can be written in the form

$$c_{j+1} = \psi(j, c_j + U_j)$$

Thus a cost function given by

$$\mathcal{H}_j(U_j) = \begin{cases} 0 & \text{if } U_j = 0 \\ \mathscr{C}_j(U_j, c_j, Q) & \text{if } U_j > 0 \end{cases}$$

can be associated to each point j and $\mathcal{H}_M = 0$ since $U_M = 0(c(\bar{l}_p) = c)$.

Denoting the new decision vector by $\mathbf{U} = [U_0 \cdots U_{M-1}]^T$, Problem (8-4-5) becomes:

$$\min [\mathcal{H}(\mathbf{U})] = \min \left[\sum_{j=0}^{M-1} \mathcal{H}_j(U_j) \right] \quad (8\text{-}4\text{-}8a)$$

subject to

$$c_{j+1} = \psi(j, c_j + U_j) \qquad j = 0, \ldots, M-1 \quad (8\text{-}4\text{-}8b)$$

$$c_j \geq \underline{c} \qquad j = 1, \ldots, M \quad (8\text{-}4\text{-}8c)$$

$$c_0 = \underline{c} \quad (8\text{-}4\text{-}8d)$$

Problem (8-4-8) is a multistage decision problem which can easily be solved by dynamic programming. The computational effort necessary to solve the problem can be greatly reduced if advantage is taken of the fact that the aerators will be located in points l_j such that $c(l_j) = \underline{c}$. In fact, let $H_h(c_h)$ be the minimum aeration cost downstream from the h-th point when c_h is the DO level at the end of the h-th interval, i.e.,

$$H_h(c_h) = \min_{\{U_j\}_{j=h}^{M-1}} \left[\sum_{j=h}^{M-1} \mathcal{H}_j(U_j) \right]$$

subject to

$$c_{j+1} = \psi(j, c_j + U_j) \qquad j = h, \ldots, M-1$$

$$c_j \geq \underline{c} \qquad j = h, \ldots, M$$

The possibility of allocating an aerator in the h-th point must be considered only if $c_h = \underline{c}$ (see Property A). Hence, the dynamic programming functional equation is

$$H_h(c_h) = \begin{cases} H_{h+1}(\psi(h, c_h)) & \text{if } c_h > \underline{c} \\ \min_{U_h} \left[\mathcal{H}_h(U_h) + H_{h+1}(\psi(h, c_h + U_h)) \right] & \text{if } c_h = \underline{c} \end{cases} \quad (8\text{-}4\text{-}9)$$

Equation (8-4-9) can be solved recursively for $h = M-1, M-2, \ldots, 0$ if the boundary condition

$$H_M(c_M) = 0$$

is taken into account. Since

$$\min [\mathcal{H}(\mathbf{U})] = H_0(\underline{c})$$

problem (8-4-8) is solved when $H_0(\underline{c})$ is computed by means of Eq. (8-4-9).

Finally, the number N_k of the aerators to be actually installed, their locations $\{l_i\}_{i=1}^{N_k}$ and their induced DO increments $\{u_i\}_{i=1}^{N_k}$, and consequently their powers π_i, are computed by backtracking, as is usually the case when dynamic programming is employed.

If the aeration cost $\mathscr{C}$ exhibits the property of economies of scale, problem (8-4-3) cannot be decomposed, but must be solved in one shot. First of all each one of the p intervals $[\underline{l}_k, \bar{l}_k]$ is discretized as before and the M_k ($k = 1,\ldots,p$) potential aeration points is determined. Then, all these points are ordered from 0 to $M = \sum_{k=1}^{p} M_k$ considering the first point of the k-th segment as the successor of the last point of the $(k-1)$-th segment. Thus, the optimal solution is obtained by applying the preceding dynamic programming algorithm (Eq. (8-4-9)) to this sequence of points.

Extension to a river basin The general case of a river basin under the assumption of piecewise constant flow rate is now considered. Obviously, in all reaches where the flow rate is constant the analysis developed for a single section can be applied, so that all the intervals $[\underline{l}_k^{(i)}, \bar{l}_k^{(i)}]$ can be found for each reach i. The new aspect is the existence of confluence points (tributaries or treatment effluents) and of diversion points where the river branches or water is derived for industrial and/or agricultural uses. First consider a diversion point. If such a point does not belong to an interval $[\underline{l}_k^{(i)}, \bar{l}_k^{(i)}]$, and if a stream standard is imposed on only one branch, there is no reason to consider the installation of an aerator immediately upstream from the diversion point, since the cost $\mathscr{C}$ is an increasing function of flow rate and the flow rate is lower downstream from the diversion. Just the contrary is true for a confluence point where two streams come together and form a larger one. Since in the two upstream branches the flow rate is lower than in the downstream one, there may be an economic advantage in installing an aerator in one of the two upstream branches (just before the confluence point) even if the corresponding DO level is greater than the standard. Thus, these confluence points must also be considered as possible points of artificial instream aeration. As a consequence, the functional equation must be suitably modified in such points to account for the possibility of locating an aerator in correspondence with any DO level. Moreover, if the cost function $\mathscr{C}(u, c, Q)$ is convex with respect to u, the optimal solution is characterized by, at most, one aerator located just upstream from the confluence point on the branch having the lower oxygen content. In fact, it can be proved that the same effect as is generated by an aerator located on the more oxygenated branch could be obtained at a lower cost by shifting it downstream from the confluence point.

To clarify the decomposition procedure, consider the example shown in Fig. 8-4-9, where four reaches are determined by the points where the flow rate has stepwise changes. Assume than in reaches 1 and 2 the DO standard is first violated in points $\underline{l}_1^{(1)}$ and $\underline{l}_1^{(2)}$, respectively, and that the segments to be considered as possible points of artificial instream aeration end in points $\bar{l}_1^{(1)}$ and $\bar{l}_1^{(2)}$. Moreover, assume that the aeration cost does not exhibit the property of economies of scale. Then, from Property C, the optimal solution turns out to be characterized by

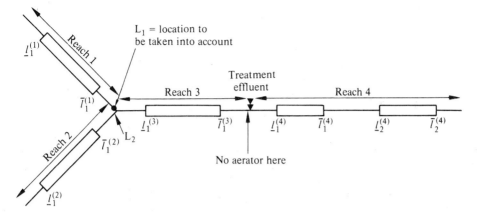

Figure 8-4-9 The optimal allocation problem for a river basin.

$c^*(\bar{l}_1^{(1)}) = c^*(\bar{l}_1^{(2)}) = \underline{c}$, so that the optimal DO levels in points L_1 and L_2 upstream from the confluence point can be a priori evaluated. Now, assume that $c^*(L_1) < c^*(L_2)$. Then from the discussion above it follows that only L_1 has to be considered as a possible aeration point. The DO concentration at the upstream end of reach 3 is dependent on the presence of an aeration device at point L_1. Thus, in order to determine the possible points for artificial aeration in reach 3, it is necessary to consider the case corresponding to the minimum value of $\underline{l}_1^{(3)}$, i.e., to assume that no aerator is present at point L_1. Finally, the position upstream from the treatment plant effluent need not be taken into account if it is certain that the DO level in the river is reduced by the deficit content of the effluent.

Once the segments of the river that are to be considered as possible points for artificial instream aeration have been determined, in the absence of economies of scale, it is possible to decide on the optimal aeration system by solving five independent subproblems. (Point L_1 must be considered with segment $[\underline{l}_1^{(3)}, \bar{l}_1^{(3)}]$.) In view of this decomposition, it is possible to apply the simple algorithm presented above, while in the presence of economies of scale the problem can be solved only by applying more complex techniques (for instance, Nonserial Dynamic Programming, see Bertelé and Brioschi, 1971).

Application to the Rhine river To show a realistic application of the techniques described, the section of the Rhine river investigated at the beginning with regard to the allocation of treatment efforts will be considered. The set of treatment plants was chosen such that compliance with an oxygen standard of 4 mg/l was guaranteed at a flow rate of $Q = 0.56MQ$, which is the flow rate exceeded with 80 percent probability. The least costly distribution of aerators which, together with those treatment plants, guarantee compliance with this standard even at a flow rate which is exceeded with 95 percent probability (low flow conditions, see Fig. 5-3-2b) will now be considered. It should be mentioned, however, that this application has to be regarded as a test for the algorithm described, and not as a real and

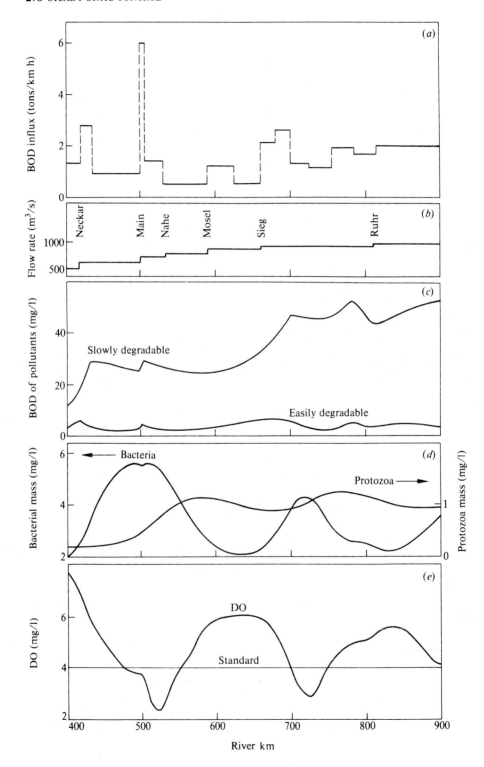

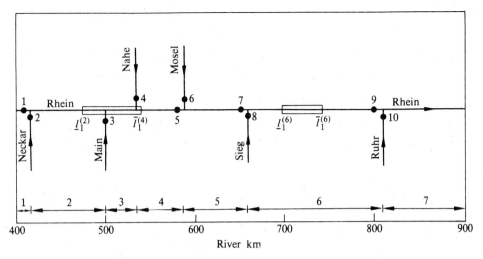

Figure 8-4-11 The seven reaches, the two segments, and the ten points to be considered as potential aeration sites.

detailed case study. The initial conditions for the problem are shown in Fig. 8-4-10. It gives the BOD discharge into the river (including the BOD load brought in by the tributaries) after the optimal solution shown in Fig. 8-4-4a has been implemented, the (reachwise constant) flow rate under low flow conditions, and the corresponding dynamics of the ecological model.

First of all, all the segments $[\underline{l}_k^{(i)}, \bar{l}_k^{(i)}]$ contained in the i-th reach ($i = 1,\ldots,7$) of the river characterized by constant flow rate (the DO concentration of all tributaries is assumed to be equal to 6 mg/l) are determined. Moreover, for each of the tributaries, two points must be considered as potential aeration points (one on the main river and one on the tributary). The final result is shown in Fig. 8-4-11, where all the points which are potential candidates for artificial aeration are indicated (ten isolated points and the two segments $[\underline{l}_1^{(2)}, \bar{l}_1^{(4)}]$ and $[\underline{l}_1^{(6)}, \bar{l}_1^{(6)}]$).

From a review of the price lists of several aeration devices, the parameters A and B of the aeration cost (Eq. 7-3-4) turned out to be 0.8 DM h/(m³ a) and 43×10^3 DM/a, i.e., for low aeration efforts (low values of u) the aeration cost exhibits the property of economies of scale. Thus, the problem cannot be decomposed into independent subproblems and the solution must be obtained in one shot. The optimal solution turned out to be characterized by the absence of aerators in all ten isolated points, while 14 aerators (total cost of 9.7×10^6 DM/a)

Figure 8-4-10 Hydrologic and biochemical characteristics of the considered steady state regime:
(a) BOD residual pattern
(b) flow rate pattern and the Rhine tributaries
(c) concentrations of easily degradable and slowly degradable pollutants
(d) concentrations of bacterial and protozoa biomasses
(e) concentration of dissolved oxygen.

280 STEADY STATE CONTROL

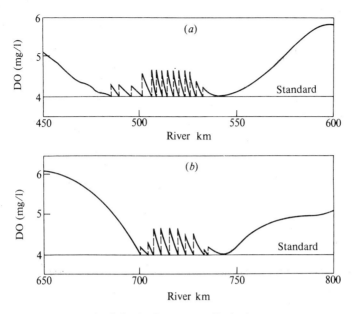

Figure 8-4-12 Optimal dissolved oxygen profiles in the two segments.

must be located in the first segment (see Fig. 8-4-12a) and 10 (total cost of 2.3×10^6 DM/a) in the second segment (see Fig. 8-4-12b). Note that the optimal oxygen profile is equal to 4 mg/l in point $\bar{l}_1^{(4)}$: this means that the same result would have been obtained by independently solving the allocation problem on the two segments. This fact and the absence of aerators on the twelve isolated points is due to the wide distances between the different tributaries and the two segments (more than ten hours of flow time between the nearest two).

A final comment Artificial instream aeration has been considered here as a substitute for advanced wastewater treatment, in order to satisfy quality standards even during extremely low flow conditions. A flow rate value $\underline{Q}$ has been arbitrarily chosen as *reference* flow value. Then the system of wastewater treatment plants was designed to meet the standards during a steady state regime characterized by a flow rate equal to $\underline{Q}$, while the aeration system was designed to cover situations of even lower flow. Both the cost $\mathscr{C}_T$ of the system of treatment plants and the cost $\mathscr{C}_A$ of the instream aeration system are dependent upon the selected reference flow rate $\underline{Q}$. More precisely, the cost $\mathscr{C}_T$ of the treatment system generally decreases as $\underline{Q}$ increases, while the aeration cost $\mathscr{C}_A$ usually increases with $\underline{Q}$. Then there exists a flow rate $\underline{Q}^0$ which corresponds to a minimum of the total cost $\mathscr{C}_T + \mathscr{C}_A$. It is therefore rational to assume this value as a reference low flow value. In order to determine $\underline{Q}^0$, a one-dimensional search procedure (see Sec. 6-1) in the flow rate space may advantageously be applied. At each step of the procedure, the costs $\mathscr{C}_T$ and $\mathscr{C}_A$ must be determined by means of the two dynamic programming algorithms presented in this section.

REFERENCES

Section 8-1

No References.

Section 8-2

Arbabi, M. and Elzinga, J. (1975). A General Linear Approach to Stream Water Quality Modelling. *Water Resour. Res.*, **11**, 191–196.

Arbabi, M., Flzinga, J., and ReVelle, C. S. (1974). The Oxygen Sag Equation: New Properties and a Linear Equation for the Critical Deficit. *Water Resour. Res.*, **10**, 921–929.

Glassey, C. R. (1973). Nested Decomposition and Multistage Linear Programs. *Management Sci.*, **20**, 282–292.

ReVelle, C. S., Dietrich, G., and Stensel, D. (1969). The Improvement of Water Quality under a Financial Constraint: A Commentary on "Linear Programming Applied to Water Quality Management." *Water Resour. Res.*, **5**, 507–513.

ReVelle, C. S., Loucks, D. P., and Lynn, W. R. (1968). Linear Programming Applied to Water Quality Management. *Water Resour. Res.*, **4**, 1–9.

Section 8-3

Beale, E. M. L. (1959). On Quadratic Programming. *Naval Research Logistics Quarterly*, **6**.

Davidson, B. and Bradshaw, R. W. (1970). A Steady State Optimal Design of Artificial Induced Aeration in Polluted Streams by the Use of Pontryagin's Minimum Principle. *Water Resour. Res.*, **6**, 383–396.

Fan, L. T., Nadkarni, R. S., and Erickson, L. E. (1971). Management of Optimum Water Quality in a Stream. *Water Research*, **5**, 1005–1021.

Fiacco, A. V. and McCormick, G. P. (1964). Computation Algorithm for the Sequential Unconstrained Minimization Technique for Nonlinear Programming. *Management Sci.*, **10**, 360–366.

Graves, G., Pingry, D. E., and Whinston, A. B. (1972). Water Quality Control: Nonlinear Programming Algorithm. *R.A.I.R.O.*, **2**, 49–78.

Hays, Jr., A. J. and Gloyna, E. F. (1972). Optimal Water Quality Management for the Houston Ship Channel. *J. San. Eng. Div., Proc. ASCE*, **98**, 195–214.

Hwang, C. L., Williams, J. L., Shojalashkari, and Fan, L. T. (1973). Regional Water Quality Management by the Generalized Gradient Method. *Water Resour. Bull.*, **9**, 1159–1181.

Nakayama, H., Sayama, Y., and Sawaragi, Y. (1975). A Generalized Lagrangian Function and Multiplier Method. *J. Optimization Theory and Appl.*, **17**, 211–227.

Pingry, D. E. and Whinston, A. B. (1973). A Regional Planning Model for Water Quality Control. In *Models for Environmental Pollution Control* (R. A. Deininger, ed.). Ann Arbor Science, Ann Arbor, Mich.

Rosen, J. B. (1960). The Gradient Projection for Nonlinear Programming: Part I. Linear Constraints. *SIAM J.*, **8**, 181–217.

Sayama, H. and Kameyama, Y. (1975). Generalized Lagrange Multiplier Method and its Application to Water Quality Management Systems (in Japanese). In *Sci. Research Program on Environmental Pollution Control*. Ministry of Education, Japan, Research Report, 21–28.

Spofford, W. O. (1973). Total Environmental Quality Management Models. In *Models for Environmental Pollution Control* (R. A. Deininger, ed.). Ann Arbor Science, Ann Arbor, Mich.

Wolfe, P. (1959). The Simplex Method for Quadratic Programming. *Econometrica*, **27**, 382–298.

Section 8-4

Bertelé, U. and Brioschi, F. (1971). *Nonserial Dynamic Programming*. Academic Press, New York.

Chang, S. and Yeh, W. W. (1973). Optimal Allocation of Artificial Aeration along Polluted Streams Using Dynamic Programming. *Water Resour. Bull.*, **9**, 985–997.

Converse, A. O. (1972). Optimum Number and Location of Treatment Plants. *J. WPCF*, **44**, 1629–1636.

Dysart II, B. C. and Hines, W. W. (1970). Control of Water Quality in a Complex Natural System. *IEEE Trans. on Systems Science and Cybernetics*, **6**, 322–329.

Fioramonti, W., Rinaldi, S., and Soncini-Sessa, R. (1974). Optimal Design of Induced Aeration in Polluted Streams Via Dynamic Programming. In *System Analysis and Modelling Approaches in Environmental Systems, Proc. IFAC-UNESCO Workshop, Zakopane, Poland, Sept. 1973*. (R. M. Dmowski, ed.). Inst. Applied Cybernetics. Polish Academy of Sciences, Warsaw, Poland.

Koivo, A. H. and Phillips, G. (1975). Optimization of Dissolved Oxygen Concentration in Polluted Streams Using Instream Aeration. *Proc. 6th IFAC Congress, Boston, Mass., August 1975*. Instrument Society of America, Pittsburgh, Pa., USA.

Liebman, J. C. and Lynn, W. R. (1966). The Optimal Allocation of Stream Dissolved Oxygen. *Water Resour. Res.*, **2**, 581–591.

Newsome, D. H. (1972). The Trent River Model: An Aid to Management. In *Modelling of Water Resources Systems*, Vol. II (A. K. Biswas, ed.). Harvest House, Montreal.

Ortolano, L. (1972). Artificial Aeration as a Substitute for Waste Water Treatment. In *Models for Managing Regional Water Quality*. (R. Dorfman et al., eds.). Harvard University Press, Cambridge, Mass.

Price, D. S., Conway, R. A., and Cheely, A. H. (1973). Surface Aerator Interactions. *J. Env. Eng. Div., Proc. ASCE*, **99**, 283–299.

Revelle, C. S., Loucks, D. P., and Lynn, W. R. (1968). Linear Programming Applied to Water Quality Management. *Water Resour. Res.*, **4**, 1–9.

Rinaldi, S. and Soncini-Sessa, R. (1978). Optimal Allocation of Artificial Instream Aerators. *J. Env. Eng. Div., Proc. ASCE*, **104**, 147–160.

Shih, C. S. (1970). System Optimization for River Basin Water Quality Management. *J. WPCF*, **42**, 1792–1804.

Stehfest, H. (1978). On the Monetary Value of an Ecological River Quality Model. Report IIASA RR-78-1, International Institute for Applied Systems Analysis, Laxenburg, Austria.

CHAPTER
NINE

UNSTEADY STATE CONTROL

9-1 GENERAL REMARKS

The need for real time river pollution control can originate from many different causes such as thermal variations, hydrological variations, load variations, and so forth. In order to adequately counteract these disturbances in real time, some of the control facilities must be such that their action (efficiency) can be varied in time. This is, indeed, what is presently done with low-flow augmentation in many instances. Real time control of wastewater treatment plants, of artificial aeration, of by-pass piping, and of cooling towers is also possible, at least in principle. Nevertheless, almost all control facilities already existing have been designed without adequate consideration of dynamic behavior or operational characteristics, so that the actual possibilities for real time control are very limited. Thus, this chapter has to be considered as a first attempt to investigate the advantages and limitations of unsteady state control in the area of river pollution. Emphasis is on methodological aspects and the main goal is to show the structural properties of the controllers under different assumptions. For this reason, the three main alternatives to consider when talking about controllers in general will now be briefly discussed.

The first and simplest control scheme is the one shown in Fig. 9-1-1 and known as *open-loop* control scheme. In this case the control function **u** is generated a

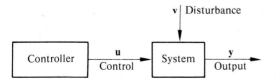

Figure 9-1-1 Open-loop controller.

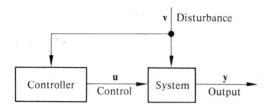

Figure 9-1-2 Feedforward controller.

priori so that the actual behavior of the system turns out to be, in general, very sensitive to the disturbance **v**, and in fact this kind of controller is never used when the main object is to lower the influence of the disturbance on the system.

A second kind of controller is the one shown in Fig. 9-1-2, known as *feedforward controller*, which differs from the open-loop controller in as far as it takes into account the actual value of the disturbance. Of course, this scheme can be considered only if the disturbance is measurable (or predictable) and when the influence of **v** on **y** is known, since otherwise no advantage can be obtained from the extra information about **v**. This is, for example, the control scheme used in low-flow augmentation: negative variations of the river flow v are detected and compensated for by releasing a certain amount u of water from a reservoir. Such a feedforward control scheme becomes more efficient when a forecast of the disturbance is available or when the cause **v*** of the disturbance itself is measureable as shown in Fig. 9-1-3. In this last case, a model describing the relationships between the primary disturbance **v*** and the disturbance **v** must be available. If, for example, v is the flow rate of the river and **v*** is rainfall, then a rainfall–runoff model for the river basin under consideration has to be developed and validated. This model will be used by the controller in order to predict the variations of v and compensate them by suitable control actions.

The third and most popular form of controller, the *feedback controller*, is shown in Fig. 9-1-4. This control scheme has three basic advantages with respect to the feedforward control scheme. First, by measuring the output of the system, the effect of any kind of disturbance can be revealed while in a feedforward scheme only the measured disturbances are compensated. Second, no particular effort is needed in modeling the disturbance, since the variations of the output with respect to a reference value are directly measured. Finally, the closed-loop structure

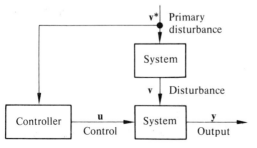

Figure 9-1-3 A more effective feedforward controller.

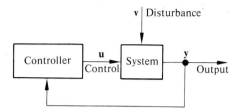

Figure 9-1-4 Feedback controller.

of the controller makes it particularly flexible; for example, the stability properties of the system can be drastically improved by means of a feedback controller (see Sec. 9-3), while no improvement can be obtained by means of a feedforward controller.

The feedback control scheme is actually very effective when the state **x** of the system is measured (i.e., when **y** = **x**), since then the determination of the best control value **u**(t) can be done without any uncertainty. On the other hand, when **y** ≠ **x** and, in particular, when **y** represents only a few components of the state vector, the controller does not receive full information about the state of the system and the selection of **u**(t) becomes more critical. In these cases it is often worth-while (see, for example, Sec. 9-3) to split the controller into two parts as shown in Fig. 9-1-5. An estimate **x̂** of the state is obtained by means of a suitable *state reconstructor* which uses measurements of the input and output of the system, and the control **u**(t) is a function of the estimate of the state, i.e.,

$$\mathbf{u}(t) = \mathbf{u}(\mathbf{\hat{x}}(t), t)$$

This function is usually called *control law* and in this case the word "controller" is used synonymously with control law while the whole feedback controller is called *regulator*.

The following two sections will deal with feedback controllers, and will show how they can be designed in view of different criteria, namely optimality and stability. In order to simplify the discussion reference will be made to the case in which no inequality constraints are imposed on the control variables. In the last section, on the other hand, a problem of feedforward optimal control with constraints on the control variables will be considered.

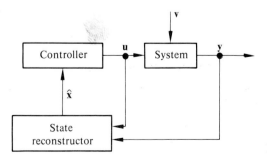

Figure 9-1-5 The structure of the regulator.

9-2 SECOND VARIATION AND ARTIFICIAL AERATION CONTROL

As a first example of unsteady state control a problem of artificial instream aeration is analyzed. This kind of problem has been extensively investigated in the last decade mainly from the point of view of optimal control. In particular, Tarassov et al., (1969) considered open-loop control laws in the presence of a single a priori known effluent source, Hullet (1974) proposed a feedback control law based on DO measurements for a river with constant cross-sectional area and constant stream velocity but with BOD sources that can vary with distance and time, while Özgören et al. (1975a) dealt with the general case of unsteady state hydrologic conditions and considered a feedback law which makes explicit use of BOD and DO measurements along the river. Sophisticated extensions of these works take into account the stochastic nature of the disturbances (see, for example, Özgören et al., 1974; and Olgac et al., 1976) and the possibility of BOD estimation from DO measurements in the context of the optimal control problem (see, for instance, Koivo and Phillips, 1975; Özgören et al., 1976).

In all these investigations the system and the performance index are assumed to be linear and quadratic, respectively, since this assumption facilitates the analysis. Unfortunately, more complex systems and/or more complex performance indices have to be used for a realistic description of such problems (see, for instance, Özgören et al., 1975b) and this is the reason why the following discussion is quite different from all others available in the literature.

In the following it will be assumed that the optimal steady state control problem has already been solved so that we will only be interested in minimizing the variations of the operational cost due to the presence of disturbances acting on the system. This problem can be formally described as a *second variation problem*, which is an optimal control problem in which the system is linear and the performance is quadratic. In general, a second variation problem for a distributed parameter system is obtained, unless dispersion is neglected, in which case the problem can be formulated along characteristic lines. Since dispersion is assumed to be negligible through this section, the main results of second variation control theory will now be briefly surveyed with reference to lumped parameter systems (see Breakwell et al., 1963). Then, the problem of optimal aeration along a river will be formulated and the implications of this theory on the structure of the optimal controller pointed out.

Second Variation Problem

Assume that a continuous, lumped parameter system of the form

$$\dot{\mathbf{x}}(\tau) = \mathbf{f}(\mathbf{x}(\tau), \mathbf{u}(\tau), \tau) \qquad \mathbf{x}(\tau_0) = \mathbf{x}_0 \qquad (9\text{-}2\text{-}1)$$

is given, together with a performance index of the kind

$$J = \int_{\tau_0}^{\tau_f} g(\mathbf{x}(\tau), \mathbf{u}(\tau), \tau) \, d\tau \qquad (9\text{-}2\text{-}2)$$

and denote by $\mathbf{x}^0(\tau)$, $\mathbf{u}^0(\tau)$ an optimal solution of this problem ($J = \min$). Under certain regularity conditions of the functions $\mathbf{f}$ and g (usually satisfied in meaningful physical problems) the optimal solution $\mathbf{x}^0(\tau)$, $\mathbf{u}^0(\tau)$ satisfies the first order necessary condition for optimality. This condition is equivalent to the necessary condition shown in Sec. 6-2 for the minimum of a function $J(\mathbf{z})$ subject to an equality constraint $\mathbf{g}(\mathbf{z}) = \mathbf{0}$, i.e., the first derivative of the Lagrangian function $[J(\mathbf{z}) + \lambda^T \mathbf{g}(\mathbf{z})]$ is equal to zero at the minimum. Here the Lagrangian function is, loosely speaking, substituted by the so-called *Hamiltonian function*

$$H(\mathbf{x}(\tau), \mathbf{u}(\tau), \lambda(\tau), \tau) = g(\mathbf{x}(\tau), \mathbf{u}(\tau), \tau) + \lambda^T(\tau)\mathbf{f}(\mathbf{x}(\tau), \mathbf{u}(\tau), \tau) \quad (9\text{-}2\text{-}3)$$

where the "multiplier" vector $\lambda(\tau)$ is called *adjoint variable* or *costate* of the system and the necessary condition for optimality is that the first derivative H_u of the Hamiltonian with respect to $\mathbf{u}$ (a row vector if $\mathbf{u}$ is a vector) is equal to zero at the optimum, i.e.,

$$H_u^0 = H_u(\mathbf{x}^0(\tau), \mathbf{u}^0(\tau), \lambda^0(\tau), \tau) = \mathbf{0} \quad (9\text{-}2\text{-}4)$$

where the optimal costate $\lambda^0(\tau)$ satisfies the differential equation

$$\dot{\lambda}(\tau) = -H_x(\mathbf{x}^0(\tau), \mathbf{u}^0(\tau), \lambda(\tau), \tau) \quad (9\text{-}2\text{-}5)$$

with given *final* condition $\lambda^0(\tau_f) = \mathbf{0}$. The interested reader can find proof of this and the following basic results in any book on optimal control such as Athans and Falb (1966), Bryson and Ho (1975), and Anderson and Moore (1971).

As in the simpler case of minimization of a function, once a solution $(\mathbf{x}^*(\tau), \mathbf{u}^*(\tau), \lambda^*(\tau))$ satisfying the necessary conditions given by Eqs. (9-2-1), (9-2-3)–(9-2-5) is obtained it is necessary to check if this satisfies some sufficient condition for optimality. While in the case of a function simply the positivity of its second derivative is checked, for optimal control problems the sufficient conditions are expressed in terms of positivity of the second derivative of the Hamiltonian. More precisely, if $(\mathbf{x}^*(\tau), \mathbf{u}^*(\tau), \lambda^*(\tau))$ is a triplet of functions satisfying Eqs. (9-2-1), (9-2-3)–(9-2-5) and

$$H_{uu}^*(\tau) = H_{uu}(\mathbf{x}^*(\tau), \mathbf{u}^*(\tau), \lambda^*(\tau), \tau) > 0 \quad (9\text{-}2\text{-}6)$$

(> 0 means positive definite) and there exists a symmetric solution $\mathbf{P}(\tau)$ of the Riccati equation

$$\dot{\mathbf{P}}(\tau) = -\mathbf{P}(\tau)\mathbf{A}(\tau) - \mathbf{A}^T(\tau)\mathbf{P}(\tau) + \mathbf{P}(\tau)\mathbf{B}(\tau)\mathbf{R}(\tau)^{-1}\mathbf{B}^T(\tau)\mathbf{P}(\tau) - \mathbf{Q}(\tau) \quad (9\text{-}2\text{-}7)$$

satisfying the boundary condition

$$\mathbf{P}(\tau_f) = \mathbf{0} \quad (9\text{-}2\text{-}8)$$

where

$$\mathbf{A}(\tau) = \mathbf{f}_x^*(\tau) - \mathbf{f}_u^*(\tau)[H_{uu}^*(\tau)]^{-1}H_{xu}^*(\tau)^T \quad \mathbf{B}(\tau) = \mathbf{f}_u^*(\tau) \quad (9\text{-}2\text{-}9)$$

$$\mathbf{Q}(\tau) = H_{xx}^*(\tau) - H_{xu}^*(\tau)[H_{uu}^*(\tau)]^{-1}H_{xu}^*(\tau)^T \quad \mathbf{R}(\tau) = H_{uu}^*(\tau) \quad (9\text{-}2\text{-}10)$$

then $(\mathbf{x}^*(\tau), \mathbf{u}^*(\tau), \lambda^*(\tau))$ is a local optimal solution of problem (9-2-1)–(9-2-2).

288 UNSTEADY STATE CONTROL

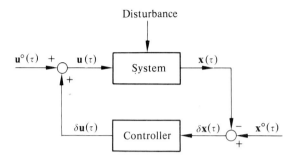

Figure 9-2-1 Second variation control scheme.

Now suppose such an optimal solution has been found, i.e., a solution $(\mathbf{x}^0(\tau), \mathbf{u}^0(\tau), \lambda^0(\tau))$ satisfying both the necessary and sufficient conditions outlined above. This optimal solution is, in general, a function of the initial state $\mathbf{x}_0$ so that a *perturbation* $\delta\mathbf{x}_0$ of the initial state implies a perturbation $(\delta\mathbf{x}^0(\tau), \delta\mathbf{u}^0(\tau), \delta\lambda^0(\tau))$ of the optimal solution. In practice, real systems are subject to uncertainty and, in particular, their initial state is rarely equal to the so-called *nominal* initial state $\mathbf{x}_0$; moreover, even if this is the case the disturbances acting on the system during the time period $[\tau_0, \tau_f]$ will induce deviations from the optimal nominal state $\mathbf{x}^0(\tau)$. Thus, the application of the nominal control $\mathbf{u}^0(\tau)$ will not give rise to the best solution of the problem. A natural countermeasure to this undesirable situation is to try to modify in real time the control variable by adding a perturbation $\delta\mathbf{u}(\tau)$ to the nominal control function $\mathbf{u}^0(\tau)$ in such a way that the new control function $[\mathbf{u}^0(\tau) + \delta\mathbf{u}(\tau)]$ is the optimum or "nearly optimum" one for the current situation. Obviously, $\delta\mathbf{u}(\tau)$ must depend upon the deviation $\delta\mathbf{x}(\tau)$ of the state from the nominal value $\mathbf{x}^0(\tau)$, so that the control scheme which has to be considered is the feedback control scheme shown in Fig. 9-2-1.

Of course, if we pretend that the perturbation $\delta\mathbf{u}(\tau)$ generated by the feedback controller minimizes the performance index, the original problem has to be solved for different initial conditions and for any initial instant of time τ, so that no particular advantage would be obtained from the feedback control scheme shown in Fig. 9-2-1. However, let us assume that the perturbations $\delta\mathbf{u}(\tau)$ and $\delta\mathbf{x}(\tau)$ are so small that they can reasonably be approximated by considering the linearized system

$$\dot{\delta\mathbf{x}}(\tau) = \mathbf{F}(\tau)\delta\mathbf{x}(\tau) + \mathbf{G}(\tau)\delta\mathbf{u}(\tau) \qquad (9\text{-}2\text{-}11)$$

where

$$\mathbf{F}(\tau) = \mathbf{f}_\mathbf{x}^0(\tau) \qquad \mathbf{G}(\tau) = \mathbf{f}_\mathbf{u}^0(\tau)$$

Thus, consistently, we can approximate the performance index with its second variation (the first variation is zero because we are in the neighborhood of an optimal solution) and the resulting problem is that the perturbation $\delta\mathbf{u}(\tau)$ must be determined such as to minimize the following quadratic performance index

$$\delta^2 J = \int_{\tau_0}^{\tau_f} [\delta\mathbf{x}^T \mathbf{Q}\, \delta\mathbf{x} + 2\delta\mathbf{x}^T \mathbf{S}\, \delta\mathbf{u} + \delta\mathbf{u}^T \mathbf{R}\, \delta\mathbf{u}]\, d\tau \qquad (9\text{-}2\text{-}12)$$

where $\mathbf{Q}(\tau)$ and $\mathbf{R}(\tau)$ are as in Eq. (9-2-10) and

$$\mathbf{S}(\tau) = H^0_{\mathbf{xu}}(\tau) \tag{9-2-13}$$

Problem (9-2-11)–(9-2-12) is the second variation problem associated with problem (9-2-1)–(9-2-2) and is a "linear-quadratic" optimal control problem, the solution of which is linear and of the kind

$$\delta\mathbf{u}(\tau) = \mathbf{K}(\tau)\,\delta\mathbf{x}(\tau) \tag{9-2-14}$$

where the *gain matrix* $\mathbf{K}(\tau)$ can be obtained by solving a suitable Riccati equation. It is interesting to note that the Riccati equation which needs to be solved to determine $\mathbf{K}(\tau)$ is exactly Eq. (9-2-7). In fact, it can be proved that

$$\mathbf{K}(\tau) = -\mathbf{R}(\tau)^{-1}(\mathbf{G}^T(\tau)\mathbf{P}(\tau) + \mathbf{S}^T(\tau)) \tag{9-2-15}$$

where $\mathbf{S}(\tau)$ is given by Eq. (9-2-13) and $\mathbf{P}(\tau)$ is the solution of Eq. (9-2-7). In other words, if the sufficient conditions for optimality have been used in solving the nominal optimal control problem no extra work is needed to correct in real time the optimal nominal solution in order to get a nearly optimum solution despite the disturbances.

Application to Optimal Real Time Aeration

Consider a finite stretch of a river with a pipeline diffuser with spargers along the length of the pipeline (see Sec. 7-3). Moreover, suppose that each sparger is hooked up with some central on-line computer so that the rate $u(t, l)$ appearing in the DO equation of the model is the control variable. For example, if the Streeter-Phelps plug flow model is used we have

$$\frac{\partial b}{\partial t} + v\frac{\partial b}{\partial l} = -k_1 b + v_b \tag{9-2-16a}$$

$$\frac{\partial c}{\partial t} + v\frac{\partial c}{\partial l} = -k_1 b + k_2(c_s - c) + v_c + u \tag{9-2-16b}$$

where v_b and v_c are distributed sources of BOD and DO and u is the control variable. Thus, a meaningful optimal control problem could be to minimize the total cost of operation with reference to a particularly significant steady state condition, and then to adjust the control function by using the second variation control law shown in Fig. 9-2-1. The nominal steady state conditions are specified by the three following functions

$$v = v(l) \qquad v_b = v_b(l) \qquad v_c = v_c(l)$$

The first one of these functions defines the characteristic lines $l = l(t_0, t)$ which are the solutions of the equation (see Eq. (3-1-19))

$$\frac{dl}{d\tau} = v(l, t_0 + \tau)$$

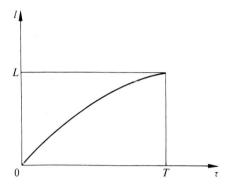

Figure 9-2-2 The reference characteristic line.

with $\tau = t - t_0$. Since under hydrological steady state conditions v depends on l only, all characteristic lines can be obtained by shifting in time the "reference" characteristic line shown in Fig. 9-2-2. This characteristic line defines a one to one relationship between distance l and flow time τ and the model can be rewritten in the lumped form

$$\dot{b}(\tau) = -k_1 b(\tau) + v_b(\tau) \tag{9-2-17a}$$

$$\dot{c}(\tau) = -k_1 b(\tau) + k_2(c_s - c(\tau)) + v_c(\tau) + u(\tau) \tag{9-2-17b}$$

(As usual, for the sake of notational simplicity the same symbols are used for the functions of τ in Eq. (9-2-17) and for the functions of l and t in Eq. (9-2-16).) The corresponding performance index is

$$J = \int_0^T g(\mathbf{x}(\tau), u(t)) \, d\tau \tag{9-2-18}$$

where $\mathbf{x}(\tau) = [b(\tau) \; c(\tau)]^T$ and g is the "cost function" taking into account fixed and operational cost of the plants and the deviations of the state of the system from a desired reference level. Problem (9-2-17)–(9-2-18) with the nominal initial condition $\mathbf{x}(0) = \mathbf{x}_0$ is a formal description of the optimal steady state control problem and its solution can be found by using one of the techniques outlined in the preceding chapter. The solution of this problem is the optimal rate of artificial aeration as a function of flow time, which can be immediately transformed into the optimal steady state control $u^0(l)$ by means of the reference characteristic line shown in Fig. 9-2-2.

If disturbances are acting on the system the state will be different from the optimal steady state so that the use of a second variation feedback control would be justified. This means that it is necessary to formulate and solve the second variation problem associated with the nominal problem (9-2-17)–(9-2-18), the solution of which will be in the form given by Eq. (9-2-14), namely

$$\delta u(\tau) = \mathbf{K}(\tau) \, \delta \mathbf{x}(\tau)$$

9-2 SECOND VARIATION AND ARTIFICIAL AERATION CONTROL

Since the control u is a scalar the matrix $\mathbf{K}(\tau)$ is a two-dimensional row vector

$$\mathbf{K}(\tau) = [k'(\tau)\ k''(\tau)] \equiv \mathbf{k}^T(\tau)$$

and the two functions $k'(\tau)$ and $k''(\tau)$ can be determined by solving a two-dimensional Riccati equation. Moreover, since system (9-2-17) is linear the second derivatives of the Hamiltonian function are the second derivatives of the cost function g, i.e.,

$$H_{uu} = g_{uu} \qquad H_{ux} = g_{ux} \qquad H_{xx} = g_{xx}$$

Finally, it is important to notice that the optimal control law given by Eq. (9-2-14) derived along the characteristics can easily be implemented since if it is interpreted in terms of distances then it follows that each point l is characterized by a time-invariant row vector $\mathbf{k}^T(l)$. This means that the control law is time-invariant and that deviations $\delta \mathbf{x}(l, t)$ from the optimal nominal solution imply a local perturbation

$$\delta u(l, t) = \mathbf{k}^T(l)\, \delta \mathbf{x}(l, t)$$

of the optimal instream aeration.

The discretization in space of this result is shown in Fig. 9-2-3 and is of particular interest since it gives a basis for the discussion that follows in the next section. Each reach i has its own *local controller* $\mathbf{k}_i^T$ characterized by two constant parameters k'_i and k''_i, and the local variation $\delta u_i(t)$ of the aeration rate is a linear combination of the local variations of BOD and DO $(\delta u_i(t) = k'_i\, \delta b_i(t) + k''_i\, \delta c_i(t))$. The parameters of these controllers are computed in *one shot* by solving the Riccati equation (9-2-7) which is, in this case, a 2×2 matrix differential equation and is therefore relatively easy to solve.

The simple problem considered in this section can be generalized to more complex situations. In particular, nonlinear ecological models could be considered for the description of the river quality and the structure of the optimal controller would still be the one shown in Fig. 9-2-3. Moreover, one could easily make reference to unsteady state situations and in particular to periodic nominal conditions (v_b and v_c can be periodic in time) and still the second variation approach could be applied. Of course in this case both the optimal solution u^0 and the matrix of the controller would be periodic functions of time. Finally, the case of

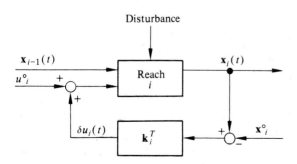

Figure 9-2-3 The local second variation controller of reach i.

a river network has also been considered by Olgac et al., (1975) and the main result is that the local controllers which are on a branch of the river are still independent one from the other. Nevertheless, any controller is connected with all other local controllers which are on other branches and are at the same distance (in flow time) from the confluence point. These controllers exchange among themselves all available information so that a disturbance on a particular branch also induces a perturbation in the artificial aeration rate of some other branch of the network. Unfortunately, the pattern of these connections depends upon the hydrology and must in general be adapted to each situation, so that a supervisory controller on the top of local controllers is needed.

9-3 POLE ASSIGNMENT AND EFFLUENT DISCHARGE CONTROL

The preceding section showed that local controllers must be used along the river if the problem is being considered from the point of view of optimal control theory. Each one of these controllers is associated with a reach and counteracts the disturbances taking place in the reach itself and no information is exchanged between the controllers (see Fig. 9-2-3).

As far as the structure of the solution is concerned, the same result would be obtained if the most traditional approach of process control engineering were to be followed. In fact, in multivariable systems in which p output variables must be kept close to a desired reference value the traditional approach is to use p control variables, each one associated with one and only one output variable through a feedback controller, as shown in Fig. 9-3-1.

For complex systems the first and sometimes most important step of the design is the *input–output assignment*, i.e., the selection of the best control variable for each output. This is often done on a heuristic basis and usually the control variable associated with an output variable is the one which has the strongest influence on it. For example, if the control variables are the effluent discharges of p wastewater treatment plants along the river and the dissolved oxygen concentration is required to be kept close to a reference value the river could be divided into exactly p reaches and the discharge u_i of the i-th treatment plant

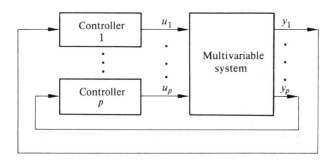

Figure 9-3-1 The classical multivariable control system.

9-3 POLE ASSIGNMENT AND EFFLUENT DISCHARGE CONTROL 293

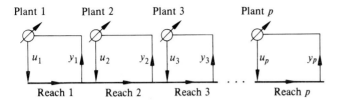

Figure 9-3-2 Independent local controllers along a river.

made sensitive to the variations of the dissolved oxygen y_i at the end of the i-th reach, as shown in Fig. 9-3-2. This solution, which seems to be very natural, can indeed be justified only if the reaches are sufficiently long. In fact, if reaches are too short, the sensitivity of the dissolved oxygen y_i with respect to u_{i-1} is greater than that with respect to u_i (see Sec. 4-1) so that the control scheme shown in Fig. 9-3-3 turns out to be preferable.

Once the input–output assignment problem has been solved the p local controllers must be suitably designed. The most simple way of solving this problem is to decompose it into the p independent control problems obtained by freezing for each pair (u_i, y_i) of adjoined variables all variables of the system with the exception of u_i and y_i. Of course, this approach can give rise to very deceptive results since it corresponds to disregarding the interactions between the subsystems. For example, some cases have been reported in the literature where this way of designing the local controllers has caused the instability of the whole system. Nevertheless, if the assignment problem has been well solved it should be expected that these interactions are relatively weak and, therefore, negligible, and this is why this approach has given satisfactory results in the past. In particular, it is important to notice that the scheme shown in Fig. 9-3-2 is exceptionally safe from the point of view of stability, since the cascade of p stable systems is always a stable system (see Sec. 1-2).

Even if this approach gives rise to the same control scheme shown in the

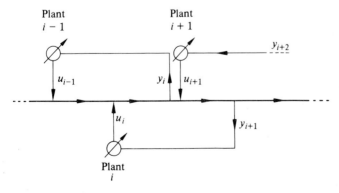

Figure 9-3-3 A more complex control structure.

preceding section the design of the local controllers is carried out in a substantially different way: a one shot procedure (solution of the Riccati equation) is used if we start from an optimality criterion, while p very simple feedback control problems must be solved if we follow the traditional approach of process control.

As far as the design of each single controller is concerned, it is necessary to select the criterion and then to find a suitable computational procedure to obtain the numerical values of the parameters characterizing the controller. Among the criteria the most popular one is *stability* which, loosely speaking, consists in selecting the controller in such a way that the resulting system has a satisfactory dynamic behavior. In the case of a river, for example, it is known that perturbations of BOD and/or DO from their nominal values due to some disturbance need days to disappear. In other words, the natural dynamics of the system are quite slow and the aim of the local controllers should be to better them in such a way that the ability of the system to absorb disturbances is improved. This problem can be easily formalized, at least in the framework of linear systems, as the problem of determining the controller such that the corresponding closed-loop system has a prescribed set of eigenvalues. Since the eigenvalues of a linear system are also called *poles* of the system this problem is known as *pole assignment*. The advantage of this criterion with respect to others is that the possibility of solving the problem, i.e., the possibility of fixing the eigenvalues at will, is related to a basic property of the system, namely *complete controllability*. Thus, a simple controllability test can reveal structural pathologic situations in which no controller can be found that gives rise to a satisfactory behavior of the closed-loop system. In these cases the proposed structure of the controller should be rejected as not being well posed and substantially changed, for example, by modifying the input–output assignment. In the following, we will show how this controllability test can be applied to river pollution problems and how satisfactory controllers can be designed by pole assignment technique.

Controllability and Pole Assignment

Consider a continuous, time-invariant, linear system of the form

$$\dot{\mathbf{x}}(t) = \mathbf{F}\mathbf{x}(t) + \mathbf{G}\mathbf{u}(t) \tag{9-3-1}$$

where $\mathbf{u}$ and $\mathbf{x}$ are the input and state vectors of dimension m and n respectively and $\mathbf{F}$ and $\mathbf{G}$ are matrices of suitable order. Then, let the controller be simply described by a constant "gain" matrix $\mathbf{K}$, i.e., let

$$\mathbf{u}(t) = -\mathbf{K}\mathbf{x}(t) \tag{9-3-2}$$

From Eqs. (9-3-1) and (9-3-2) we obtain

$$\dot{\mathbf{x}}(t) = (\mathbf{F} - \mathbf{G}\mathbf{K})\mathbf{x}(t) \tag{9-3-3}$$

which is the description of the resulting closed-loop system. Thus, the problem of pole assignment can be formulated as follows: given a system described by

Eq. (9-3-1) determine the control law (the matrix **K** in Eq. (9-3-2)) in such a way that the eigenvalues of the matrix (**F** − **GK**) describing the motion of the resulting system (9-3-3) are equal to a prescribed set of eigenvalues. Since the eigenvalues of the matrix (**F** − **GK**) are the same as those of its transpose (**F**^{T} − **K**^{T}**G**^{T}) then the problem is exactly the same as the Luenberger state reconstructor problem described in Sec. 5-4. In that case, in fact, a matrix **L** had to be chosen in such a way that the matrix (**F** − **LH**) had prescribed eigenvalues. Thus, the results described in Sec. 5-4 can be directly transformed by substituting the matrices **F** and **H** by **F**^{T} and **G**^{T}. In particular, under this transformation the observability matrix **O** given by Eq. (5-4-5) becomes the matrix

$$\mathbf{C} = [\mathbf{G} \ \mathbf{FG} \ldots \mathbf{F}^{n-1}\mathbf{G}] \qquad (9\text{-}3\text{-}4)$$

and the main result of state reconstruction theory is transformed into the following: the eigenvalues of the closed-loop system (9-3-3) can be fixed at will if and only if the matrix **C** given by Eq. (9-3-4) is of rank n. This matrix **C** is called *controllability matrix* and it can be shown that its rank is equal to n if and only if any point of the state space can be reached from the origin in finite time by applying a suitable control function **u**(·) to the system. For this reason, systems satisfying this property (rank **C** = n) are called *completely controllable systems* and the final conclusion is that complete controllability is equivalent to pole assignability.

River Pollution Control Problems and Pole Assignment

Returning to river pollution control problems, assume that each reach of the river is described by a lumped parameter system of the form

$$\dot{\mathbf{x}}(t) = \mathbf{f}(\mathbf{x}(t), \mathbf{u}(t), \mathbf{v}(t))$$

where **x** is the state vector and **u** and **v** are the control and the disturbance acting on the reach. Moreover, suppose that a reference steady state $(\bar{\mathbf{u}}, \bar{\mathbf{v}}, \bar{\mathbf{x}})$ satisfying the equation

$$\mathbf{f}(\bar{\mathbf{x}}, \bar{\mathbf{u}}, \bar{\mathbf{v}}) = 0$$

is given and that the aim of the local feedback controller is to modify the stability properties of this equilibrium state. Of course, it may be expected that after implementing the feedback controller the deviations from the desired reference state will be quite small, since perturbations due to disturbances will be quickly damped. Consistently, it may be assumed that the behavior of the system is sufficiently well described by the linearized system (see Sec. 1-2)

$$\delta\dot{\mathbf{x}}(t) = \mathbf{F}\,\delta\mathbf{x}(t) + \mathbf{G}\,\delta\mathbf{u}(t) + \mathbf{H}\,\delta\mathbf{v}(t) \qquad (9\text{-}3\text{-}5)$$

where the three constant matrices **F**, **G**, **H** are given by

$$\mathbf{F} = \left[\frac{\partial \mathbf{f}}{\partial \mathbf{x}}\right]_{\bar{\mathbf{x}},\bar{\mathbf{u}},\bar{\mathbf{v}}} \qquad \mathbf{G} = \left[\frac{\partial \mathbf{f}}{\partial \mathbf{u}}\right]_{\bar{\mathbf{x}},\bar{\mathbf{u}},\bar{\mathbf{v}}} \qquad \mathbf{H} = \left[\frac{\partial \mathbf{f}}{\partial \mathbf{v}}\right]_{\bar{\mathbf{x}},\bar{\mathbf{u}},\bar{\mathbf{v}}}$$

If the disturbance **v** is assumed to be unmeasurable, the most natural solution (see Sec. 9-1) is to determine the variations $\delta \mathbf{u}$ of the control vector on the basis of the deviations of the state from its reference value, i.e.,

$$\delta \mathbf{u}(t) = \mathbf{K}\, \delta \mathbf{x}(t)$$

Thus, this is the case described above and it can therefore be concluded that if the linearized system (9-3-5) is completely controllable with respect to $\delta \mathbf{u}$ the eigenvalues of the reach can be fixed by suitably selecting the matrix **K**.

The determination of the matrix **K** is relatively complex for multi-input systems (see, for example, Heymann, 1968) while for single-input systems the problem becomes so simple that it can often be solved by hand.

Examples

Three very simple examples are now given in order to show how the technique outlined above can be applied. In all these examples the reach is assumed to be described by a suitable CSTR model (see Sec. 4-3).

Instream aeration control Assume the reach contains some kind of artificial aeration device and let $u(t)$ be its rate of aeration. Thus the system is described by the following equations

$$\dot{b}(t) = -a_{11} b(t) + v_b(t)$$

$$\dot{c}(t) = -a_{21} b(t) - a_{22} c(t) + u(t) + v_c(t)$$

where $v_b(t)$ takes into account the BOD coming into the reach from upstream and from uncontrollable sources and $v_c(t)$ is the oxygenation rate due to the DO of the upstream reach and to sources and sinks of oxygen within the reach which are not accounted for by the other terms. This system is already linear so that the matrices **F** and **G** can be immediately obtained and are given by

$$\mathbf{F} = \begin{bmatrix} -a_{11} & 0 \\ -a_{21} & -a_{22} \end{bmatrix} \qquad \mathbf{G} \equiv \mathbf{g} = \begin{bmatrix} 0 \\ 1 \end{bmatrix}$$

Thus the controllability matrix **C** is the 2×2 matrix

$$\mathbf{C} = [\mathbf{g} \quad \mathbf{Fg}] = \begin{bmatrix} 0 & 0 \\ 1 & -a_{22} \end{bmatrix}$$

and since it is not full rank the system is not completely controllable. Therefore, by means of a feedback control law of the kind

$$\delta u(t) = -k'\, \delta b(t) - k''\, \delta c(t)$$

it is not possible to fix at will the two eigenvalues of the closed-loop system. Actually, in this very simple case, this fact is quite obvious and becomes even more transparent if the block-diagram representation of the closed-loop system shown in Fig. 9-3-4 is considered (notice that the BOD block is not in the loop).

9-3 POLE ASSIGNMENT AND EFFLUENT DISCHARGE CONTROL

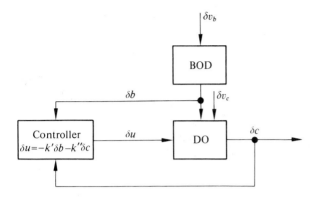

Figure 9-3-4 The local controller in the case of artificial instream aeration.

Treatment plant efficiency control Assume that every reach receives one major controlled effluent discharge from a sewage treatment facility. Moreover, assume that the BOD content of the discharge of the treatment plant can be varied by using some form of variable treatment of the sewage effluent. Although the exact method of implementing such kind of control is still open to speculation we will now show by means of our controllability analysis that this kind of intervention is of potential interest. In fact, under the above assumptions the reach is described by the following CSTR model

$$\dot{b}(t) = -a_{11}b(t) + u(t) + v_b(t) \tag{9-3-6a}$$

$$\dot{c}(t) = -a_{21}b(t) - a_{22}c(t) + v_c(t) \tag{9-3-6b}$$

so that

$$\mathbf{F} = \begin{bmatrix} -a_{11} & 0 \\ -a_{21} & -a_{22} \end{bmatrix} \quad \mathbf{G} \equiv \mathbf{g} = \begin{bmatrix} 1 \\ 0 \end{bmatrix}$$

and

$$\mathbf{C} = [\mathbf{g} \ \mathbf{Fg}] = \begin{bmatrix} 1 & -a_{11} \\ 0 & -a_{21} \end{bmatrix}$$

The controllability matrix $\mathbf{C}$ is full rank since $a_{21} \neq 0$; thus, the eigenvalues of the closed-loop system can be fixed at will by means of a feedback control law of the form

$$\delta u(t) = -k' \ \delta b(t) - k'' \ \delta c(t)$$

The computation of k' and k'' is actually very simple. In fact, the matrix $(\mathbf{F} - \mathbf{GK})$ is given by

$$\begin{bmatrix} -a_{11} & 0 \\ -a_{21} & -a_{22} \end{bmatrix} - \begin{bmatrix} k' & k'' \\ 0 & 0 \end{bmatrix} = \begin{bmatrix} -(a_{11} + k') & -k'' \\ -a_{21} & -a_{22} \end{bmatrix}$$

and its characteristic polynomial is

$$\Delta_{\mathbf{F}-\mathbf{GK}}(\lambda) = \lambda^2 + (a_{11} + a_{22} + k')\lambda - a_{21}k'' + (a_{11} + k')a_{22}$$

so that if the eigenvalues of the closed-loop system are chosen to be equal to λ_1 and λ_2 the parameters k' and k'' have to be found so that

$$\Delta_{\mathbf{F}-\mathbf{GK}}(\lambda) = (\lambda - \lambda_1)(\lambda - \lambda_2)$$

The solution of this problem is guaranteed by the complete controllability of the system and in this case is given by (recall that $a_{21} \neq 0$ implies complete controllability of system (9-3-6))

$$k' = -(a_{11} + a_{22} + \lambda_1 + \lambda_2)$$

$$k'' = -\frac{1}{a_{21}}(a_{22} + \lambda_1)(a_{22} + \lambda_2)$$

Detention tank control The effluent flow rate of the discharge can also be considered as the control variable of the reach. This can be realized by storing the sewage treated in detention tanks to some standard level and then by varying the amount of water flowing from the tank into the river as required by the control law. This time, the control variable u appears in both equations since by varying the volumetric flow rate of the effluent discharge the rate of oxygenation due to the discharge itself is varied. Thus, the model is given by

$$\dot{b}(t) = -a_{11}b(t) + u(t) + v_b(t)$$
$$\dot{c}(t) = -a_{21}b(t) - a_{22}c(t) + \alpha u(t) + v_c(t)$$

where α is the ratio of the nominal steady state concentrations of DO and BOD in the detention tank and u is proportional to the effluent flow rate. The matrices **F** and **G** are now given by

$$\mathbf{F} = \begin{bmatrix} -a_{11} & 0 \\ -a_{21} & -a_{22} \end{bmatrix} \quad \mathbf{G} \equiv \mathbf{g} = \begin{bmatrix} 1 \\ \alpha \end{bmatrix}$$

so that the controllability matrix

$$\mathbf{C} = [\mathbf{g} \quad \mathbf{Fg}] = \begin{bmatrix} 1 & -a_{11} \\ \alpha & -a_{21} - a_{22}\alpha \end{bmatrix}$$

turns out to be full rank if

$$\alpha \neq \frac{a_{21}}{a_{11} - a_{22}} \qquad (9\text{-}3\text{-}7)$$

From Eq. (4-3-11) it follows that

$$\frac{a_{21}}{a_{11} - a_{22}} = \frac{k_1}{k_1 - k_2}$$

and this expression can never be positive and smaller than one at the same time. On the other hand, the ratio α between DO and BOD concentration in the detention tank is positive and certainly smaller than one. Thus, Eq. (9-3-7) is always satisfied and this means that a feedback control law of the kind

$$\delta u(t) = -k'\, \delta b(t) - k''\, \delta c(t)$$

can be considered as a good way of modifying the dynamic behavior of the reach.

Linear Regulators

In Sec. 9-1 it was already stated that feedback control schemes are sometimes difficult to realize in practice because some of the state variables could be difficult to measure in real time. This is, for example, the case for BOD which is usually one of the components of the state vector. The solution suggested by linear system theory is to use a *linear regulator* constituted by a linear state reconstructor in cascade with a linear controller as shown in Fig. 9-3-5.

The order of the resulting system is $2n$ (the state reconstructor is a system of order n) and its characteristic polynomial $\Delta(\lambda)$ can be proved to be given by

$$\Delta(\lambda) = \Delta_{F-LH}(\lambda)\, \Delta_{F-GK}(\lambda)$$

namely, by the product of the two characteristic polynomials which were considered when dealing with state reconstructors and controllers. The main consequence of this fact is that if the system is completely controllable and completely observable the $2n$ eigenvalues of the closed-loop system can be fixed at will by suitably selecting the regulator $(\mathbf{K}, \mathbf{L})$. Since examples of state reconstructors (see Chapter 5) and controllers have already been presented no further detailed description of this matter is given at this point.

Industrial Regulators

A somewhat different approach consists in using the industrial regulators which are usually constituted by a linear dynamical single-input single-output system. Among these regulators perhaps the most popular one is the so-called *P.I. regulator*, the output of which is a linear combination of the input variable and of its integral (P. and I. stand, in fact, for proportional and integral (action)).

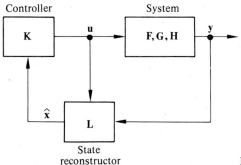

Figure 9-3-5 The structure of a linear regulator.

Thus, a P.I. regulator has the following transfer function

$$M_{P.I.}(s) = \alpha + \frac{\beta}{s}$$

Moreover, if such a regulator is used in a control scheme it will guarantee that its input is zero in steady state conditions. Thus, if one of the state variables of a system has to be kept close to a reference value it is worth while using the deviation of this variable as the input of a P.I. regulator. For example, if the DO concentration at the end of a reach must be kept constant in time use of the deviation δc as the input of the P.I. regulator is recommended. Of course, the parameters α and β of the regulator must be determined in such a way that the dynamic behavior of the closed-loop system is satisfactory, and again this could be done by referring to the eigenvalues of the resulting system (see, for example, Young and Beck, 1974). Unfortunately, the relationships between this problem and the controllability properties of the system are not so clear as in the case of state feedback controllers. Thus, no particular insight into the problem can be obtained by analyzing the controllability properties of the system and this is actually the reason why linear feedback state controllers or linear regulators should be preferred to industrial regulators, at least in principle. Nevertheless, industrial regulators have been extensively used with great success in other fields, especially in the case in which the model is unknown or partially unknown, so that it will be surprising if the use of this methodology in river pollution control does not emerge in the future.

9-4 FEEDFORWARD EMERGENCY CONTROL

A problem relevant for many rivers is the fact that with a certain, though very low probability an extraordinarily high effluent discharge may occur as a consequence of some accident. This section discusses how to cope with such an *emergency* situation.

Let the BOD effluent discharges along the whole river be measured continuously in time and assume that these BOD discharges are at their optimal steady state level when working in normal conditions. Now, suppose a very high BOD discharge takes place at point l_0 and at time t_0 as shown in Fig. 9-4-1. Then $b(l_0, t_0)$ will deviate greatly from the optimal steady state. If, after the accident the control variables are reset to the optimal steady state values, this high BOD impulse will propagate downstream and will generate very low values of DO somewhere. Thus, in order to minimize the effect of the accident, unsteady state control of the discharges is needed during the time interval $[t_0, t_f]$, where t_f is the time of arrival of the impulse at the final point l_f.

The control scheme discussed in this section is a *feedforward control* scheme like the one shown in Fig. 9-4-2. Once an *impulsive disturbance* is detected this information is transmitted to all the controllers downstream, and the controllers can start to act. If there is no dispersion, each effluent discharge should be shut

9-4 FEEDFORWARD EMERGENCY CONTROL

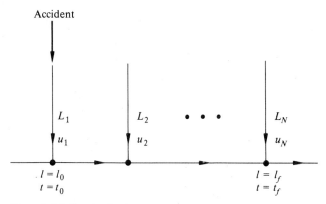

Figure 9-4-1 Sketch of a river.

down, if possible, when the BOD impulse is passing through the corresponding discharge point (the control action of each controller is just instantaneous). However, if dispersion is taken into account, each controller should act dynamically before and after the BOD peak transitates. The *perfect and instantaneous compensation* of the disturbance is technically unfeasible, since it would require BOD to be extracted from the river. Therefore, the solution of the problem must result from the trade-off between the benefit obtained by compensation and the cost of the compensation itself.

The river quality model used here is the discrete-time distributed-lag model which has already been used in Sec. 5-4 (see Eq. (5-4-13))

$$b_i(t+1) = \alpha_i b_i(t) + \frac{Q_{i-1}}{V_i} \sum_{j=0}^{\theta} \phi_{i-1}(j) b_{i-1}(t-j) + (1 - u_i(t)) \frac{L_i}{V_i} \quad (9\text{-}4\text{-}1a)$$

$$d_i(t+1) = \alpha'_i b_i(t) + \beta_i d_i(t) + \frac{Q_{i-1}}{V_i} \sum_{j=0}^{\theta} \phi'_{i-1}(j) b_{i-1}(t-j)$$

$$+ \frac{Q_{i-1}}{V_i} \sum_{j=0}^{\theta} \psi_{i-1}(j) d_{i-1}(t-j) + \frac{D_i}{V_i} \quad (9\text{-}4\text{-}1b)$$

$$i = 1, 2, \ldots, N$$

$$t = 0, 1, \ldots, T-1$$

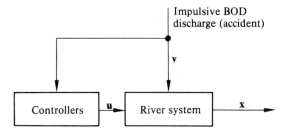

Figure 9-4-2 Feedforward emergency control system.

where $u_i(t)$ is the percentage of removal of BOD load L_i, $T = (t_f - t_0)/\Delta t$, and Δt is the elementary time interval. Equation (9-4-1) can be written as Eq. (5-4-23) in the vector form

$$\mathbf{x}(t+1) = \mathbf{F}_0 \mathbf{x}(t) + \mathbf{F}_1 \mathbf{x}(t-1) + \cdots + \mathbf{F}_\theta \mathbf{x}(t-\theta) + \mathbf{G}\mathbf{u}(t) + \mathbf{v}$$
$$t = 0, 1, \ldots, T-1 \qquad (9\text{-}4\text{-}2)$$

where $\mathbf{u}$, $\mathbf{G}$, and $\mathbf{v}$ are given by

$$\mathbf{u}(t) = \begin{bmatrix} u_1(t) & u_2(t) & \cdots & u_N(t) \end{bmatrix}^T$$

$$\mathbf{G} = \begin{bmatrix} g_1 & 0 & \cdots & 0 \\ 0 & g_2 & \cdots & 0 \\ \vdots & \vdots & & \vdots \\ 0 & 0 & \cdots & g_N \end{bmatrix} \qquad \mathbf{g}_i = \begin{bmatrix} -\dfrac{L_i}{V_i} \\ 0 \end{bmatrix} \qquad i = 1, 2, \ldots, N$$

$$\mathbf{v} = \begin{bmatrix} \mathbf{v}_1 \\ \mathbf{v}_2 \\ \vdots \\ \mathbf{v}_N \end{bmatrix} \qquad \mathbf{v}_1 = \begin{bmatrix} \dfrac{Q_0}{V_1} b_0^* + \dfrac{L_1}{V_1} \\ \dfrac{Q_0}{V_1} d_0^* + \dfrac{D_1}{V_1} \end{bmatrix} \qquad \mathbf{v}_i = \begin{bmatrix} \dfrac{L_i}{V_i} \\ \dfrac{D_i}{V_i} \end{bmatrix} \qquad i = 2, \ldots, N$$

b_0^* and d_0^* are the optimal steady state values of BOD and DO deficit in the upstream end point of the first reach, and $\mathbf{F}_0, \mathbf{F}_1, \ldots, \mathbf{F}_\theta$ are the same as in Eq. (5-4-23). The initial conditions of system (9-4-2) are assumed to be

$$\mathbf{x}(t) = \mathbf{x}^* = \begin{bmatrix} b_1^* & d_1^* & b_2^* & d_2^* & \cdots & b_N^* & d_N^* \end{bmatrix}^T$$
$$t = -\theta, -\theta+1, \ldots, -1$$
$$\mathbf{x}(0) = \begin{bmatrix} b_1(0) & d_1^* & b_2^* & d_2^* & \cdots & b_N^* & d_N^* \end{bmatrix}^T$$

where b_i^* and d_i^* are the optimal steady state values of BOD and DO deficit in each reach, and $b_1(0)$ is the extraordinary high BOD value detected in the first reach at $t = 0$. The constraints on the control variables are given by

$$0 \le u_i(t) \le \bar{u}_i \qquad (9\text{-}4\text{-}3)$$

where $\bar{u}_i$ is the maximum possible percentage of removal of BOD in the i-th reach. The performance index to be minimized is

$$J = \frac{1}{2} \sum_{t=0}^{T-1} \sum_{i=1}^{N} \left\{ w_{1i}(b_i(t+1) - b_i^*)^2 + w_{2i}(d_i(t+1) - d_i^*)^2 + m_i(u_i(t) - u_i^*)^2 \right\}$$

or in vector form

$$J = \frac{1}{2} \sum_{t=0}^{T-1} \left\{ \|\mathbf{x}(t+1) - \mathbf{x}^*\|_{\mathbf{W}}^2 + \|\mathbf{u}(t) - \mathbf{u}^*\|_{\mathbf{M}}^2 \right\} \qquad (9\text{-}4\text{-}4)$$

where $\mathbf{u}^*$ is the steady state control value and $\|\mathbf{a}\|_{\mathbf{Q}}^2$ stands for $\mathbf{a}^T \mathbf{Q} \mathbf{a}$. In Eq. (9-4-4) the first term is a measure of the degradation of the river quality and the second term represents the cost associated with supplementary treatment of the effluents. If we take care of both BOD and DO deficit deviations from the steady state, then $w_{1i} > 0$, $w_{2i} > 0$ for all i, and $\mathbf{W}$ is a positive definite diagonal matrix with w_{1i} and w_{2i} on its diagonal. If, on the other hand, we take care only of DO deficit deviations from the steady state, $w_{1i} = 0$, $w_{2i} > 0$, then $\mathbf{W}$ becomes a positive semi-definite diagonal matrix. The matrix $\mathbf{M}$ is a positive definite diagonal matrix with m_i on its diagonal. It may be expected that during an emergency period the BOD, the DO deficit, and the control (treatment) level will be greater than their steady state values, and this justifies the use of the quadratic performance index given by Eq. (9-4-4).

The overall *optimal emergency control* problem is then stated as follows: given the system (9-4-2) and the constraints (9-4-3), find the optimal control sequence $u(0), u(1), \ldots, u(T-1)$ which minimizes the performance index (9-4-4).

This problem is a linear-quadratic optimal control problem like those considered in Sec. 9-2. Unfortunately, in this case there are inequality constraints on the control variables, and therefore the optimal control cannot be obtained by solving (discrete) Riccati equations. Thus, this problem must be solved as a quadratic programming problem (equality constraints are linear since the model is linear and the inequality constraints on the control variables are also linear, while the performance index is quadratic) and as already said in Sec. 6-2 this turns out to be possible even for relatively large-scale cases. Moreover, advantage can be taken of the particular structure of the problem by using the optimization technique recently developed by Tamura (1975) based on *Lagrange duality and decomposition* (Lasdon, 1970). Here, this optimization method is only briefly described. By defining the new vector

$$\mathbf{z}_t = [\mathbf{x}^T(t) \quad \mathbf{u}^T(t-1)]^T$$

the optimal control problem can be converted into the following quadratic programming problem $(\mathbf{z} = [\mathbf{z}_1^T \quad \mathbf{z}_2^T \cdots \mathbf{z}_T^T]^T)$

$$\min_{\mathbf{z}} J = \min_{\mathbf{z}} [f_1(\mathbf{z}_1) + f_2(\mathbf{z}_2) + \cdots + f_T(\mathbf{z}_T)]$$

subject to

$$\mathbf{P}_1 \mathbf{z}_1 + \mathbf{P}_2 \mathbf{z}_2 + \cdots + \mathbf{P}_T \mathbf{z}_T = \mathbf{q}_0$$

$$\mathbf{R}_1 \mathbf{z}_1 \leq \mathbf{q}_1 \quad (9\text{-}4\text{-}5)$$

$$\mathbf{R}_2 \mathbf{z}_2 \leq \mathbf{q}_2$$

$$\vdots$$

$$\mathbf{R}_T \mathbf{z}_T \leq \mathbf{q}_T$$

where the functions f_t, the vectors $\mathbf{q}_t$, and the matrices $\mathbf{P}_t$ and $\mathbf{R}_t$ can easily be derived from Eqs. (9-4-2)–(9-4-4). The *dual problem* of problem (9-4-5) is

$$\max_{\lambda} \phi(\lambda) \quad (9\text{-}4\text{-}6)$$

where

$$\phi(\lambda) = \min_{z} \{L(z, \lambda): R_t z_t \le q_t, t = 1, 2, \ldots, T\}$$

and

$$L(z, \lambda) = J + \lambda^T \left[\sum_{t=1}^{T} P_t z_t - q_0 \right] = \sum_{t=1}^{T} \{f_t(z_t) + \lambda^T P_t z_t\} - \lambda^T q_0$$

This dual problem (9-4-6) has no constraints, and the dual objective function $\phi(\lambda)$ is a concave function. Therefore, the dual objective function has a unique maximum point and if the value $\phi(\lambda)$ and the gradient $\phi_\lambda(\lambda)$ can be computed at each point λ^*, the dual problem (9-4-6) can be easily solved by means of some gradient technique (see Sec. 6-2). To evaluate the dual function at $\lambda = \lambda^*$, we must solve the problem

$$\min_{z} L(z, \lambda^*) = \min_{z} \left[\sum_{t=1}^{T} \{f_t(z_t) + \lambda^{*T} P_t z_t\} - \lambda^{*T} q_0 \right]$$

subject to

$$R_t z_t \le q_t \qquad t = 1, 2, \ldots, T$$

This minimization problem can be decomposed into the following T independent simpler subproblems ($t = 1, 2, \ldots, T$)

$$\min_{z_t} \left[f_t(z_t) + \lambda^{*T} P_t z_t \right] \tag{9-4-7}$$

subject to

$$R_t z_t \le q_t$$

Thus, if the optimal solution of problem (9-4-7) is $z_t = z_t^*$, we have

$$\phi(\lambda^*) = L(z^*, \lambda^*)$$

$$\phi_\lambda(\lambda)|_{\lambda=\lambda^*} = \sum_{t=1}^{T} P_t z_t^* - q_0 \tag{9-4-8}$$

Since any gradient technique needs the information of Eq. (9-4-8), the gradient technique

$$\lambda^{i+1} = \lambda^i + \alpha_i s^i \qquad i = 1, 2, \ldots \tag{9-4-9}$$

can be applied to maximize the dual objective function $\phi(\lambda)$ starting from an initial crude estimate λ^1, where in each iterative step (i.e., for $\lambda^* = \lambda^i$) it is necessary to solve T independent small minimization problems (9-4-7) to obtain $z_t = z_t^i$, $t = 1, 2, \ldots, T$. In Eq. (9-4-9) s^i is the search direction computed by using $\phi_\lambda(\lambda^i)$, and α_i is the step length (see Sec. 6-2). The optimization procedure described above is, therefore, a *two-level hierarchical optimization* procedure as shown in Fig. 9-4-3. Thus, we can conclude that if an effective procedure for solving Eq. (9-4-7) is available, the dual problem is easily solved and at the same

9-4 FEEDFORWARD EMERGENCY CONTROL

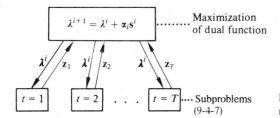

Figure 9-4-3 Two-level hierarchical optimization scheme.

time the optimal solution of the primal problem is obtained. In the case considered, since Eq. (9-4-4) is quadratic, it is easy to solve problem (9-4-7). Moreover, it can be proved (see Tamura, 1975) that if **W** is positive definite, problem (9-4-7) can be further reduced to a one-dimensional quadratic problem.

An example with $N = 4$ and related to the River Cam (see Sec. 5-5) is now presented to show the nature of the feedforward optimal control (Tamura, 1974). The values of the parameters appearing in Eqs. (9-4-2)–(9-4-4) are

$k_1 = 0.32$ days^{-1} $k_2 = 0.20$ days^{-1} $\Delta t = 0.5$ days

$Q_0 = 75 \times 10^3$ m$^3/\Delta t$ $Q_1 = 82.5 \times 10^3$ m$^3/\Delta t$

$Q_2 = 90 \times 10^3$ m$^3/\Delta t$ $Q_3 = 97.5 \times 10^3$ m$^3/\Delta t$

$Q_4 = 105 \times 10^3$ m$^3/\Delta t$

$b_i^* = 5$ mg/l $d_i^* = 3$ mg/l $V_i = 150 \times 10^3$ m^3

$\bar{u}_i = 0.8$ $L_i = 300$ kg/Δt $D_i = -12.75$ kg/Δt

$\phi_i(0) = \psi_i(0) = 0.15$ $\phi_i(1) = \psi_i(1) = 0.7$

$\phi_i(2) = \psi_i(2) = 0.15$ $\phi_i'(0) = \phi_i'(1) = \phi_i'(2) = 0$ for all i

The initial BOD value $b_1(0)$ is assumed to be

$$b_1(0) = 15 \text{ mg/l}$$

and the optimal solution is shown in Fig. 9-4-4, where $c_s = 10$ mg/l. Although the control level is changed only twice a day ($\Delta t = 0.5$ days) in each reach, the effects of the accident are well modulated in the second, third, and fourth reach of the river. The control sequences have some *anticipatory nature*, that is high control (treatment) levels appear before the situation becomes too bad. This kind of control action cannot be achieved by local feedback controllers such as those described in the previous sections. Thus, the conclusion can be drawn that when relevant deviations from the nominal steady state do occur, as in the case of a high accidental release of BOD, the most suitable structure of real time control seems to be that of a feedforward emergency control of the kind shown in this section.

The anticipatory action in the downstream reaches is an indication, however, that the performance index given by Eq. (9-4-4) may not be as realistic as originally

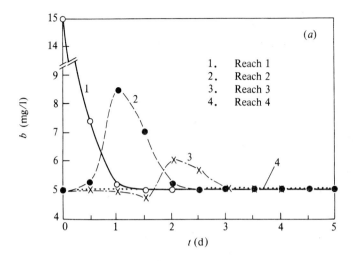
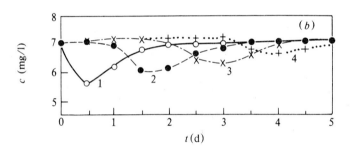
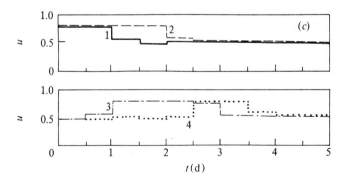

Figure 9-4-4 Optimal BOD–DO trajectories and control sequences: (a) BOD (b) DO (c) treatment level and BOD effluent discharge.

thought (Stehfest, 1978): the optimal control may create an improvement of the nominal water quality just before the pollution peak arrives, and an intimation of it can indeed be observed in Fig. 9-4-4. These deviations ought to be considered beneficial, but the performance index (9-4-4) scores them as detrimental. A non-quadratic performance index would not introduce too many troubles, however, because it would at least be convex.

The only serious limitation to the method seems to be the dimensionality of the optimization problem which needs to be solved. For a large number of reaches a promising scheme for obtaining suboptimal controls may be the reach-wise computation proposed by Singh (1975), which requires less computational effort. In this scheme the control problem is solved reach by reach starting from upstream and the optimal output of reach $(i - 1)$ is handled as the exogenous input of reach i when the optimal control problem for reach i is solved.

REFERENCES

Section 9-1

No References.

Section 9-2

Anderson, B. D. D. and Moore, J. B. (1971). *Linear Optimal Control.* Prentice Hall, London.
Athans, M. and Falb, P. L. (1966). *Optimal Control: An Introduction to the Theory and its Applications.* McGraw-Hill, New York.
Breakwell, J. V., Speyer, J. L., and Bryson, A. E. (1963). Optimization and Control of Non Linear Systems Using the Second Variation. *SIAM Journal on Control,* 1, 192–223.
Bryson, A. E. and Ho, Y. C. (1975). *Applied Optimal Control.* John Wiley, New York.
Hullett, W. E. (1974). Optimal Estuary Aeration: an Application of Distributed Parameter Control Theory. *Applied Mathematics and Optimization,* 1, 20–63.
Koivo, A. J. and Phillips, G. R. (1975). Optimal Estimation and Control of Polluted Stream Variables. *Proc. 6th IFAC Congress, Boston, Mass., August 1975.* Instrument Society of America, Pittsburgh, Pa., USA.
Olgac, N. M., Cooper, C. A., and Longman, R. W. (1975). Optimal Control of Artificial Aeration in River Networks. *Industry Oriented Conference and Exhibit. Milwaukee, Wisconsin, Oct. 1975.* Instrument Society of America, Pittsburgh, Pa., USA.
Olgac, N. M., Cooper, C. A., and Longman, R. W. (1976). Optimal Allocation of Measurement and Control Resources with Application to River Depollution. *IEEE Trans. on Systems, Man and Cybernetics,* 6, 377–384.
Özgören, M. K., Cooper, C. A., and Longman, R. W. (1976). Optimal Control of a Partially Time Lagged River Aeration System Subject to Random Disturbances. *Proc. Joint Automatic Control Conf., West Lafayette, Ind., July 1976.*
Özgören, M. K., Longman, R. W. and Cooper, C. A. (1974). Stochastic Optimal Control of Artificial River Aeration. *AACC Proc. Joint Automatic Control Conf., Austin, Texas, June 1974.* The University of Texas, Austin, Tex., USA.
Özgören, M. K., Longman, R. W., and Cooper, C. A. (1975a). Optimal Control of a Distributed Parameter, Time-Lagged River Aeration System. *Journal of Dynamic Systems, Measurements and Control,* ASME Trans. series G, 97, 166–171.

Özgören, M. K., Longman, R. W., and Cooper, C. A. (1975b). Application of Lie Transform Based Canonical Perturbation Methods to the Optimal Control of Bilinear Systems. *Proc. AAS/AIAA Astrodynamics Specialist Conf., Nassau, Bahamas, July 1975.* AAS Publications Office, P.O. Box 746, Tarzana, Ca., U.S.A.

Tarassov, V. J., Perlis, H. J., and Davidson, B. (1969). Optimization of a Class of River Aeration Problems by the Use of Multivariable Distributed Parameter Control Theory. *Water Resour. Res.*, **5**, 563–573.

Section 9-3

Heymann, M. (1968). Comments on Pole Assignment in Multi-Input Controllable Linear Systems. *IEEE Trans. on Automatic Control*, **13**, 748–749.

Young, P. and Beck, B. (1974). The Modelling and Control of Water Quality in a River System. *Automatica*, **10**, 455–468.

Section 9-4

Lasdon, L. S. (1970). *Optimization Theory for Large Systems.* Macmillan, New York.

Singh, M. G. (1975). River Pollution Control. *Int. J. Systems Sci.*, **6**, 9–21.

Stehfest, H. (1978). Some Remarks on Real-Time Control of Dissolved Oxygen in Rivers. In *Recent Developments in Real-Time Forecasting and Control of Water Resources Systems, Proc. IIASA Workshop, 18–20 Oct. 1976.* John Wiley, New York.

Tamura, H. (1974). A Discrete Dynamic Model with Distributed Transport Delays and its Hierarchical Optimization for Preserving Stream Quality. *IEEE Trans. on Systems, Man and Cybernetics*, **4**, 424–431.

Tamura, H. (1975). Decentralized Optimization for Distributed-Lag Models of Discrete Systems. *Automatica*, **11**, 593–602.

CHAPTER
TEN

MANAGEMENT OF THE RIVER BASIN

10-1 TASKS OF RIVER QUALITY MANAGEMENT

Although the problems analyzed in Chapters 5, 8, and 9 are often characterized by the presence of a decision maker, they should not be qualified as management problems since this term usually refers to much more complex situations. Being very pragmatic, management problems could be defined as the problems which must be solved by any kind of *control agency* (C.A.), or groups of them, responsible for the use of a resource in a well defined area. Thus, management problems can be very different as far as their spatial and temporal scale is concerned since it is possible to have municipal, regional, and state C.A.s, which can deal with short, medium, and long term planning. Moreover, regarding water quality, a great variety of problems occurs if the interactions between water quality and other environmental aspects are taken into account. Following Kneese and Bower (1968) it can be said that the main task of a C.A. is to implement all possible relevant measures to improve water quality in the river basin. In doing this, particular care must be devoted to all interrelations with *water resources development*, *land use management*, and other environmental aspects such as *soil* and *air pollution*. Finally, the socio-economic attributes of the problem cannot be neglected; in particular, the C.A. should internalize the major externalities associated with waste discharges and provide an opportunity for affected parties to participate in the decision-making process.

If management problems are analyzed in detail it can be recognized that their solution consists, in general, of the following four sequential steps.

1. decomposition of the problem into subproblems
2. solution of the subproblems
3. determination of the best compromise
4. implementation

The first step is possibly the most important and delicate one and is usually influenced by the particular responsibilities of the C.A., and by the decision and information structures of the system. Decomposition techniques have been developed in the context of large scale mathematical programming (see, for instance, Himmelblau, 1973), but they are not easy to apply to such complex cases as water resources problems. For this reason, environmental management problems are usually transformed into several simpler subproblems by using very pragmatic decomposition criteria. Thus, for example, river quality control problems are decomposed over time into independent subproblems (for instance, unsteady state control problems can be solved independently from steady state control problems). Another very simple way of decomposing large problems is to subdivide the river basin into smaller river basins. Almost all the problems discussed in the preceding chapters make reference to a simple stretch of river and are therefore to be thought of as single subproblems of a larger problem. Sometimes the solution of the problem can be obtained by a very simple aggregation of all the suboptimal solutions, but more often some coordination is needed as shown, for example, at the end of Sec. 9-2 for the problem of optimal aeration of a river network. Finally, when complex environmental management problems must be solved a third possible decomposition consists of considering all main components like land use, water pollution, urbanization, industrialization, as different compartments of the problem. Each one of these compartments can be considered as an independent problem and is often simple enough to be formulated and solved as a mathematical programming problem.

The second step (solution of subproblems) proves to be the most simple one and is certainly the one on which most has been written. The estimation problems presented in Chapter 5 and the control problems discussed in Chapters 8 and 9 are significant examples of this phase of the solution of a complex management problem.

After having solved the subproblems one is usually left with the problem of determining the best compromise solution (third step on page 309). Indeed, the most significant characteristic of a management problem is the fact that often a supercriterion must be devised in order to choose among different alternatives, which are sometimes quite difficult to compare. Vector optimization, better known as *multiobjective programming*, is the field of mathematics which deals with these kind of problems. Since multiobjective programming recently became popular in water resources management (see Cohon and Marks, 1975), the main ideas and results of this technique will be surveyed in the next section. The problem of determining the *best compromise* is, in general, formulated as a *secondary optimization problem* (determination of the best solution among a set of optimal solutions). A great variety of problems of this kind could be described in order to point out the main characteristics of these secondary optimization problems (trade-off between different attributes, uncertainty, subjective evaluation).

One example of this kind is the selection of the "best" solution among various problem solutions which have been obtained with different river quality models. To be more specific, suppose that a decision maker of a river authority

10-1 TASKS OF RIVER QUALITY MANAGEMENT

has to solve the problem posed at the beginning of Sec. 8-4, namely to find the least costly allocation of wastewater treatment effort on a river such that certain river quality standards are satisfied (see Stehfest, 1978). He might be given by two consultants the two solutions shown in Fig. 8-4-4, and have the task of selecting one of the two plans for realization. Obviously, it is not sufficient just to check how well the underlying models fit actual measurements. If a model has a sufficient number of parameters the fit can always be made perfect, but the model may give very inaccurate results when applied to circumstances which are considerably different from the ones for which the parameters have been estimated (see Sec. 5-1). Hence, the decision maker will also include in his choice the confidence he places in the concept of the model (see Sec. 1-3). This confidence is certainly subjective, and may be influenced, for instance, by the reputation of the model builders. Another aspect to be included are the costs associated with each solution and the costs to be incurred if the solution selected turns out to be wrong. Although the choice involves subjective judgements, the integration of all the aspects of the choice may be formalized, and, to a certain extent, made lucid by means of decision theoretical arguments.

As for the primary problems discussed in Sec. 8-4, minimization of cost seems to be a reasonable objective for the secondary optimization, but because of the uncertainty as to which model is better we can only speak of expected cost. For simplicity in notation, we look upon the two models as one aggregate model, which contains a parameter z which can assume two values z^1 and z^2. This aggregate model may be imagined as the union of the two models plus a linear output transformation, which contains the parameter z. Depending on the value of z the output variables of the aggregate model are equal to those variables of one model or the other on which standards are imposed. (Of course, a variable on which a standard is imposed must be contained in either model.) The secondary optimization problem appears now as the problem of estimating the parameter z. (Obviously, this reasoning may be generalized to more than two models, and there is no restriction as to the type of the models.) If it is assumed a priori that one of the two models correctly describes the real behavior of the system, the expectation of the total cost is given by

$$\mathscr{C}(z', z^1)p(z^1 | \mathbf{Y}) + \mathscr{C}(z', z^2)p(z^2 | \mathbf{Y}) \qquad (10\text{-}1\text{-}1)$$

where $\mathscr{C}(z', z^i)$ denotes the cost to be incurred if z^i is the "right" value and the model characterized by z' is used, and $p(z^i | \mathbf{Y})$ is the subjective probability for z^i being the "right" value given set $\mathbf{Y}$ of observations. Expression (10-1-1) is nothing else than expression (5-1-8) in the case when z can attain only two values. The secondary optimization consists in minimizing this expression over $z' \in \{z^1, z^2\}$. The set $\mathbf{Y}$ comprises all informations on which the decision maker decides on the two models. The probability $p(z^i | \mathbf{Y})$ could be resolved according to Bayes' formula into

$$p(z^i | \mathbf{Y}) = \frac{p(z^i | \mathbf{Y} \cap \mathbf{Y}_0)p(\mathbf{Y}_0 | z^i)}{p(\mathbf{Y}_0)}$$

where $\mathbf{Y}_0$ is the set of observations of the variables on which standards are imposed. The a priori probability $p(z^i | \mathbf{Y} \cap \mathbf{Y}_0)$ would have to be quantified subjectively by the decision maker (see, for example, Raiffa, 1968), while the other probabilities in principle could be calculated from the models. In view of the other uncertainties, however, it seems more appropriate for the decision maker to estimate $p(z^i | \mathbf{Y})$ directly; for the sake of notational simplicity $p(z^i | \mathbf{Y})$ will be denoted by p_i in the following. The cost $\mathscr{C}(z', z^i)$ to be incurred if $z' = z^i$ is obviously the cost $\mathscr{C}'$ of the optimal control S' belonging to the model characterized by z' (see Sec. 8-4). For $z' \neq z^i$ it is reasonable to assume that the costs for realizing the solution belonging to z' have to be incurred, too, because planning and construction of the treatment plants have to begin long before it becomes manifest that the wrong model was relied upon. Then these costs have to be corrected for the fact that either the standard is not met or too many treatment plants have been built. A reasonable correction would be to add the costs necessary to make the solution S' meet the standards given that z^i is the "right" value. This would be consistent with the maxim behind standard setting, which says that no damage occurs if the standard is met, while the damage is very high if it is violated. The corrections can be determined by means of the same optimization procedure which was used to find the optimal solutions among which one has to choose. If, in the case of $z' \neq z^i$, the solution S' meets the standards even with z^i being the "right" value, the correction might be negative, if shutting down of treatment plants is considered (which saves operational costs). If Δ_{ik} denotes the cost of modifying the optimal control S^i such that it meets the standards if value z^k is used, minimization of expression (10-1-1) means comparing the two expressions

$$\mathscr{C}^1 + \Delta_{12}(1 - p_1) \qquad (10\text{-}1\text{-}2a)$$

and

$$\mathscr{C}^2 + \Delta_{21} p_1 \qquad (10\text{-}1\text{-}2b)$$

If the first is smaller than the second, the decision maker should choose solution S^1 in order to minimize the expected cost; otherwise S^2 should be chosen. The two expressions (10-1-2a) and (10-1-2b) as functions of p_1 are shown in Fig. 10-1-1. Obviously, the decision maker need not assign a specific value to p_1, but only has to specify whether p_1 is smaller or greater than the abscissa of the intersection point of the two lines representing the expressions (10-1-2a) and (10-1-2b). In the case of the example given in Sec. 8-4 the total annual costs are (DM = German Marks)

$$\mathscr{C}^1 = 1.080 \times 10^9 \text{ DM/a} \qquad \Delta_{12} = 0.102 \times 10^9 \text{ DM/a}$$

and

$$\mathscr{C}^2 = 1.162 \times 10^9 \text{ DM/a} \qquad \Delta_{21} = 0.017 \times 10^9 \text{ DM/a}$$

where index 1 refers to the Streeter-Phelps model and index 2 to the ecological model. Hence, the control S^2 does not satisfy the standards if the Streeter-Phelps model is the "right" one, even though the treatment effort for S^2 is considerably

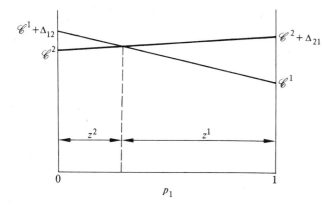

Figure 10-1-1 Choice between two optimal control strategies, which are based on two different river quality models.

higher than for S^1. Shutting down of treatment plants is not considered. The abscissa of the intersection point is in this case about 0.17.

The scheme described may be generalized to cover other options, which may be different even qualitatively. It could be, for example, that the consultants who formulated the two different solutions of the optimal control problem, propose (as is often the case) to take more measurements on the river in order to get a better validated model. Hence, beside realization of S^1 or S^2, the decision maker has the third option of first taking more measurements at cost $\mathscr{C}_m$. He can assume that the model which emerges from the new measurements will give the control which is optimal in reality, but he does not yet know anything about the costs. In this case, the following judgemental probabilities of the decision maker have to be assessed:

p_1 = probability for z^1 giving an aggregate model output which agrees with reality within desired accuracy

p_2 = probability for z^2 giving an aggregate model output which agrees with reality within desired accuracy

p_3 = probability that the model which describes reality adequately differs from the two models offered to the decision maker by more than the prescribed accuracy

The three probabilities add up to one if the two given models differ by more than the prescribed accuracy. For the costs $\mathscr{C}^3$ of realizing the optimal control for the third, yet unknown model, the only reasonable assumption which can be made is

$$\mathscr{C}^3 = \frac{p_1 \mathscr{C}^1 + p_2 \mathscr{C}^2}{p_1 + p_2}$$

i.e., $\mathscr{C}^3$ is the mean of $\mathscr{C}^1$ and $\mathscr{C}^2$ weighted with the probabilities for the corresponding models. Similarly, for the cost Δ_{13} to be incurred if optimal control

S^1 is realized and neither z^1 nor z^2 gives the "right" model, the only reasonable assumption is $\Delta_{13} = \Delta_{12}$. Analogously, $\Delta_{23} = \Delta_{21}$ must be assumed. Hence, minimization of the expected cost means searching for the smallest one among the following three expressions

$$\mathscr{C}^1 + \Delta_{12}(1 - p_1) \tag{10-1-3a}$$

$$\mathscr{C}^2 + \Delta_{21}(1 - p_2) \tag{10-1-3b}$$

$$\mathscr{C}_m + \frac{p_1 \mathscr{C}^1 + p_2 \mathscr{C}^2}{p_1 + p_2} \tag{10-1-3c}$$

Figure 10-1-2 shows, in analogy to Fig. 10-1-1, expressions (10-1-3a) to (10-1-3c) for a fixed value of $p_1 + p_2 = 1 - p_3$ as a function of $p_1/(p_1 + p_2)$.

The same approach as the one just described can be used for selecting the best plan among different water resources development plans involving economic, environmental, social, and technical considerations. Any single plan can be thought of as the optimal solution within a certain class of possible plans and the secondary optimization problem consists of giving a grade to each alternative proposal, thus determining the best one. A certain number of *attributes* such as cost of the plan, water quality, land and forest use, recreation potential, social impact, and others are selected to indicate the degree to which the objectives are achieved in the river basin. A utility function is assessed over these attributes taking into account the manager's subjective evaluation of the possible consequences of the many alternatives (see Sec. 10-2). This utility function explicitly indicates the preference trade-offs among attributes and yields an overall rating of the desirability of the different plans. Interesting case studies of this kind can be found in the literature (see, for instance, David and Duckstein, 1975; Keeney et al., 1976), but they are not described in detail in this chapter since in these studies river pollution plays only a minor role.

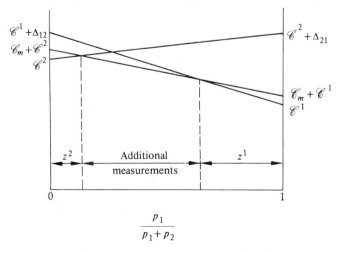

Figure 10-1-2 Choice among three options for realizing a river sanitation program.

In the above examples, the solutions among which the manager has to select are particular optimal solutions computed in the second step of the management problem. Sometimes this is not possible since the second step becomes computationally unfeasible, so that it is necessary to solve the problem in an approximate way. However, if the second and third steps are not so rigidly separated some reduction of the computational effort is often obtained.

These facts can be better clarified by means of a simple example. For this reason, consider the problem of designing a measurement network in a river basin. By means of the network samples are taken in time and space of the output variables of the system, which are, in general, different from the state variables of the system. Nevertheless, if a model of the river basin is available the state of the system can be estimated as shown in Chapter 5. Assume that given a network (i.e., given n instruments and their locations) the method of obtaining the best estimate of the state is known. Thus, for any given number of instruments the problem of optimal location of the measurement points can be solved (at least in principle). If the cost of the network is determined by the number of measurement points and the efficiency of the network is described by a simple scalar indicator, called *estimation error* (for example, the trace of the error covariance matrix as proposed by Yu and Seinfeld, 1973), a series of points in a two-dimensional space is obtained (see Fig. 10-1-3). The determination of each one of these points requires the solution of an n dimensional optimization problem (optimal allocation of n instruments). Of course, the estimation error decreases as the cost of the network increases so that the next step consists in finding the best compromise between these two attributes. This is very simple to do if *indifference curves* are known in the space of the attributes. If, for example, each unit of the estimation error has a fixed price P, the indifference curves are straight lines as shown in Fig. 10-1-3 and the best compromise solution can be found by determining the point of the "optimal" curve for which the tangent is an indifference curve.

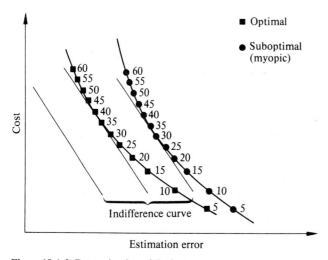

Figure 10-1-3 Determination of the best measurement network.

Unfortunately, the procedure just outlined cannot be followed in general since it cannot be carried out in a reasonable time. In this case the computational effort can be reduced if the second and third step of the problem are interrelated as follows. Suppose the manager knows from his experience that a network with eighty instruments is certainly too costly. Thus he has to search the best compromise in the interval of uncertainty $(0, 80)$ and this can be done without solving eighty optimal location problems. In fact, one can test in the middle of the interval of uncertainty (*bisection procedure*). By determining the best networks with forty and forty-one instruments it is possible to check if the decrement in the estimation error is smaller or greater than the decrement the manager is willing to pay at the price P. In the case shown in Fig. 10-1-3, the decrement in the estimation error obtained by adding one instrument to a network of forty instruments is smaller than the decrement the manager is willing to pay and this means that the best compromise is in the range $(0, 40)$. Thus, one can solve the two subproblems $n = 20$ and $n = 21$ and then $n = 30$ and $n = 31$, and finally the best compromise is obtained when solving the problems with $n = 35$ and $n = 36$.

It is worth-while noting that the use of this bisection procedure requires the use of the preference structure (the price P) at each iteration. This is the equivalent to saying that the manager does participate in the phase of the solution of the subproblems and, indeed, his knowledge is used to eliminate some of the subproblems at each step of the procedure until the best compromise is found. If this reduction of dimensionality is not yet sufficient to make the problem solvable, the procedure must be further simplified, and a well-known way of doing this is to use the *myopic optimization criterion*. Following this criterion implies solving a sequence of very simple allocation problems until a satisfactory network is obtained. More precisely, the first problem to be solved is that of the allocation of a single instrument ($n = 1$). Then, the second instrument is allocated leaving the first one fixed, so that the second subproblem ($n = 2$) is still a one-dimensional problem. Continuing in the same way a "suboptimal" (myopic) sequence of points in the space of the attributes is obtained, which can be used instead of the optimal sequence, to determine the best compromise solution, which is characterized by a smaller number of instruments (thirty instead of thirty-five in Fig. 10-1-3) and, of course, by a higher estimation error. This is actually the price which has to be paid in order to make the problem solvable. In this case the bisection method cannot be used, but the knowledge of the price P can be used to stop the iterative procedure as soon as the best compromise is obtained.

Finally, the fourth step of a management problem, which is that of implementation, is now briefly discussed. If all economic, social, and technological constraints have been taken into account when solving the preceding steps of the problem, this final step does not present any particular difficulty. Unfortunately, quite often, some constraints are not taken into account when solving the problem so that the implementation of the best compromise is not always an easy task. For example, when determining the best water resources development plan it can be assumed that there are no particular technological and economic constraints for the plan. Nevertheless, after solving the problem the manager might recognize that the plan cannot be realized immediately since the required effort is too high.

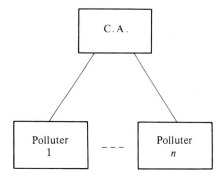

Figure 10-1-4 Decentralized decision structure.

Thus, a new problem arises, namely the problem of the determination of the sequence of actions to be undertaken in order to realize the plan in minimum time or, more generally, in the best possible way. Optimal *sequencing problems* of this kind are very common in water resources management because of the time needed to construct reservoirs, channels, networks of pipelines and so forth and because of the yearly budget limitations that such plans must usually satisfy. A particular but quite interesting example of a sequencing problem related to river pollution control will be analyzed in detail in Sec. 10-3.

The problem of implementation of the best compromise is certainly very complex in systems characterized by a decentralized decision structure. Indeed, during the first three steps of a management problem it is usually assumed that a unique decision maker maximizes the total social benefit associated with the use of the resource. How the C.A. can realize this maximum social benefit in a decentralized system of the kind shown in Fig. 10-1-4 is not at all a simple problem. Effluent charges (taxes) or standards are the classical tools for obtaining a predetermined compromise solution in a non-cooperative system as shown in Fig. 10-1-4, where each polluter is maximizing his own profit. Some of the basic notions of the Theory of Games can be used to deal with such complex situations and in particular to analyze the tendency of forming coalitions which can take advantage of the economies of scale in wastewater treatment plants. In Sec. 10-4 this kind of problem will be analyzed in detail and particular attention will be paid to the role the C.A. can play in stabilizing these coalitions. Studies of this kind can only be considered as crude and initial investigations of a problem as complex as that of the use of the resources in a river basin. Possible extensions are related to the stochastic nature of the problems and to more realistic decision structures like the *multilevel structure* shown in Fig. 10-1-5.

Economic analysis, and also Sec. 10-4, traditionally refers to the static case, while industrialization and urbanization are typically dynamic phenomena. Thus, a different approach for analyzing the effects of charges imposed on polluters is needed, even at the expense of theoretical rigor. For this reason, the last section of this chapter illustrates an example of a dynamic taxation scheme which can be applied to any developing region and shows how the industrialization of the region can be reasonably kept under control by the C.A.

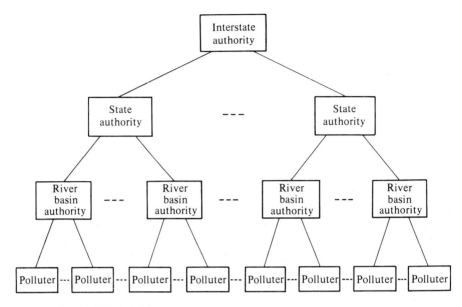

Figure 10-1-5 Multilevel decision structure.

10-2 MULTIOBJECTIVE PROGRAMMING IN WATER QUALITY MANAGEMENT

Vector optimization, or multiobjective programming, refers to decision problems characterized by more than a single objective function. The mathematical theory of such problems was first introduced by Kuhn and Tucker (1951) and Koopmans (1951). On the other hand, the interest that so-called management science has shown towards these techniques is relatively new. In particular, in the field of river pollution management, multiobjective programming has only been applied on an exercise level up to now. For this reason, only the main ideas and results of this branch of mathematical programming (see Cohon and Marks, 1975) will be covered in this section.

In a crude but descriptive way, a multiobjective programming problem can be formulated as follows:

$$\max_{\mathbf{z}} \left[\mathbf{J}(\mathbf{z}) \right] = \max_{\mathbf{z}} \left[J_1(\mathbf{z}) \quad J_2(\mathbf{z}) \quad \ldots \quad J_p(\mathbf{z}) \right] \qquad (10\text{-}2\text{-}1)$$

subject to

$$\mathbf{z} \in \mathbf{Z} \qquad (10\text{-}2\text{-}2)$$

where the functions $J_i(\mathbf{z})$, $i = 1, \ldots, p$, are the p different *objectives* (or goals, or *attributes*) of the problem and $\mathbf{Z}$ is the feasibility set usually described by a finite set of equality and inequality constraints. In particular, the attributes may represent the net benefits of p water users affected by the decisions of a control agency. Equation (10-2-1) expresses, in a compact form, that each objective must be maximized; this operation is clearly impossible without information about *preferences* which provides a rule for combining the objective functions.

A very particular preference structure which is, nevertheless, quite important to mention, is the *lexicographic* one, which basically says that each objective is infinitely more important than the following (in the order 1, 2, ..., p). Thus, under this preference structure the problem can obviously be solved step by step in the following way. First solve the problem

$$\max_{\mathbf{z}} [J_1(\mathbf{z})] \tag{10-2-3}$$

subject to

$$\mathbf{z} \in \mathbf{Z} \tag{10-2-4}$$

If this problem has a unique solution, then this is the solution of the whole problem. Otherwise, indicate by $\mathbf{Z}^{(1)}$ the set of all solutions of problem (10-2-3)-(10-2-4) and solve the problem

$$\max_{\mathbf{z}} [J_2(\mathbf{z})]$$

subject to

$$\mathbf{z} \in \mathbf{Z}^{(1)}$$

and continue until the end or until one of the problems has a unique solution. Although problems of this kind seem to be exceptional they have quite often been considered in the literature (possibly because they are simple to solve). The reader will, for example, recognize that the parameter estimation problem discussed in Sec. 5-2, which is characterized by the performance index (5-2-5), is a problem of this kind with two objectives if $\lambda = 1$, and in fact the solution procedure is in two steps. Another example of this kind will be shown in the next section.

In general, the preference structure is not so rigid since the attributes $J_1, \ldots, J_p$ are of comparable importance. Fortunately, a relatively small set of "good" solutions can in general be associated with any multiobjective programming problem. These solutions are called *noninferior solutions* (or *Pareto optimal solutions*) and are exactly defined as follows: a feasible solution $\mathbf{z}^*$ is said to be a noninferior solution if there exists no other feasible solution $\mathbf{z}$ such that

$$J_i(\mathbf{z}) > J_i(\mathbf{z}^*)$$

for some i and

$$J_j(\mathbf{z}) \geq J_j(\mathbf{z}^*)$$

for all $j \neq i$. This means that by perturbing a noninferior solution no attribute can be improved without lowering at least one of the others. The set $\mathbf{Z}^*$ of all noninferior solutions can be represented in the decision space, but a more attractive representation is obtained in the space of the attributes where the set $\mathbf{J}(\mathbf{Z}^*)$ can be immediately obtained from the set $\mathbf{J}(\mathbf{Z})$ as shown in Fig. 10-2-1.

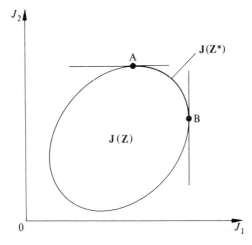

Figure 10-2-1 Noninferior solutions J(Z*) in the space of the attributes.

Even if the set of noninferior solutions is available, the problem still remains of determining the best one, called the *best compromise solution*. Thus, for the moment the multiobjective programming problem can be considered to consist of the following two separate subproblems

(a) determination of noninferior solutions
(b) determination of the best compromise

The solution of problem (a) was first given by Kuhn and Tucker (1951) and then interpreted and discussed by many authors (see, for example, Zadeh, 1963), and the main conclusion is that the set **Z*** of all noninferior solutions can be computed (under suitable assumptions on **J** and **Z** (for example, convexity of **J** and **Z**)) by solving the following mathematical programming problem for all possible combinations of the nonnegative weights w_i with $\Sigma_i w_i = 1$:

$$\max_{\mathbf{z}} \sum_i w_i J_i(\mathbf{z}) \qquad (10\text{-}2\text{-}5)$$

subject to

$$\mathbf{z} \in \mathbf{Z} \qquad (10\text{-}2\text{-}6)$$

This method, called *weighting method*, is not very efficient from a computational point of view and can become practically unfeasible for high values of p, since the number of solutions of the scalarized problem required to identify completely or even approximately the noninferior set **Z*** increases exponentially with the number of objectives. Nevertheless, this method is very simple to apply, particularly in the case in which the attributes are linear and the feasibility set **Z** is described by linear equality and inequality constraints, since Eqs. (10-2-5) and (10-2-6) then constitute a linear programming problem. It should be mentioned that if the attributes J_i correspond to various interest groups involved, the weights w_i may be related to the political influence of the i-th group (see Dorfman, 1972).

10-2 MULTIOBJECTIVE PROGRAMMING IN WATER QUALITY MANAGEMENT

A kind of dual problem of the scalarized problem described by Eqs. (10-2-5) and (10-2-6) is the following

$$\max_{\mathbf{z}} \left[J_i(\mathbf{z}) \right] \qquad (10\text{-}2\text{-}7)$$

subject to

$$\mathbf{z} \in \mathbf{Z} \qquad (10\text{-}2\text{-}8)$$

$$J_j(\mathbf{z}) \geq \underline{J}_j \qquad j \neq i \qquad (10\text{-}2\text{-}9)$$

which can be proved to give noninferior solutions $\mathbf{z}^*$. Thus, the set $\mathbf{Z}^*$ can be generated by parametrically varying the lower bounds $\underline{J}_j$ of the attributes J_j in Eq. (10-2-9). This method, called *constraint method*, has the same advantages and disadvantages as the weighting method. The only (minor) difference is that the constraint method is perhaps more intuitive in the field of river pollution, since it is easier to have some a priori knowledge or guess of the lower bounds $\underline{J}_j$ of the attributes than it is of the relative weighting factors w_i (see also Sec. 8-1). Moreover, meaningful values of the bounds $\underline{J}_j$ can be easily computed by solving the p optimization problems

$$\max_{\mathbf{z}} \left[J_j(\mathbf{z}) \right]$$

subject to

$$\mathbf{z} \in \mathbf{Z}$$

and this is also a reason to be in favor of the constraint method.

Other methods exist for the determination of the noninferior set $\mathbf{Z}^*$ (see, for example, Beeson and Meisel, 1971; and Reid and Vemuri, 1971), but they are not described here since their applicability to river pollution management problems seems to be questionable.

As far as the determination of the best compromise is concerned, the first approach to be mentioned is the *utility function* approach. This method assumes that a utility function $u(\mathbf{J})$ is known, which quantifies the preferences among the points in the attribute space, and whose expectation is the guide for decisions under uncertainty (Keeney and Raiffa, 1977). Hence, if there is no uncertainty about $\mathbf{J}(\mathbf{z})$ the best compromise is obtained by solving the problem

$$\max_{\mathbf{J}} \left[u(\mathbf{J}) \right]$$

subject to

$$\mathbf{J} \in \mathbf{J}(\mathbf{Z}^*)$$

as shown in Fig. 10-2-2 for a two-dimensional case. In other words, the best compromise solution is obtained as the point at which the noninferior set and a contour of equal utility are tangent. Since utility functions are monotonically increasing with any objective J_i (due to the problem definition (10-2-1)) the best compromise solution can be determined in one shot by the following mathe-

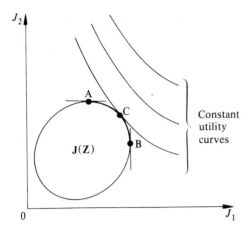

Figure 10-2-2 The use of utility functions: the line AB is the set $J(Z^*)$ and the point C is the best compromise solution.

matical programming problem in the decision space:

$$\max_{z} \left[u(J(z)) \right]$$

subject to

$$z \in Z$$

thus avoiding the explicit determination of all noninferior solutions. If there is uncertainty as to the values of the attributes the whole feasibility set Z has to be considered anyway for maximizing the expected utility $E[u]$. The main drawback of this approach is that the utility function, which is in general highly subjective, has to be assessed by means of relatively intricate interrogation procedures (Keeney and Raiffa, 1976). That is the reason why the utility function approach has still to be proven as a tool which really facilitates decision making on river pollution problems (for an application of this approach to a pollution control problem of the Rhine river see Ostrom and Gros, 1975).

As should be obvious from the explanation of Fig. 10-2-2, for determining the best compromise it suffices to know the curves of constant utility, rather than the full utility function, if there are no uncertainties about the attribute values. These so-called *indifference curves* are much easier to determine, since only ordinal comparisons have to be made. In other words, for the assessment of the indifference surfaces the decision maker only has to fix preferences between decision alternatives, while for the utility function he has to give also a quantitative rating of his preferences.

Another approach consists in fixing a priori the *optimal weights* w_i^* of the objectives (as it is done in classical cost-benefit analysis) and solving the following problem:

$$\max_{z} \sum_{i} w_i^* J_i(z)$$

subject to

$$z \in Z$$

This problem is of the form given by Eqs. (10-2-5) and (10-2-6) and its solution z^* is therefore a noninferior solution. Moreover, in this way the best compromise solution is obtained without explicitly computing all noninferior solutions, a fact which can be considered as a relevant advantage from a computational point of view. The method is completely equivalent to the previous one if hyperplanes of a given direction are used as indifference surfaces. Although only one scalarized problem must be solved, in general, a careful sensitivity analysis with respect to the optimal weights is desirable when this method is used.

Other approaches like the *surrogate worth trade-off method* (see Haimes et al., 1975), and *Electre method* (see Roy, 1971) have been proposed and some of them have been used in water resources management problems (see, for example, Haimes et al., 1975). Among these other methods, the most interesting ones are those which rely on the progressive articulation of preferences, i.e., those methods in which as soon as a noninferior solution has been obtained it is presented to the decision maker, who reacts on it and allows whoever is solving the problem to search in the decision space in a relatively efficient way. If one of these methods is applied it is not possible to distinguish any more between the two phases (a) and (b) since the decision maker is dynamically involved with his preference structure (not specified in analytical terms) during all the solution process. One method of this kind which is worth mentioning here is *Semops* (see Monarchi et al., 1973), which has been successfully applied to a river pollution management problem of realistic dimensions. Applications of this type will certainly be more and more frequent in the future and will help to give a final value judgement to multiobjective programming in river pollution management.

10-3 A SEQUENCING PROBLEM

This section is devoted to the presentation of a particular *sequencing* problem which can be thought as part of the solution of a complex multiobjective optimization problem.

Sequencing problems are frequently encountered when dealing with water resources development plans (for example, the determination of the best sequence in which a set of dams must be built). In all these problems the set of items to be built or the set of actions to be sequentially undertaken is given and is usually predetermined as the optimal solution of a *primary optimization* problem. The sequencing problem can therefore be considered as a *secondary optimization problem*. A careful analysis of all such problems shows that in the great majority of cases the whole problem is characterized by two objectives and that the decomposition into primary and secondary phase is often arbitrary and can be rationally justified only when the first objective is much more important than the second one. In fact, under this assumption the multiobjective optimization problem must be solved following the *lexicographic* approach discussed in the preceding section and the result is that the problem is decomposed into two subproblems, which must be solved in cascade. The solution of the first subproblem gives the set of actions to be undertaken, while the second subproblem consists in ordering such a set.

Consider the case of a river basin for which it is necessary to satisfy a certain number of quality standards within a given period of time T. Assume that all the characteristics of the river and the location and intensity of all the main (N) pollution sources in the time-period $[0, T]$ are known. Thus, the problem is that of determining the dimension of the treatment plants and the order in which they must be built. One objective, certainly very important and already discussed in the preceding chapters, is the total *cost* $\mathscr{C}$ of the operation, which can be assumed to be given by the cost of the N treatment plants, i.e.,

$$\mathscr{C} = \sum_{i=1}^{N} \mathscr{C}_i \qquad (10\text{-}3\text{-}1)$$

A second objective, which is also quite important, in particular when the river is highly polluted and the planning period T is relatively long, is the mean value P over the period $[0, T]$ of a suitable *pollution index* $\mathscr{P}(t)$, i.e.,

$$P = \frac{1}{T} \int_0^T \mathscr{P}(t)\,\mathrm{d}t \qquad (10\text{-}3\text{-}2)$$

The problem of defining a pollution index for a river basin is certainly not a new one and many suggestions can be found in the literature. The formulation of such an index can be done in two steps. First, a *water quality indicator* is defined and then this indicator is suitably integrated over the entire river basin. The first step is without doubt the more difficult one to accomplish, since the water quality indicator should take into account the composite influence of significant physical and chemical parameters and the different uses of the water. Unfortunately, the soundest proposals are so complex and detailed that they cannot be used for solving problems of the kind considered here, since they would require the use of models far more sophisticated than those that have been validated so far. For example, the water quality index described by Brown et al., (1972) takes into account the following eleven parameters: dissolved oxygen, fecal coliforms, pH, 5-day BOD, nitrate, phosphate, temperature, turbidity, total solids, toxic elements, pesticides, and there is no model that can predict all these variables at one time. Therefore, it is necessary to select so compact an indicator of water quality that any standard river quality model allows the computation of this measure. Fortunately, there are no significant alternatives in making this choice since all reasonable models of water quality have only one variable in common, namely the dissolved oxygen concentration. For these reasons, the pollution index used is the total amount of oxygen missing in the river basin with respect to the ideal conditions of fully saturated water, i.e.,

$$\mathscr{P} = \int_{\mathscr{L}} A(l)d(l)\,\mathrm{d}l \qquad (10\text{-}3\text{-}3)$$

where $\mathscr{L}$ is the set of spatial coordinates defining the river basin and $A(l)$ and $d(l)$ are, respectively, the cross-sectional area and the oxygen deficit at point l. The index $\mathscr{P}$ is, in general, time-varying, since the hydrologic and environmental conditions vary seasonally and the total pollution load entering the river is usually increasing in the period $[0, T]$.

This pollution index, first proposed by Liebman (see Kneese and Bower, 1971, pages 94–95), is opposed to other indices such as *"the mileage out of standards"* (Deininger, 1965) or *"maximum deviation from the stream standards"* (Revelle et al., 1969). The reason it is preferred is that it takes into account the global situation of the basin, since each point of the river gives its contribution to the total deficit; by contrast, the maximum deviation from the stream standards refers to only one point.

The multiobjective programming problem

$$\min \begin{bmatrix} \mathscr{C} & P \end{bmatrix}$$

is very difficult to solve and indeed no serious indications can be found in the literature on this topic. Nevertheless, if the cost $\mathscr{C}$ of the treatment plants is supposed to be of primary importance, the lexicographic approach can be followed, and the problem splits into two subproblems. The first subproblem has already been dealt with in Chapter 8 and gives the optimal set **X** of wastewater treatment plants to be built in the planning period $[0, T]$. The order in which the plants are built is assumed not to influence the cost $\mathscr{C}$ of treatment, while it does influence the value P of the second attribute of the problem. Thus, the second subproblem consists of selecting that schedule (i.e., that sequence of plants) which minimizes P.

Given a schedule one can compute the value of P simply by simulation; thus, in principle, the problem could be solved by determining P for each possible schedule and then by selecting that one which gives the minimum value of P. Unfortunately, this naive approach cannot be followed in practice since the number of schedules is $N!$ and the time required for simulation would be tremendously high even for very simple cases (for example, if the number of plants is 10 the model must be run more than 10^6 times). Therefore, it is necessary to develop a simpler algorithm for the determination of the optimal schedule. In the field of *Combinatorial Optimization* (see, for example, Rinaldi, 1975) the algorithms are often subdivided into two classes (simple and complex) depending upon the number of elementary operations required to obtain the solution. If such a number is of the order of a polynomial in the variable N identifying the dimension of the problem the algorithm is said to be *polynomial-bounded* and is considered a simple algorithm. Thus, if it is assumed that one simulation of the model over the entire planning period $[0, T]$ corresponds to an elementary operation, the exhaustive algorithm mentioned above is not polynomial-bounded, since the number of elementary operations it requires is of the order of $N!$. Unfortunately, no polynomial-bounded algorithm can be devised for our problem unless some relatively strong simplifying assumptions are made (see page 327). Nevertheless, a polynomial-bounded algorithm which gives a reasonably good suboptimal solution can easily be determined. This can be accomplished by approaching the problem with the so-called *myopic criterion* which has been extensively used in the past, in particular in control problems (e.g., minimum time control) and in mathematical programming problems (e.g., steepest descent method). With this criterion each decision is selected by looking only at its "local" effect.

In order to be more specific in dealing with our case, let us assume that the plants are built in series and that the time Δ_i for constructing the plant i is proportional to its cost $\mathscr{C}_i$, i.e.,

$$\Delta_i = \frac{\mathscr{C}_i}{\mathscr{C}} T \qquad (10\text{-}3\text{-}4)$$

(notice that this assumption means that there is a continuous and constant effort in plant construction during the planning period). Moreover, let t_i be the time at which the construction of the plant i is initiated, so that $(t_i + \Delta_i)$ is the time at which the plant i starts working. Thus, at $(t_i + \Delta_i)$ the pollution index $\mathscr{P}$ is suddenly lowered by a quantity $\mathscr{R}_i$ as shown in Fig. 10-3-1. A measure of the momentary effect of constructing the plant i on the pollution index is the ratio between $\mathscr{R}_i$ and Δ_i or, in view of Eq. (10-3-4), the ratio

$$\eta_i = \frac{\mathscr{R}_i}{\mathscr{C}_i} \qquad (10\text{-}3\text{-}5)$$

which is called *efficiency* of the plant and is expressed in mg of oxygen per dollar or in equivalent units. Thus, applying the myopic criterion means that each time the construction of a plant is accomplished the next plant to be built must be the one with the highest efficiency. Unfortunately, the efficiency η_i of a plant depends upon the time at which the plant is built and upon the set of plants which have

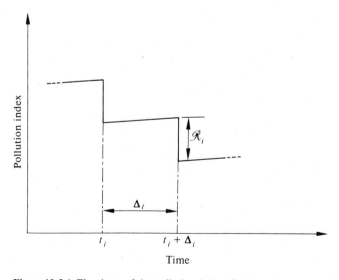

Figure 10-3-1 The shape of the pollution index $\mathscr{P}$ during the construction period of plant i.

already been built until that time. This implies that the optimal myopic schedule can only be determined by simulating the effects of all the plants not yet built each time a decision must be taken. For deciding which one is the first plant to be built N pairs of simulations are necessary (it is necessary to determine for all i the pollution index at time Δ_i with and without plant i). The second plant of the sequence can be determined with $(N-1)$ pairs of simulations, the third with $(N-2)$ pairs and so forth. Thus, the total number of simulations is of the order of N^2, which proves that the myopic procedure is polynomial-bounded.

Consider the possibility of further simplifying the procedure by making some suitable assumptions. First of all suppose that the hydrologic and environmental conditions are constant and that there is no trend on the pollution load during the planning period. This implies that the efficiency of a plant does not depend upon the time at which the plant is built. Moreover, assume that the efficiency of a plant is also independent of the presence of other plants in the river basin. This implies that at each instant of time t the pollution index $\mathscr{P}(t)$ can be given the expression

$$\mathscr{P}(t) = \mathscr{P}_0 - \sum_{i \in \mathbf{X}_t} \mathscr{R}_i \qquad (10\text{-}3\text{-}6)$$

where $\mathscr{P}_0$ is the initial value of $\mathscr{P}$, $\mathbf{X}_t$ is the set of plants operating at time t and $\mathscr{R}_i$ is the contribution of plant i. The property expressed by Eq. (10-3-6) is referred to below as the *additivity property*.

Various conditions can be given under which Eq. (10-3-6) holds, depending on the model and on the spatial variability of the parameters involved in it. The simplest case is that of a basin constituted by a uniform and semiinfinite channel in which the integral of the distributed load along the river is finite, so that all significant variables describing the system go to zero for $l \to \infty$ because of self-purification (photosynthesis processes are neglected). In fact, first assume that the river is described by the Streeter-Phelps model (see, for instance, Eq. (8-2-1))

$$\frac{d}{dl} b(l) = -\frac{k_1}{v} b(l) + \frac{1}{Q} \left\{ v_b(l) + \sum_{i \in \mathbf{X}_t} u_i \delta(l - l_i) + \sum_{i \notin \mathbf{X}_t} U_i \delta(l - l_i) \right\} \qquad (10\text{-}3\text{-}7a)$$

$$\frac{d}{dl} d(l) = \frac{k_1}{v} b(l) - \frac{k_2}{v} d(l) \qquad (10\text{-}3\text{-}7b)$$

where $v_b(l)$ is the BOD load distributed along the river, u_i is the amount of BOD discharged by the i-th plant, U_i is the amount of BOD produced by the i-th pollution source, l_i is the spatial coordinate of the i-th discharge, and δ is the impulse function. Since Eq. (10-3-7) is linear, its solution depends linearly on the boundary conditions $b(0)$ and $d(0)$, and on the amount u_i of BOD discharged by each plant. Moreover, the integral of $v_b(l)$ is finite, so that the pollution index $\mathscr{P}$ is well defined and is a linear functional of the deficit $d(l)$. Therefore, Eq. (10-3-6) a priori follows, with

$$\mathscr{R}_i = K(U_i - u_i) \qquad (10\text{-}3\text{-}8)$$

where U_i is the BOD load of the i-th plant. The constant K in Eq. (10-3-8) does not depend upon l_i since a given amount $(U_i - u_i)$ of BOD removed will have an effect on the index $\mathcal{P}$ which is independent of the location l_i of the treatment plant due to the infinite length of the river.

The additivity property holds also for the case in which the river is described by a higher order nonlinear model of the kind

$$\frac{d}{dl}\mathbf{z}(l) = -\frac{1}{v}\mathbf{f}(\mathbf{z}(l), d(l)) + \frac{1}{Q}\left\{\mathbf{v}_b(l) + \sum_{i \in X_t}\mathbf{u}_i\delta(l-l_i) + \sum_{i \notin X_t}\mathbf{U}_i\delta(l-l_i)\right\} \quad (10\text{-}3\text{-}9a)$$

$$\frac{d}{dl}d(l) = \frac{\boldsymbol{\alpha}^T}{v}\mathbf{f}(\mathbf{z}(l), d(l)) - \frac{k_2}{v}d(l) \quad (10\text{-}3\text{-}9b)$$

where $\mathbf{z}(l)$ can be looked upon as a suitable n-th order vector describing the various stages in the degradation of the organic pollutants, $\mathbf{f}$, $\mathbf{v}_b(l)$, $\mathbf{u}_i$ and $\mathbf{U}_i$ are n-th order vectors, and $\boldsymbol{\alpha}^T$ is an n-th order row vector of conversion factors. In fact, solving Eq. (10-3-9a) with respect to $\mathbf{f}$ and substituting in Eq. (10-3-9b), one obtains

$$d = \frac{v}{k_2}\left[-\boldsymbol{\alpha}^T\frac{d}{dl}\mathbf{z} - \frac{d}{dl}d + \frac{\boldsymbol{\alpha}^T}{Q}\mathbf{v}_b + \frac{\boldsymbol{\alpha}^T}{Q}\sum_{i \in X_t}\mathbf{u}_i\delta(l-l_i) + \frac{\boldsymbol{\alpha}^T}{Q}\sum_{i \notin X_t}\mathbf{U}_i\delta(l-l_i)\right] \quad (10\text{-}3\text{-}10)$$

from which $(vA = Q)$

$$\mathcal{P}(t) = A\int_0^\infty d(l)\,dl = \frac{Q}{k_2}\left[\boldsymbol{\alpha}^T\mathbf{z}(0) + d(0) - \lim_{l\to\infty}\boldsymbol{\alpha}^T\mathbf{z}(l) - \lim_{l\to\infty}d(l)\right.$$
$$\left. + \frac{\boldsymbol{\alpha}^T}{Q}\int_0^\infty\mathbf{v}_b(l)\,dl + \frac{\boldsymbol{\alpha}^T}{Q}\sum_{i \in X_t}\mathbf{u}_i + \frac{\boldsymbol{\alpha}^T}{Q}\sum_{i \notin X_t}\mathbf{U}_i\right]$$

follows. If we confine ourselves to the biochemical degradation processes, the two limits in the preceding expression are zero under the assumption that the integral of $\mathbf{v}_b$ is finite; the final formula is then

$$\mathcal{P}(t) = \frac{Q}{k_2}\left[\boldsymbol{\alpha}^T\mathbf{z}(0) + d(0) + \frac{\boldsymbol{\alpha}^T}{Q}\int_0^\infty\mathbf{v}_b(l)\,dl + \frac{\boldsymbol{\alpha}^T}{Q}\sum_{i \in X_t}\mathbf{u}_i + \frac{\boldsymbol{\alpha}^T}{Q}\sum_{i \notin X_t}\mathbf{U}_i\right] \quad (10\text{-}3\text{-}11)$$

which can be compared with Eq. (10-3-6), thus giving

$$\mathcal{R}_i = \frac{1}{k_2}\boldsymbol{\alpha}^T(\mathbf{U}_i - \mathbf{u}_i)$$

Again $\mathcal{R}_i$ is independent of l_i.

The structure of Eq. (10-3-9) is so general that it contains as particular cases all models discussed in this book; therefore, the next thing we have to do is to relax the assumptions of the channel being infinite and uniform. Thus, suppose that the river is described by a linear model of the kind

$$\frac{d\mathbf{x}(l)}{dl} = \mathbf{F}(l)\mathbf{x}(l) + \mathbf{G}(l)\left[\mathbf{v}_b(l) + \sum_{i \in X_t}\mathbf{u}_i\delta(l-l_i) + \sum_{i \notin X_t}\mathbf{U}_i\delta(l-l_i)\right] \quad (10\text{-}3\text{-}12)$$

where $\mathbf{x}(l)$ is an n-th order vector, $\mathbf{F}(l)$ and $\mathbf{G}(l)$ are matrices of suitable order and $0 \leq l \leq L$. Since the deficit $d(l)$ is certainly one of the components (for example the last one) of the vector $\mathbf{x}(l)$, for the sake of simplicity in notation, a row vector $\mathbf{h}^T$ can be introduced such that

$$A(l)d(l) = \mathbf{h}^T(l)\mathbf{x}(l)$$

where

$$\mathbf{h}^T(l) = [0 \quad 0 \quad \ldots \quad A(l)] \qquad (10\text{-}3\text{-}13)$$

The solution of Eq. (10-3-12) is

$$\mathbf{x}(l) = \mathbf{\Phi}(0, l)\mathbf{x}(0) + \int_0^l \mathbf{\Phi}(\xi, l)\mathbf{G}(\xi)\left[\mathbf{v}_b(\xi) + \sum_{i \in X_t} \mathbf{u}_i \delta(\xi - l_i) + \sum_{i \notin X_t} \mathbf{U}_i \delta(\xi - l_i)\right] d\xi$$

$$= \mathbf{\Phi}(0, l)\mathbf{x}(0) + \int_0^l \mathbf{\Phi}(\xi, l)\mathbf{G}(\xi)\mathbf{v}_b(\xi)\, d\xi + \sum_{i \in X_t} \mathbf{\Phi}(l_i, l)\mathbf{G}(l_i)\mathbf{u}_i$$

$$+ \sum_{i \notin X_t} \mathbf{\Phi}(l_i, l)\mathbf{G}(l_i)\mathbf{U}_i$$

where the $n \times n$ matrix $\mathbf{\Phi}(\xi, l)$ is the transition matrix (see Sec. 1-2). From Eqs. (10-3-3) and (10-3-13) the following is obtained

$$\mathscr{P}(t) = \int_0^L \mathbf{h}^T(l)\mathbf{\Phi}(0, l)\mathbf{x}(0)\, dl + \int_0^L \mathbf{h}^T(l) \int_0^l \mathbf{\Phi}(\xi, l)\mathbf{G}(\xi)\mathbf{v}_b(\xi)\, d\xi\, dl$$

$$+ \sum_{i \in X_t} \int_0^L \mathbf{h}^T(l)\mathbf{\Phi}(l_i, l)\, dl\, \mathbf{G}(l_i)\mathbf{u}_i + \sum_{i \notin X_t} \int_0^L \mathbf{h}^T(l)\mathbf{\Phi}(l_i, l)\, dl\, \mathbf{G}(l_i)\mathbf{U}_i$$

which has the form of Eq. (10-3-6) with

$$\mathscr{R}_i = \int_0^L \mathbf{h}^T(l)\mathbf{\Phi}(l_i, l)\, dl\, \mathbf{G}(l_i)(\mathbf{U}_i - \mathbf{u}_i) \qquad (10\text{-}3\text{-}14)$$

Equation (10-3-14) for the quality improvement $\mathscr{R}_i$ shows that even in the case in which $\mathbf{u}_i$ is a scalar, the coefficient $\mathscr{R}_i/(\mathbf{U}_i - \mathbf{u}_i)$ is, in general, dependent upon l_i, and this turns out to be true also for uniform but finite channels. In other words, in a uniform river two plants characterized by the same BOD removal give rise to the same improvement of the pollution index only if they are located sufficiently far upstream. This fact explains why the total biodegradable load proposed by Deininger (1965) as a pollution index, differs from Liebman's index even in the simple case of a uniform finite channel described by the Streeter-Phelps model. Finally, it is worthwhile noticing that with linear models Eq. (10-3-6) holds also for the cases in which some of the plants are located on tributaries of the main river (this result follows immediately from the linearity of the model). On the contrary, Eq. (10-3-6) does not hold for a finite, non-uniform channel described by a nonlinear model.

In summary, the additivity property holds for linear models under very general conditions, while for nonlinear models we can only say that there is a tendency for this property to be satisfied if the river basin is approximately uniform and if the amount of biodegradable matter going out of the river basin is small enough.

If there are no time-varying factors and the additivity property holds, the solution of the myopic procedure is very simple. In fact, under these assumptions, the efficiencies η_i are characteristics of the single plants and therefore do not have to be recomputed at each step of the procedure. The result is that the plants are built in the order of decreasing efficiency, a very intuitive criterion. Notice that the computational effort required by the algorithm has been reduced since only $(N + 1)$ simulations are now necessary to determine the efficiencies of the N plants. But the most interesting fact is that the schedule obtained by means of the myopic procedure is the optimal schedule, i.e., the schedule minimizing the pollution index given by Eq. (10-3-2) (of course, under the simplifying assumptions described above). The proof is very simple. Assume by contradiction, that the optimal sequence contains at least one pair of successive plants (i, k) with

$$t_k = t_i + \Delta_i \qquad \eta_i < \eta_k$$

This situation is illustrated in Fig. 10-3-2a where the dashed area $\mathscr{A}_{ik}$ given by

$$\mathscr{A}_{ik} = (\mathscr{R}_k + \mathscr{R}_i)(\Delta_k + \Delta_i) - \mathscr{R}_i\Delta_k$$

represents the contribution of the plants (i, k) to the performance index P (the mean value of $\mathscr{P}$). If the two plants (i, k) are inverted in the sequence, the situation shown in Fig. 10-3-2b is obtained, where the dashed area $\mathscr{A}_{ki}$ is given by

$$\mathscr{A}_{ki} = (\mathscr{R}_k + \mathscr{R}_i)(\Delta_k + \Delta_i) - \mathscr{R}_k\Delta_i$$

The assumption $\eta_i < \eta_k$ implies (see Eq. (10-3-4))

$$\mathscr{R}_i\Delta_k < \mathscr{R}_k\Delta_i$$

from which it follows that

$$\mathscr{A}_{ik} > \mathscr{A}_{ki}$$

Thus, the performance index P is lowered when the plants (i, k) are inverted in the sequence and this contradicts the assumption of the sequence being optimal.

Application to the Rhine River

For a realistic application of the techniques described above, the section of the Rhine River discussed in Secs. 5-3 and 8-4 was chosen. Two models, namely the Streeter-Phelps model and the ecological model described in Sec. 5-3, were available, as well as the optimal solutions of the primary problem (see Sec. 8-4). The problem now under consideration is one of finding the sequence in which the treatment plants required by the solution in Fig. 8-4-4a should be built. The total treatment efforts per reach, which are shown in Fig. 8-4-4a, are associated

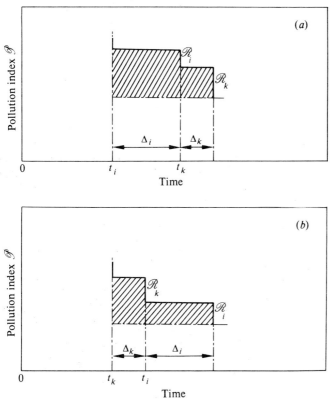

Figure 10-3-2 The shape of the pollution index $\mathscr{P}$ during the construction of the pair of plants (i, k):
(a) plant i is built first
(b) plant k is built first.

with single treatment plants in such a way that 15 treatment plants result. In practice, this would require unrealistically large treatment plants, but for illustrating the differences between various solutions of the sequencing problem, a small number of treatment plants is more appropriate. The plants in each reach were assumed to be uniformly distributed over the reach. For the sake of simplicity, the wastewater production was assumed to be constant during the planning period and equal to the values used for the primary optimization. The time T within which the plants had to be installed was chosen to be five years.

If the ecological model is used, the question arises whether the additivity property, which holds exactly in case of an infinite, homogeneous river, is approximately satisfied for the river section investigated. Numerical calculations showed that the $\mathscr{R}_i$'s depend strongly on the presence of the other treatment plants; only for small subsets X_t is Eq. (10-3-6) approximately fulfilled. Thus, the solution of the problem was determined following the general myopic procedure described above. The result of the first step of the procedure is shown in the third and fourth columns of Table 10-3-1. Plant 2 is characterized by the highest

Table 10-3-1 Costs $\mathscr{C}_i$, improvements $\mathscr{R}_i$, and efficiencies η_i of the single plants when none, 3, 7, and 11 plants are working in the river basin (computed with the ecological model)

Plant i	Cost $\mathscr{C}_i$ (10^6 DM/a)	Selection of the 1st plant		Selection of the 4th plant		Selection of the 8th plant		Selection of the 12th plant	
		Improvement $\mathscr{R}_i(t\,O_2)$	Efficiency η_i	Improvement $\mathscr{R}_i(t\,O_2)$	Efficiency η_i	Improvement $\mathscr{R}_i(t\,O_2)$	Efficiency η_i	Improvement $\mathscr{R}_i(t\,O_2)$	Efficiency η_i
1	27.9	8.09	0.290	6.25	0.224	3.01	0.107	4.15	0.149
2	114.8	56.39	0.491	—	—	—	—	—	—
3	114.8	56.39	0.491	36.47	0.318	—	—	—	—
4	27.9	6.72	0.240	5.02	0.180	2.59	0.093	4.04	0.145
5	57.4	17.57	0.306	11.45	0.199	5.35	0.093	13.16	0.299
6	50.3	15.50	0.308	10.21	0.203	4.85	0.097	—	—
7	14.0	1.36	0.097	0.84	0.060	0.40	0.029	—	—
8	71.8	8.20	0.115	11.37	0.158	19.20	0.267	1.33	0.095
9	143.6	27.20	0.189	31.49	0.219	45.35	0.316	—	—
10	129.2	24.43	0.188	28.32	0.219	40.65	0.314	—	—
11	86.2	26.89	0.312	28.01	0.325	—	—	—	—
12	79.0	24.63	0.311	25.65	0.324	—	—	—	—
13	107.0	39.23	0.364	40.19	0.373	—	—	—	—
14	79.0	29.68	0.375	—	—	—	—	—	—
15	57.4	21.55	0.374	—	—	—	—	—	—

efficiency and, therefore must be built first. Then the procedure was applied to the $(N-1)$ remaining plants and plant 14 turned out to be the second one in the sequence (notice that η_{14} in the fourth column of Table 10-3-1 is not the second highest efficiency). In the remaining columns of Table 10-3-1, the improvements $\mathscr{R}_i$ and the corresponding efficiencies η_i are shown under the assumption that 3, 7, and 11 plants have already been installed according to the myopic criterion. Continuing in the same way, the optimal myopic sequence was determined and the corresponding value of the performance P was 738 tons of oxygen ($t\,O_2$).

If the Streeter-Phelps model is used Eq. (10-3-6) is satisfied because the pollution load during the planning period is assumed to be constant, and hence the problem can be solved with only N simulations. The corresponding results are shown in Table 10-3-2 from which the optimal sequence can be derived in one shot simply by putting the plants in the order of decreasing efficiencies (2, 3, ..., 7). The corresponding value of the performance P (computed with the ecological model) turned out to be $777t\,O_2$, which is not very different from the preceding value.

Evaluating the results of this illustrative example, one could say that, considering the model uncertainties, it is sufficient to install the plants in the order given by their efficiencies η_i computed by means of the Streeter-Phelps model. However, in cases where the differences among plant costs are more relevant and nonlinear effects are dominant, it could be that the general myopic procedure gives rise to better results.

Table 10-3-2 Costs $\mathscr{C}_i$, improvements $\mathscr{R}_i$, and efficiencies η_i of the plants (computed with the Streeter-Phelps model)

Plant i	Cost $\mathscr{C}_i$ (10^6 DM/a)	Improvement $\mathscr{R}_i(t\,O_2)$	Efficiency η_i
1	27.9	7.55	0.270
2	114.8	60.90	0.530
3	114.8	60.90	0.530
4	27.9	7.76	0.278
5	57.4	29.02	0.506
6	50.3	25.39	0.505
7	14.0	3.29	0.235
8	71.8	34.05	0.474
9	143.6	67.07	0.467
10	129.2	60.36	0.468
11	86.2	38.70	0.448
12	79.0	35.48	0.449
13	107.0	46.71	0.434
14	79.0	31.60	0.400
15	57.4	22.99	0.401

10-4 TAXATION SCHEMES: STATIC ANALYSIS

As already pointed out in Sec. 10-1, one of the most important management problems is how to implement a "good" solution once it has been selected. Three main alternatives are available:

(a) The prescription of *effluent standards* by legislative regulation, i.e., the prescription of upper bounds on the discharge of residuals into the environment for each unit operating in the system
(b) *Subsidies* (*bribes*) for the construction of treatment plants, both in the form of tax concessions or direct government grants
(c) *Effluent charges*, i.e., taxes depending on the amount of pollutant discharged.

The last solution rests on the concept that the waste discharger should bear the damages his activities impose on all other parties using the same resource (the river).

A wide discussion has taken place during the last ten years on the relative advantages and disadvantages of these three alternatives (see, among others, Kamien et al., 1966; Kneese and Bower, 1968; and Mäler, 1974). Nowadays, most resource economists are in favor of effluent charges, which are considered as one of the most important tools of effective water quality management. Thus, the present section (as well as the next one) deals with effluent charges, while the other solutions are sometimes referenced only for comparison.

The existing contributions to water quality control by means of effluent charges (taxes) can be subdivided into two groups. The first group follows the academic tradition and deals with the effects induced by, and the properties of different taxation schemes. Static situations are analyzed since only in this case can the social welfare be easily defined. On the other hand, the second group considers the dynamic characteristics of the system, where firms are entering and leaving the economy, and technology is evolving. The scope of this second category is to point out heuristic rules for "good" taxation schemes. In the present section a particular but very interesting aspect of the static case, namely the problem of aggregation of different polluters into groups, is considered, while in the next section a simulation study on the dynamic case is presented.

Many opportunities and problems faced by individual decision units can be better dealt with or exploited by group behavior. One example is the problem of dealing with water pollution, and the opportunity presented is for a joint arrangement to treat and transport wastes. Regional or areawide wastewater treatment systems offer economic and environmental benefits to wastewater dischargers. Economic benefits arise from economies of scale and the opportunity to develop a comprehensive, consistent wastewater treatment system which minimizes redundant capacities. Environmental benefits arise from the increased reliability of larger, better funded systems and the opportunity to move effluents to discharge points with minimal adverse impact.

Along with the potential benefits come the problems of how to organize efficiently the regional system including an agreement on how the benefits should be distributed. One mechanism for allocating benefits is through a charging structure based on the services of a regional authority. Depending on the particular form of the charging structure and the administrative control exercised by a central coordinating agency, different distributions of profits and benefits result. Certain charging rules may unduly favor one class of users as against another.

The central problem investigated in the present section will be the question of what allocation of benefits among the participants will reinforce group adhesion. A charging rule will be referred to as stable if once stated there are no incentives for the users of the service to reject the regional plan. Such an incentive to reject would exist if a smaller group of participants could conceive of a plan that would allow for greater rewards. The desirability of central treatment of water depends on the existence of economies of scale. Given the technical condition, such as economies of scale over the relevant range of demand, the potential for a profitable collaboration among users exists. Depending on the degree of economies of scale, including their absence, there are greater options for choosing stable charges.

Description of the Parties

In general, reference will be made to the scheme of Fig. 10-4-1 where the first block represents the pollution units called *firms* (or producers) and the second and third blocks are the *network of treatment plants* and the *environment* (i.e., the river). The variable U_i ($i = 1, \ldots, n$) represents the mass flow rate of pollutant from firm i to the network of treatment plants, while the variable u_j ($j = 1, \ldots, p$) represents the mass flow rate of pollutant discharged by the j-th effluent of the network into the environment.

Each firm i is characterized by a *profit* $D_i(U_i)$, i.e., the profit for producing an amount of goods which corresponds to a production U_i of pollutant.

The network of treatment plants is characterized by a *cost* $\mathscr{C}$ which is, in general, a function of the input and output vectors $\mathbf{U} = [U_1 \ldots U_n]^T$ and $\mathbf{u} = [u_1 \ldots u_p]^T$, i.e.,

$$\mathscr{C} = \mathscr{C}(\mathbf{U}, \mathbf{u}) \qquad (10\text{-}4\text{-}1)$$

Given a class of networks (i.e., given the structure of the network) the cost $\mathscr{C}$ is assumed to be the one corresponding to the least cost solution, i.e., Eq. (10-4-1)

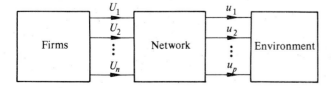

Figure 10-4-1 Structure of the system.

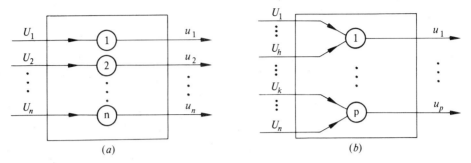

Figure 10-4-2 Two particular structures of the treatment network (each circle represents a wastewater treatment plant):
(a) completely disaggregated network
(b) partially disaggregated network.

represents the cost of the cheapest network in the class. For example, if the structure of the network is the one described in Fig. 10-4-2a (completely disaggregated network: one firm–one plant–one effluent) then

$$\mathscr{C} = \sum_{i=1}^{n} \mathscr{C}_i(U_i, u_i)$$

where $\mathscr{C}_i(U_i, u_i)$ is the cost of the cheapest treatment plant which transforms U_i into u_i (see Sec. 7-1), i.e., in this particular case the cost of the network is the sum of the costs of individual treatment plants. On the other hand, when groups y_j of firms (see Fig. 10-4-2b) jointly take care of their waste, the cost of the treatment network is

$$\mathscr{C} = \sum_{j=1}^{p} \mathscr{C}_j\left(\sum_{i \in y_j} U_i, u_j\right)$$

Finally, if there is no restriction on the structure of the network, and transportation costs are of minor importance, then $\mathscr{C}$ only depends upon the total input and the total output, i.e.,

$$\mathscr{C} = \mathscr{C}\left(\sum_{i=1}^{n} U_i, \sum_{j=1}^{p} u_j\right)$$

The third component of our system, the environment is characterized by a function

$$E = E(\mathbf{u})$$

which is the sum, in monetary terms, of all possible *damages* to society (health, vegetation, goods, esthetics, etc.).

At this point, two possible institutional arrangements can be considered: either the waste treatment network is owned and operated by the firms, or by a regional authority from now on called the *Control Agency*. The setting involves an exchange between the producers in the region and the Control Agency. This

10-4 TAXATION SCHEMES: STATIC ANALYSIS

exchange consists of a transfer of waste discharge, eventually treated, and of funds from the producers to the regional authority. The amount of waste discharge to be transferred to the treatment network is based on achieving the greatest net regional profit, where profit is measured by private gains by firms, environmental damage, and treatment cost.

If the Control Agency is taking care of the construction and the operation of the network (see Fig. 10-4-3a), a charge depending upon the amount U_i of pollutant is imposed on each firm i, while if the producers themselves take care of the treatment (Fig. 10-4-3b) a charge is set on each output u_j of the network of treatment plants. These two cases can be formally described as a unique case and a common theory can be developed for both (see Rinaldi et al., 1977). Nevertheless, for the sake of simplicity, only the case in which the Control Agency is the owner of the treatment network is discussed in this section (the interested reader can make reference to the original paper even for the proofs of some of the properties presented in the following).

The Control Agency, whose aim is to maximize the regional profit, is characterized by a *cost function*

$$C(\mathbf{U}) = \min_{\mathbf{u}} \left[\mathscr{C}(\mathbf{U}, \mathbf{u}) + E(\mathbf{u}) \right]$$

From now on the set of producers is called $\mathscr{N}$, i.e.,

$$\mathscr{N} = \{1, 2, \ldots, n\}$$

while

$$\overline{\mathscr{N}} = \{0, 1, 2, \ldots, n\} = \mathscr{N} \cup \{0\}$$

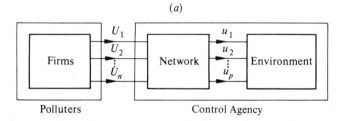

(a)

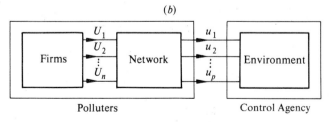

(b)

Figure 10-4-3 The two possible ways of taxing:
(a) charges on U_i
(b) charges on u_j.

denotes the set of all parties (firms and Control Agency). Consistently, if **y** is a subset of the parties (containing, or not, the Control Agency) then

$$\bar{\mathbf{y}} = \begin{cases} \mathbf{y} & \text{if } \mathbf{y} \text{ contains the Control Agency} \\ \mathbf{y} \cup \{0\} & \text{otherwise} \end{cases}$$

while

$$\underline{\mathbf{y}} = \begin{cases} \mathbf{y} & \text{if } \mathbf{y} \text{ does not contain the Control Agency} \\ \mathbf{y} - \{0\} & \text{otherwise} \end{cases}$$

Notice that $\underline{\mathbf{y}} \subseteq \mathbf{y} \subseteq \bar{\mathbf{y}}$ and either $\underline{\mathbf{y}} = \mathbf{y}$ or $\bar{\mathbf{y}} = \mathbf{y}$.

Moreover, given a set **y** of producers, $\mathbf{U}^\mathbf{y}$ denotes the vector $\{U_i\}$ with $i \in \mathbf{y}$ and for the sake of simplicity in notation the vector $\mathbf{U}^\mathbf{y}$ is also defined for sets **y** containing the Control Agency as $\mathbf{U}^\mathbf{y} = \mathbf{U}^{\underline{\mathbf{y}}}$. The aggregated profit function $D_\mathbf{y}(\mathbf{U}^\mathbf{y})$ for any set **y** of producers can be defined as

$$D_\mathbf{y}(\mathbf{U}^\mathbf{y}) = \sum_{i \in \mathbf{y}} D_i(U_i) \tag{10-4-2}$$

while for sets **y** containing the Control Agency we write

$$D_\mathbf{y}(\mathbf{U}^\mathbf{y}) = D_{\underline{\mathbf{y}}}(\mathbf{U}^{\underline{\mathbf{y}}})$$

Similarly, given a set **y** of producers the cost $C_\mathbf{y}(\mathbf{U}^\mathbf{y})$ is defined as the cost characterizing the Control Agency in the case in which only the producers of the set **y** are present in the system, i.e.,

$$C_\mathbf{y}(\mathbf{U}^\mathbf{y}) = C(\mathbf{U}^*) \tag{10-4-3}$$

where the i-th component of $\mathbf{U}^*$ is zero if $i \notin \mathbf{y}$ and U_i otherwise. Again $C_\mathbf{y}(\mathbf{U}^\mathbf{y})$ can be written instead of $C_\mathbf{y}(\mathbf{U}^{\underline{\mathbf{y}}})$ in the case **y** containing the Control Agency.

The Characteristic Function

In order to understand and analyze the problem of supporting financially a regional system based on overall planning, some of the concepts from the Theory of Games will be used, in particular the notion of characteristic function. Given a *system* (i.e., a set $\mathcal{N}$ of producers, their demand functions, the Control Agency, and its cost function) we are interested in the maximum net profit attainable by any subset **y** of the parties. This net profit, denoted by $V(\mathbf{y})$ is called the *characteristic function*.

In order to completely specify this function it is necessary to define the legal conditions governing the regional system. It will be assumed that the regional authority has been given legal responsibility for all discharges. Based on a damage function, it must make appropriate payments for compensation. Dischargers must obtain an agreement with the regional authority for a certain discharge level and without agreement, no discharge is possible. Reflecting these conditions, the characteristic function is defined to have the following properties. First, the value of a coalition without the regional authority as a member is zero. Second, a coalition with only the Control Agency as a member will have zero value since there is no discharge taking place. Thus the only coalitions with potentially positive value are those that contain both the Control Agency and at least one producer.

Formally, the characteristic function $V(\mathbf{y})$ is defined on all subsets $\mathbf{y}$ of $\bar{\mathcal{N}}$ as follows

$$V(\mathbf{y}) = \begin{cases} 0 & \text{if } \mathbf{y} = \underline{\mathbf{y}} \text{ or } \mathbf{y} = \{0\} \\ \max_{\mathbf{U}^y} [D_y(\mathbf{U}^y) - C_y(\mathbf{U}^y)] & \text{if } \mathbf{y} = \bar{\mathbf{y}} \neq \{0\} \end{cases} \quad (10\text{-}4\text{-}4)$$

where $D_y(\mathbf{U}^y)$ and $C_y(\mathbf{U}^y)$ are defined as in Eqs. (10-4-2) and (10-4-3). In particular, $V(\bar{\mathcal{N}})$ is the maximum profit attainable by the set $\bar{\mathcal{N}}$, also called *maximum social benefit*.

In the following, $\mathbf{U}^y$ will denote that particular vector which solves the optimization problem (10-4-4).

The reader accustomed to Game Theory must notice that it is not assumed a priori that the characteristic function V is superadditive ($V(\bar{\mathbf{x}}) + V(\bar{\mathbf{y}}) \leq V(\bar{\mathbf{x}} \cup \bar{\mathbf{y}})$ for all sets $\underline{\mathbf{x}}$ and $\underline{\mathbf{y}}$ of producers such that $\underline{\mathbf{x}} \cap \underline{\mathbf{y}} = \emptyset$) as is usually the case in Game Theory. Superadditivity can be a priori inferred only if the option always exists for groups to act separately if their joint action would not lead to an improvement of their total profit. However, in the present case this option never exists, since the environmental damage can never be decomposed into the sum of damages each one imputable to a given group.

Changes in the characteristic function V indicate how profit depends on the coalition structure. Thus consider $\mathbf{y}_1$ and $\mathbf{y}_2$ distinct collections of all parties $\bar{\mathcal{N}}$ where the central authority is a member of, say $\mathbf{y}_1$. Then $V(\mathbf{y}_2) = 0$ and $V(\mathbf{y}_1) \geq 0$. Suppose $V(\mathbf{y}_1) \leq V(\mathbf{y}_1 \cup \mathbf{y}_2)$. In that case, the addition of the members of $\mathbf{y}_2$ to the group consisting of $\mathbf{y}_1$ improves the total profit. This would mean that the increase in environmental damage and treatment costs is less than the additional profit induced by $\mathbf{y}_2$. The reverse condition, $V(\mathbf{y}_1) > V(\mathbf{y}_1 \cup \mathbf{y}_2)$, reflects a situation where the additional members from $\mathbf{y}_2$ lower the total profits. Under this circumstance, the environmental damages and treatment costs are extremely high. One would expect strong resistance to having the additional members added to the region.

An important attribute of characteristic functions, which is not assumed to be a priori satisfied in the following, is convexity.

Definition of convexity A characteristic function is convex if

$$V(\mathbf{x} \cup \{i\}) - V(\mathbf{x}) \leq V(\mathbf{y} \cup \{i\}) - V(\mathbf{y}) \quad (10\text{-}4\text{-}5)$$

for all $\mathbf{x} \subseteq \mathbf{y} \subseteq \bar{\mathcal{N}}$ and for all $i \in \bar{\mathcal{N}} - \mathbf{y}$.

Roughly speaking it can be said that the satisfaction of this condition means that the overall cooperation is profitable. Thus, when the characteristic function is convex, it would be expected that charges exist such that once stated there are no incentives for the producers to reject the regional plan.

The Taxation Scheme

A set of charges paid by the firms to the Control Agency will be associated with any vector of discharge **U**. The charges have a variety of purposes; first, they are incentive to the firms to develop production processes which generate less waste discharge, secondly, the charges are used by the Control Agency to pay for its costs including compensation for environmental damages, and lastly, the charges should be supportive of the regional system.

The charge on U_i, denoted by $\tau_i(U_i)$, will often be referred to as *tax* in what follows. When the i-th producer is charged an amount $\tau_i(U_i)$ his *private benefit* is given by

$$B_i(U_i) = D_i(U_i) - \tau_i(U_i) \qquad i \in \mathcal{N} \qquad (10\text{-}4\text{-}6)$$

while the benefit accruing to the Control Agency is

$$B_0(\mathbf{U}) = \sum_{\mathcal{N}} \tau_i(U_i) - C(\mathbf{U})$$

It is assumed that each producer is *profit maximizing* in the sense that he selects his discharge U_i^* by maximizing Eq. (10-4-6), i.e.,

$$B_i^* = \max_{U_i} \left[B_i(U_i) \right] = B_i(U_i^*)$$

Then the corresponding benefit for the Control Agency is given by

$$B_0^* = B_0(\mathbf{U}^*)$$

The sum of all private benefits B_i^* and of the benefit B_0^* of the Control Agency is the *social benefit*; obviously, if the taxes $\tau_i(\cdot)$ are suitably selected the social benefit will reach its maximum $V(\mathcal{N})$.

We are now in the position to formally define a taxation scheme.

Definition of taxation scheme A *taxation scheme* is a set of rules that, given a system, generates a set of taxes $\tau_i(\cdot)$, $i \in \mathcal{N}$.

Some examples of taxation schemes may clarify this definition.

Example 1 The following two rules define a taxation scheme.

1. Compute $V(\mathcal{N})$ and $U_i^{\mathcal{N}}$.

2. Determine $\tau_i(\cdot)$, $i \in \bar{\mathcal{N}}$ so that

    (i) $U_i^* = U_i^{\bar{\mathcal{N}}}$ $i \in \bar{\mathcal{N}}$

    (ii) $B_i^* = \dfrac{1-\alpha}{n} V(\bar{\mathcal{N}})$ $0 \leq \alpha \leq 1$ $i \in \bar{\mathcal{N}}$

Condition (i) means that the taxes are such that the producers by solving their own problems will maximize the social benefit since (i) implies

$$\sum_{\bar{\mathcal{N}}} B_i^* = V(\bar{\mathcal{N}})$$

Condition (ii) means that the producers divide equally a part of the total benefit. Moreover, if $\alpha = 0$ there is no benefit for the Control Agency, while if $\alpha = 1$ there is no benefit for the producers.

Point 2 above makes sense only if the functions $\tau_i(\cdot)$ satisfying conditions (i) and (ii) can actually be found. Obviously, if the profit functions $D_i(\cdot)$ are concave then such functions $\tau_i(\cdot)$ exist and are characterized by

$$\left.\frac{d\tau_i}{dU_i}\right|_{U_i^{\bar{\mathcal{N}}}} = \left.\frac{dD_i}{dU_i}\right|_{U_i^{\bar{\mathcal{N}}}} \qquad \tau_i(U_i^{\bar{\mathcal{N}}}) = D_i(U_i^{\bar{\mathcal{N}}}) - B_i^*$$

The first condition means that the marginal charge equals the marginal profit and implies proposition 2(i). The second condition, which follows from Eq. (10-4-6), leads to proposition 2(ii). The functions $\tau_i(\cdot)$ may not be differentiable everywhere as in the case of bulk or two part tariffs. However, for our purpose it is sufficient to assume that $\tau_i(\cdot)$ is locally differentiable at $U_i^{\bar{\mathcal{N}}}$.

Example 2 (*Lexicographic taxation scheme*) The following two rules define a taxation scheme.

1. Given an ordering $i \to \omega(i)$ in $\bar{\mathcal{N}}$ compute

$$V(\mathbf{x}_k) \quad \text{for} \quad \mathbf{x}_k = \{i : \omega(i) \leq k\} \qquad k = 0, 1, \ldots, n$$

(note that the computation of the last term $V(\mathbf{x}_n)$ gives the vector $\mathbf{U}^{\bar{\mathcal{N}}}$).

2. Determine $\tau_i(\cdot)$, $i \in \bar{\mathcal{N}}$ so that

    (i) $U_i^* = U_i^{\bar{\mathcal{N}}}$ $i \in \bar{\mathcal{N}}$

    (ii) $B_i^* = V(\mathbf{x}_{\omega(i)}) - V(\mathbf{x}_{\omega(i)-1})$ $i \in \bar{\mathcal{N}}$, $V(\mathbf{x}_{-1}) = 0$

As in the preceding example, we have $\sum_{\bar{\mathcal{N}}} B_i^* = V(\bar{\mathcal{N}})$ (this follows from (i)). The sense of condition (ii) is as follows: a producer gets a benefit equal to the improvement he generates in the total benefit when he enters the system following the order ω. If $\omega(0) = 0$ the Control Agency has no benefit, while if $\omega(0) = n$ we have the scheme of Example 1 with $\alpha = 1$.

Each example shows that a taxation scheme generates different taxes $\tau_i(\cdot)$ and different benefits B_i^* when applied to different systems. For this reason $B_i^{\bar{y}}$, $i \in \bar{y}$ will indicate the benefit accruing to the i-th party when the set of producers in the system is $\mathbf{y}$. Moreover, we define $B_i^{\mathbf{y}} = 0$, $i \in \mathbf{y}$ since the producers alone cannot make profit. Therefore, given a set $\mathbf{y} \subseteq \mathcal{N}$ either $B_i^{\mathbf{y}} = B_i^{\bar{\mathbf{y}}}$ if $\mathbf{y} = \bar{\mathbf{y}}$ or $B_i^{\mathbf{y}} = B_i^{\mathbf{y}} = 0$ if $\mathbf{y} = \underline{\mathbf{y}}$.

Attributes of Taxation Schemes

Three fundamental attributes can characterize taxation schemes, namely *acceptability*, *efficiency*, and *stability*.

The acceptability of a taxation scheme corresponds to the fact that the benefits of all parties are non-negative. More precisely, we have the following definition.

Definition of acceptability A taxation scheme is *acceptable*, for the set $\bar{\mathcal{N}}$, if the corresponding vector of benefits $\mathbf{B}^{\bar{\mathcal{N}}} = (B_0^{\bar{\mathcal{N}}}, B_1^{\bar{\mathcal{N}}}, \ldots, B_n^{\bar{\mathcal{N}}})$ is non-negative, i.e.,

$$B_i^{\bar{\mathcal{N}}} \geq 0 \qquad \forall \, i \in \bar{\mathcal{N}}$$

The notion of efficiency is directly related to the definition of characteristic function. In brief, a taxation scheme is said to be efficient when the solution it generates (through the profit maximization of the producers) is characterized by the maximum total benefit.

Definition of efficiency A taxation scheme is *efficient* with respect to $\bar{\mathcal{N}}$ if

$$\sum_{\mathbf{y}} B_i^{\mathbf{y}} = V(\mathbf{y}) \qquad \forall \, \mathbf{y} \subseteq \bar{\mathcal{N}}$$

The two preceding examples are examples of acceptable and efficient taxation schemes.

The literature on "optimal taxing" (see, for instance, Upton 1968; Hass 1970; Ferrar 1973) has dealt extensively with the problem of selecting charges such that the total cost of treatment is minimized while a given water quality standard is satisfied. If the environmental damage $E(\mathbf{u})$ is defined to be zero when the standard is met, and infinity when it is not, the classical problem is reduced to the determination of a particular efficient taxation scheme.

Finally, stability is defined as follows.

Definition of stability A taxation scheme is *stable*, with respect to $\bar{\mathcal{N}}$, if

$$\sum_{\mathbf{y}} B_i^{\bar{\mathcal{N}}} \geq \sum_{\mathbf{y}} B_i^{\mathbf{y}} \qquad \forall \, \mathbf{y} \subseteq \bar{\mathcal{N}}$$

That is a stable taxation scheme exists when all subset $\mathbf{y}$ of $\bar{\mathcal{N}}$ take advantage of the coalition with the remaining parties.

10-4 TAXATION SCHEMES: STATIC ANALYSIS

If a taxation scheme is efficient with respect to $\overline{\mathcal{N}}$ then

$$\sum_y B_i^y = V(\mathbf{y}) \qquad \forall \, \mathbf{y} \subseteq \overline{\mathcal{N}}$$

Therefore, in this case, the condition of stability can be modified as follows: an efficient taxation scheme is stable with respect to $\overline{\mathcal{N}}$ if

$$\sum_y B_i^{\bar{y}} \geq V(\mathbf{y}) \qquad \forall \, \mathbf{y} \subseteq \overline{\mathcal{N}}$$

Two of the preceding notions (efficiency and stability) are used to give the following definition.

Definition of core The set of the vectors $\mathbf{B}^{\bar{y}}$ of benefits generated by all the taxation schemes which are efficient and stable with respect to $\overline{\mathcal{N}}$ is called the *core* of $\overline{\mathcal{N}}$.

A simple and visual representation of the core can be given in the case $n = 2$, i.e., when there are only two producers in the system. In Fig. 10-4-4 the three-

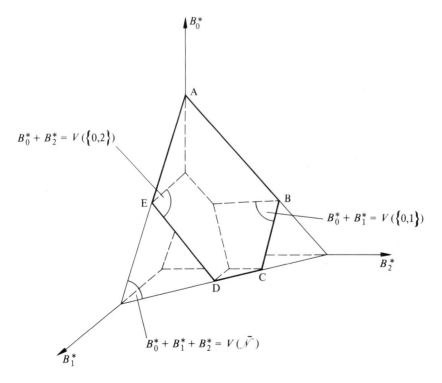

Figure 10-4-4 The space of the benefits and the core (ABCDE).

dimensional space of benefits (B_0^*, B_1^*, B_2^*) is shown, together with the three planes

$$B_0^* + B_1^* = V(\{0,1\})$$
$$B_0^* + B_2^* = V(\{0,2\})$$
$$B_0^* + B_1^* + B_2^* = V(\{0,1,2\})$$

These planes are characterized by the fact that they contain all the vectors of benefits generated by efficient taxation schemes. More precisely, if an efficient taxation scheme is applied to $\overline{\mathcal{N}}$ the corresponding vector of benefits belongs to the last plane, while if it is applied to, for instance, $\{0,1\}$ the corresponding vector of benefits lies on the intersection of the first plane with the plane $B_2^* = 0$. One can notice from the figure that $V(\{0,1,2\})$ is greater than $V(\{0,1\})$ and $V(\{0,2\})$, which means that, in this case, the characteristic function is convex (see Eq. (10-4-5)). Moreover, the vectors **B*** corresponding to the points of the polyhedron ABCDE, are such that

$$\sum_{\overline{\mathcal{N}}} B_i^* = V(\overline{\mathcal{N}}) \qquad \sum_y B_i^* \geq V(\mathbf{y}) \qquad \mathbf{y} \subseteq \overline{\mathcal{N}}$$

and there are no other points satisfying these relationships. Since these are the conditions of efficiency and stability it means that the core is contained in the convex polyhedron ABCDE. Moreover, it is possible to prove (see Rinaldi et al., 1977) that there exist taxation schemes generating all the points of the polyhedron ABCDE, so that we can conclude that the core is the convex polyhedron ABCDE.

The following three properties can usefully be employed (an example is given below) to analyze particular situations. The first two properties are very simple, while the third is a suitable reformulation of an important result proved by Shapley (1971).

Property 1 *A taxation scheme which is stable and efficient with respect to $\overline{\mathcal{N}}$ is also acceptable with respect to $\mathcal{N}$.*

Property 2 *If the profit functions $D_i(\cdot)$ are concave functions, then a necessary and sufficient condition for the existence of a stable and efficient taxation scheme is*

$$V(\overline{\mathcal{N}}) \geq V(\mathbf{y}) \qquad \forall \, \mathbf{y} \subseteq \overline{\mathcal{N}}$$

Property 3 *If the characteristic function is convex the core exists and is a convex polyhedron. Moreover, the lexicographic taxation schemes are stable and efficient and generate all the vectors of benefits, which are vertices of the core.*

Properties of the Model

In order to analyze the existence of stable and efficient taxation schemes in regional environmental management, it is necessary to postulate some structural properties for the functions characterizing the units we called firms, treatment network and environment.

It is assumed that the profit function $D_i(\cdot)$ has the following properties:

(i) $D_i(0) = 0$

(ii) $\dfrac{dD_i}{dU_i} > 0 \qquad U_i > 0$

(iii) $\dfrac{d^2 D_i}{dU_i^2} < 0 \qquad U_i > 0$

Assumptions (i) and (ii) only state that no production implies no profit and that more production implies more profit. Assumption (iii), namely the fact that the marginal profit is a decreasing function of U_i, is usually satisfied for sufficiently high values of U_i, i.e., for sufficiently large firms. On the other hand, small firms are sometimes characterized by increasing marginal profits because of the economies of scale in the technology of production. This means that the theory developed in the following can only be applied to the cases when the firms exploiting the common resource are so large that their marginal profit cannot be increased by increasing the amount of goods produced. On the other hand, point (iii), namely the fact that the profit function $D_i(\cdot)$ is concave, cannot be relaxed since it implies that efficient taxation schemes can be levied on U_i (see the comment in Example 1, page 341).

If the treatment network is completely disaggregated, it can be assumed that the cost $\mathscr{C}_i$ of each plant exhibits the properties shown in Sec. 7-1. If there are no constraints on the structure of the network, the cost of the treatment network is a function $\mathscr{C}(U_+, u_+)$ of the total input $U_+ = \sum_i U_i$ and the total output $u_+ = \sum_j u_j$, which will be assumed to satisfy an important property referred to as "*economies of scale*"

$$\mathscr{C}(U'_+ + \Delta, u'_+ + \delta) - \mathscr{C}(U'_+, u'_+) \geq \mathscr{C}(U''_+ + \Delta, u''_+ + \delta) - \mathscr{C}(U''_+, u''_+) \qquad (10\text{-}4\text{-}7)$$

for

$$U'_+ < U''_+ \qquad u'_+ < u''_+ \qquad \delta < \Delta$$

This property is a natural extension of the property of economies of scale of the cost of a single treatment plant. Thus, if it is assumed that

$$u'_+ = \alpha U'_+ \qquad u''_+ = \alpha U''_+ \qquad \delta = \alpha \Delta$$

Eq. (10-4-7) becomes

$$\mathscr{C}(U'_+ + \Delta, \alpha(U'_+ + \Delta)) - \mathscr{C}(U'_+, \alpha U'_+) \geq \mathscr{C}(U''_+ + \Delta, \alpha(U''_+ + \Delta)) - \mathscr{C}(U''_+, \alpha U''_+)$$

from which it follows that the function $\mathscr{C}(U_+, \alpha U_+)$ is concave (see Sec. 7-1). If the structure of the network of the treatment plant is somehow constrained, the property of economies of scale should be formulated in a more general way, by substituting $U'_+, u'_+, U''_+, u''_+, \Delta$, and δ in Eq. (10-4-7) with the vectors $\mathbf{U}', \mathbf{u}', \mathbf{U}'', \mathbf{u}''$, $\boldsymbol{\Delta}$, and $\boldsymbol{\delta}$ ($\boldsymbol{\Delta} > 0, \boldsymbol{\delta} > 0$).

The main feature to be taken into account when discussing the damages produced by the users of an environmental resource is the so-called *congestion*

346 MANAGEMENT OF THE RIVER BASIN

effect. At some low level of use an additional use of the resource may practically generate no surplus of damage. A point is reached, however, where an additional user will cause others to incur additional costs or suffer disutilities associated with congestion (see Kneese, 1971, for details). This property can be given the following very general form

$$E(\mathbf{u}' + \boldsymbol{\delta}) - E(\mathbf{u}') \leq E(\mathbf{u}'' + \boldsymbol{\delta}) - E(\mathbf{u}'') \qquad (10\text{-}4\text{-}8)$$

where $\mathbf{u}'$ $\mathbf{u}''$ and $\boldsymbol{\delta}$ are three non-negative vectors and $\mathbf{u}'' \geq \mathbf{u}'$. Equation (10-4-8) implies that E is convex with respect to each component u_i ($u'_j = u''_j$ and $\delta_j = 0$ for all $j \neq i$ in Eq. (10-4-8)).

A Particular Case

The cases that can be analyzed in order to detect the implications of possible institutional arrangements and different legal constrictions on the existence of the core, are many and different. The case discussed below is perhaps one of the most interesting, since it corresponds to a compromise between the scheme based purely on effluent standards and the one based only on effluent charges.

The traditional approach to pollution control has been legislative regulation of the discharges and most of the laws currently existing stipulate the allowable amount of waste each type of firm can discharge. In some environmental laws it is tacitly assumed that all the discharges satisfying a standard $\bar{u}_i$ induce a negligible environmental damage ($E(\mathbf{u}) = 0$), while other legislations do not consider $E(\mathbf{u})$ to be zero. In these cases each polluter is asked to compensate for the damage in monetary terms. Usually, each producer acts by himself and takes care of his own treatment plant, while a regional waste treatment system could often be of advantage because of the economies of scale and the possibility of reallocating the discharges. Then it is of interest to analyze the case of a regional agency which takes care of the treatment network, acquires the rights $\bar{u}_i$ of discharge owned by each firm and levies taxes on the pollutant flow rate U_i generated by each firm. To gain insights into the advantages of a well designed regional waste treatment network a comparison will be made between the case in which the treatment network is unconstrained and the case in which it is a priori assumed to be disaggregated.

When no constraints are imposed on the structure of the network the total cost of the Control Agency is

$$C(U_+) = \min_{\mathbf{u}} \left[\mathscr{C}(U_+, u_+) + E(\mathbf{u}) \right]$$

$$u_+ \leq \bar{u}_+^{\mathscr{T}} \triangleq \sum_{\mathscr{N}} \bar{u}_i$$

The last constraint means that the Control Agency can discharge up to a maximum $\bar{u}_+^{\mathscr{T}}$ given by the sum of the rights of discharge of the producers. In general, the optimal solution $\mathbf{u}^0$ is a function of U_+, i.e.,

$$\mathbf{u}^0 = \mathbf{u}^0(U_+)$$

while, if the environmental damage $E(\mathbf{u})$ is assumed to be zero, the total output u_+^0 is not dependent upon U_+, since it is obviously given by

$$u_+^0 = \bar{u}_+^{\bar{\mathcal{I}}}$$

When the Control Agency is constrained to use a completely disaggregated network it loses the right to reallocate the discharges of the firms, i.e., the total cost of the Control Agency is

$$C(\mathbf{U}) = \min_{\mathbf{u}} \left[\sum_{\mathcal{N}} \mathscr{C}_i(U_i, u_i) - E(\mathbf{u}) \right]$$

subject to

$$u_i \leq \bar{u}_i \quad \forall\, i \in \mathcal{N}$$

In this case the vector $\mathbf{u}^0$ is a function of the vector $\mathbf{U}$ (and not of U_+), while if $E(\mathbf{u}) = 0$ obviously $u_i^0 = \bar{u}_i$.

The four possible cases are now analyzed in the following order

(i) unconstrained network, $E(\mathbf{u}) = 0$
(ii) unconstrained network, $E(\mathbf{u}) \neq 0$
(iii) completely disaggregated network, $E(\mathbf{u}) = 0$
(iv) completely disaggregated network, $E(\mathbf{u}) \neq 0$

(i) Under the hypothesis $E(\mathbf{u}) = 0$, it is possible to prove (see Rinaldi et al., 1977) the following basic property.

Property 4 *If the environmental damage is negligible and the cost of the treatment network satisfies Eq. (10-4-7), then there exist efficient and stable taxation schemes.*

In fact, under the preceding assumptions, the characteristic function turns out to be convex, so that Property 4 can be inferred from Property 3. Moreover, if the strict inequality is satisfied in Eq. (10-4-7) then there exists an infinity of stable and efficient taxation schemes characterized by $B_0 = 0$ (see segment CD of Fig. 10-4-4). In other words, the Control Agency can charge the firms only for the cost of treating the waste and still have options in sharing the total benefit among the producers without generating incentives for the users of the service to reject the regional plan.

(ii) Since $E(\mathbf{u}) \neq 0$ for $\mathbf{u} > 0$ it is not possible to prove that the characteristic function is convex and therefore the existence of stable and efficient taxation schemes cannot be inferred any more by means of Property 3. On the other hand, it can be shown by means of very simple examples that the necessary and sufficient condition of Property 2 can either be satisfied or not, so that it is possible to have situations in which stable taxation schemes exist and situations in which all the schemes are unstable. But even when stable and efficient taxation schemes exist, the Control Agency is more constrained than in the preceding case $(E(\mathbf{u}) = 0)$. It is possible to prove that when the congestion effect of the environment is particularly important, taxation schemes characterized by zero profit of the Control Agency cannot be stable. This means that stability can be obtained only at the price of transforming the Control Agency into a profit corporation. More precisely the following property can be proved.

Property 5 *When the congestion effect is dominant with respect to the economies of scale (in the sense specified below), each producer strives to expel the others from the system, unless the Control Agency is to some extent a profit-making corporation ($B_0^{\bar{T}} \neq 0$).*

Let $\mathbf{u}^{\bar{T}}$ denote the output vector of the treatment network serving the system $\mathcal{N}(\mathbf{u}^{\bar{T}} = \mathbf{u}^0(U_+^{\bar{T}}))$ and $\mathbf{u}^{(i)}$ the output vector of the treatment network that would be used if the i-th producer were alone in the regional system with a pollutant production $U_i^{\bar{T}}(\mathbf{u}^{(i)} = \mathbf{u}^0(U_i^{\bar{T}}))$. Then the reduction of the treatment cost due to economies of scale is

$$\sum_{\mathcal{N}} \mathscr{C}(U_i^{\bar{T}}, u_+^{(i)}) - \mathscr{C}(U_+^{\bar{T}}, u_+^{\bar{T}})$$

while

$$E(\mathbf{u}^{\bar{T}}) - \sum_{\mathcal{N}} E(\mathbf{u}^{(i)})$$

is the increase of the environmental damage due to the congestion effect. Property 5 says that the condition

$$E(\mathbf{u}^{\bar{T}}) - \sum_{\mathcal{N}} E(\mathbf{u}^{(i)}) > \sum_{\mathcal{N}} \mathscr{C}(U_i^{\bar{T}}, u_+^{(i)}) - \mathscr{C}(U_+^{\bar{T}}, u_+^{\bar{T}}) \quad (10\text{-}4\text{-}9)$$

implies the non-existence of stable and efficient taxation schemes with $B_0^{\bar{T}} = 0$. Moreover, it can be proved that stable and efficient taxation schemes can exist only if the benefit of the Control Agency satisfies the following condition

$$B_0^{\bar{T}} \geq \left[\sum_{\mathcal{N}} V_i - V(\bar{\mathcal{N}}) \right] \Big/ (n-1) \quad (10\text{-}4\text{-}10)$$

where $V_i = V(\{0, i\})$.

In conclusion, if a sequence of cases were to be considered in which the environment is increasingly sensitive to the congestion effect cores would be obtained that became smaller and smaller. In Fig. 10-4-5 a sequence (a) (b) (c) (d) is shown; in the first two cases Eq. (10-4-9) is not satisfied, while the third figure refers to the limit case in which Eq. (10-4-9) is satisfied with the equality sign.

(iii) The implications of constraining the Control Agency to use a completely disaggregated network of treatment plants will now be analyzed. For this, consider first the case in which the environmental damage is neglected

$$E(\mathbf{u}) = 0$$

and recall that in this case $u_i^0 = \bar{u}_i$. Then

$$V(\bar{\mathcal{N}}) = \max_{\mathbf{U}} \left[\sum_{\mathcal{N}} D_i(U_i) - \sum_{\mathcal{N}} \mathscr{C}_i(U_i, \bar{u}_i) \right]$$

$$= \sum_{\mathcal{N}} \max_{U_i} \left[D_i(U_i) - \mathscr{C}_i(U_i, \bar{u}_i) \right]$$

i.e.,

$$V(\bar{\mathcal{N}}) = \sum_{\mathcal{N}} V_i \qquad (10\text{-}4\text{-}11)$$

This condition implies (easy to check) that the characteristic function is convex. Thus, Property 3 applies and the conclusion is that the core always exists. The fundamental difference with respect to the case of aggregated networks is that there now exists only one point in the core with $B_0^{\bar{\mathcal{N}}} = 0$. In fact, if $B_0^{\bar{\mathcal{N}}} = 0$ then

$$V(\bar{\mathcal{N}}) = \sum_{\bar{\mathcal{N}}} B_i^{\bar{\mathcal{N}}} = \sum_{\mathcal{N}} B_i^{\bar{\mathcal{N}}}$$

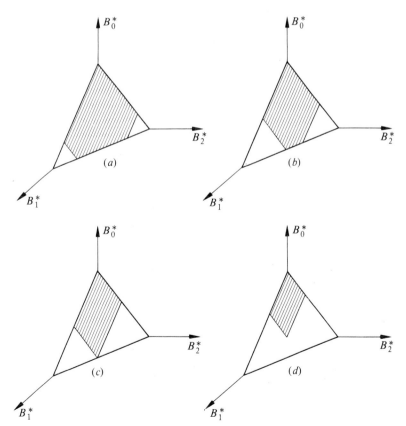

Figure 10-4-5 Smaller cores are obtained for increasing congestion effects.

350 MANAGEMENT OF THE RIVER BASIN

Hence stability and Eq. (10-4-11) imply

$$B_i^{\bar{\mathcal{N}}} = V_i$$

which proves the uniqueness of such kind of taxation scheme. This case is shown in Fig. 10-4-6 where the point $C = D$ represents the unique possibility for the Control Agency to be a non-profit corporation. This point requires a specific distribution of the benefit $V(\bar{\mathcal{N}})$ among the firms. If a different distribution is desired (see, for instance, point S of Fig. 10-4-6) this can be done only by means of an unstable taxation scheme. Nevertheless, if the benefit of the Control Agency is allowed to be positive, then the scope can be obtained by means of stable taxation schemes (see segment AR of Fig. 10-4-6).

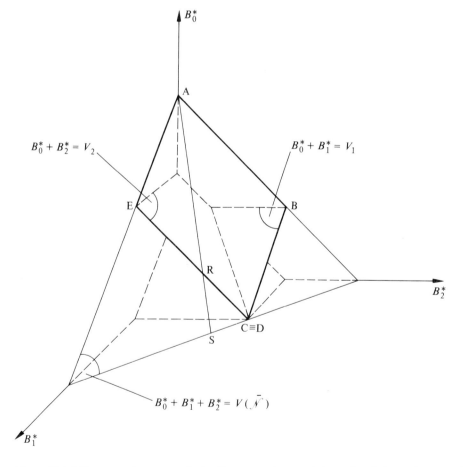

Figure 10-4-6 The core when a completely disaggregated network is used and the environmental damage is neglected.

(iv) Finally, consider the case

$$E(\mathbf{u}) \neq 0 \qquad \mathbf{u} > \mathbf{0}$$

with the Control Agency constrained to use a completely disaggregated network of treatment plants. Again, cases exist in which there are no stable and efficient taxation schemes. This can happen when there exists a subset $\mathbf{x}$ of $\mathcal{N}$ such that the variation $E(\mathbf{u}^{\bar{\mathcal{N}}}) - E(\mathbf{u}^{\bar{x}})$ of the damage produced by the set $\mathcal{N} - \mathbf{x}$ is greater than the variation $W(\bar{\mathcal{N}}) - W(\bar{\mathbf{x}})$ of the net benefit

$$W(\bar{\mathcal{N}}) = \sum_{\mathcal{N}} [D_i(U_i^{\bar{\mathcal{N}}}) - \mathcal{C}_i(U_i^{\bar{\mathcal{N}}}, u_i^{\bar{\mathcal{N}}})]$$

i.e., when

$$E(\mathbf{u}^{\bar{\mathcal{N}}}) - E(\mathbf{u}^{\bar{x}}) > W(\bar{\mathcal{N}}) - W(\bar{\mathbf{x}}) \tag{10-4-12}$$

where,

$$\mathbf{u}^{\bar{\mathcal{N}}} = \mathbf{u}^0(U^{\bar{\mathcal{N}}}) \qquad \mathbf{u}^{\bar{x}} = \mathbf{u}^0(U^{\bar{x}})$$

In fact, Eq. (10-4-12) implies

$$V(\bar{\mathcal{N}}) = W(\bar{\mathcal{N}}) - E(\mathbf{u}^{\bar{\mathcal{N}}}) < W(\bar{\mathbf{x}}) - E(\mathbf{u}^{\bar{x}}) = V(\bar{\mathbf{x}})$$

i.e.,

$$V(\bar{\mathcal{N}}) < V(\bar{\mathbf{x}})$$

which contradicts the necessary condition (see Property 2) for the existence of stable and efficient taxation schemes. On the other hand, if the inequality sign $\leq$ holds in Eq. (10-4-12) for all $\mathbf{x} \subseteq \mathcal{N}$, it can be proved that

$$V(\bar{\mathcal{N}}) \geq V(\bar{\mathbf{x}})$$

which implies the existence of the core. The only difference with the case of the unconstrained network (see point (ii)) is the uniqueness of the stable and efficient taxation scheme which assigns a zero benefit to the Control Agency, while the difference with respect to the preceding case (iii) is the possibility of the non-existence of such a particular taxation scheme. This easily follows from Eq. (10-4-10) and the condition

$$V(\bar{\mathcal{N}}) \leq \sum_{\mathcal{N}} V_i$$

which can be proved using the property of congestion effect of the environment. In Fig. 10-4-6 an example is shown with $V(\bar{\mathcal{N}}) = \sum_{\mathcal{N}} V_i$, while in Fig. 10-4-7 (notice that $V(\bar{\mathcal{N}}) < \sum_{\mathcal{N}} V_i$) the lowest point of the core (point $C \equiv D$) is characterized by

$$B_0^{\bar{\mathcal{N}}} = \left[\sum_{\mathcal{N}} V_i - V(\bar{\mathcal{N}}) \right] \bigg/ (n-1)$$

which corresponds to the bound given by Eq. (10-4-10).

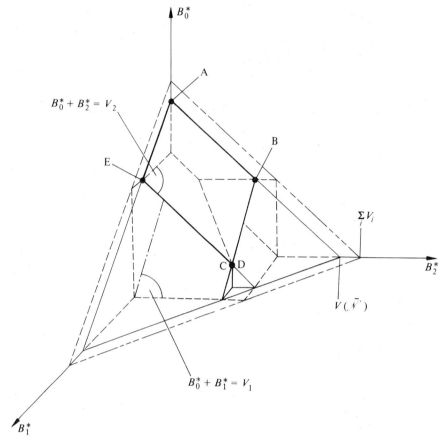

Figure 10-4-7 The core when a completely disaggregated network is used and the environmental damage is taken into account.

In order to make the analysis complete the case in which the Control Agency can reallocate the rights of discharge, even when it is forced to use a completely disaggregated network can now be considered. The cost of the Control Agency is then

$$C(\mathbf{U}) = \min_{\mathbf{u}} \left[\sum_{\mathcal{N}} \mathscr{C}_i(U_i, u_i) - E(\mathbf{u}) \right]$$

$$u_+ \leq \bar{u}_+^{\mathcal{N}}$$

The analysis of this situation could be accomplished in a way similar to the preceding one and the main conclusion is that the freedom to reallocate the rights of discharge gives rise to larger cores, i.e., the number of efficient taxation schemes is generally larger than in the case in which the Control Agency cannot reallocate the rights of discharge.

Interpretation and Implications of the Results

In summary, the preceding discussion shows that if the damages to the environment are not negligible and, if these damages must be refunded by the users of the resource, it is very unlikely that efficient and stable taxation schemes can be found if the regional authority acts as a non-profit corporation. These results also state that if the damages to the environment are neglected, or, in other words, if the total benefit of the firms instead of the social benefit is maximized, then efficiency and stability are easily obtained. This is, indeed, what happened in the history of the industrial development of the last century in almost all countries. The efficiency of the firms has been very high and there has been no friction or competition for the use of the self-purification capacity of the environment. Nevertheless, these two attributes have been obtained at a price which must be considered too high: namely, the fact that damages to society are neglected. The increased public awareness nowadays makes this solution no longer feasible. In this respect, the approach followed in this section indicates an alternative solution, since stability and efficiency can also be obtained by letting the regional authority obtain a profit from the sale of emission rights. The higher the environmental congestion the greater this profit must be. If ethical and political attitudes are against this kind of solution there is no way to maximize the social benefit without generating frictions among the producers.

10-5 TAXATION SCHEMES: DYNAMIC ANALYSIS

The static analysis carried out in the preceding section is relatively unsatisfactory, since it makes reference to an ideal situation in which all the firms have to select their treatment effort at the same time. For this reason, the studies based on static analysis are justifiable only because they give the "maximum" to be expected in such an ideal situation and the corresponding results must be considered only as a reference and guideline for selecting good empirical solutions. The real situation is characterized by firms entering and leaving the system and by evolving production and treatment technology, so that the discharge demands of the firms must be thought as time-varying. Each time a new firm enters the system or an existing one raises its discharge rate, some water user downstream from the discharge point is affected (in general adversely) and bears an *external diseconomy*. More precisely, an external diseconomy (or external cost) occurs when the decision of an economic entity (a firm) inflicts an appreciable damage (higher costs, less valuable production) on a managerially independent entity which was not a consenting party in reaching the decision (Meade, 1973). The water quality management problem can then be seen as the problem of controlling the spreading off of external costs. From this point of view, an effluent charge equal to the external diseconomies induced by the discharge appears as the most natural and efficient controlling tool (Pigou, 1929). In fact, under an effluent charge, a waste discharger has an economic incentive to take care of the damages it generates, so that it can be said that "external costs are internalized."

Many contributions which develop a theoretically oriented analysis of the static case can be found in the literature, while the dynamic case, with a decentralized decision structure and firms entering and leaving the system, has never been clearly analyzed. The available literature is mainly concerned with case studies, like the Delaware estuary study and the Genossenschaften in the Ruhr basin (see Kneese and Bower, 1968; Mäler, 1974), while many legislations are only based on heuristic proposals. In the same way, this section is not to be considered an attempt at developing a theoretical analysis of the dynamic case: its scope is definitively less ambitious. It is only attempting to give an idea of how taxation schemes can be derived from some "basic principles" and to point out the role that water quality models can play in solving these types of management problems (see Guariso et al., 1977). In particular, attention will be paid to the differences which can be obtained by using different water quality models. For this reason, the last part of this section is devoted to a simulation study of the industrial development in the Rhine river basin (Guariso and Soncini-Sessa, 1976). Different patterns of industrial waste discharges which would be induced by different taxation schemes and different ways of estimating the damage are presented. Both the Streeter-Phelps model and the ecological model given in Sec. 5-3 are used, and some important differences between the results obtained in the various cases are pointed out.

The Basic Principles

The main task of this section is to devise a *taxation scheme* (a set of rules) for calculating the charges to be imposed on each polluter such that an overall solution results which is "not too far" from the one which maximizes the total benefit. The reason why we ask for a solution "not too far" from the optimal one, instead of asking for an efficient solution (see Sec. 10-4) is that the latter is practically unfeasible in a dynamic context, since it would require the redefinition of the discharges of all units operating in the system each time a firm enters, or leaves the system, or modifies its discharge. The consequence is that it is not possible to derive a dynamic taxation scheme by solving a well posed optimization problem as in the static case. Thus, the approach will be mainly heuristic: the taxation scheme will be derived from some basic principles and some of its properties will be briefly discussed.

From an analysis of the different cases described in the literature it emerges that it is generally considered that any acceptable taxation scheme has to comply with the following three very general principles.

1. *Feasibility of control.* This practically implies that the tax must be directly connected to the discharge flow rate, since this is the only easily measurable variable.
2. *Unmodifiability of charges.* The tax cannot be increased by the control agency as long as the discharge rates are unchanged. This attribute is a prerequisite for planning with an acceptable degree of risk.

3. *No veto on any development.* This means that everybody must be allowed to enter the system at any time and at any place, provided that he is willing to pay the corresponding charge.

A Taxation Scheme Satisfying the Principles

As has already been mentioned, the Pigouvian point of view (Pigou, 1929) suggests that the charge on a particular discharge must be equal to the damage it causes to all parties using the river basin. However, even if the difficulties involved in estimating the damage are neglected, it is impossible to associate a given damage to each firm, since in general the damage is not a separable function because of the interactions and synergistic effects among the pollutants. To overcome this difficulty the Control Agency can charge each firm with the increment of the damage generated when entering the regional environmental system. More precisely, let l_i be the position (still to be selected) of the discharge of the i-th firm and let u_i be the corresponding mass flow rate of pollutants discharged. If $(i - 1)$ firms are already in the system, the locations $\{\bar{l}_j\}_{j=1}^{i-1}$ and the pollutant flow rates $\{\bar{u}_j\}_{j=1}^{i-1}$ of these firms (from now on called the *predecessors* of i) are fixed as well as the damage $\bar{E}_{i-1}$ they generate. Thus, the surplus of damage $E_i - \bar{E}_{i-1}$ generated by the i-th firm can be seen as a function of l_i and u_i. These ideas can be formally expressed as follows.

Rule 1 *The tax τ_i proposed by the Control Agency on the i-th firm is a function of l_i and u_i given by*

$$\tau_i(l_i, u_i) = E_i(l_i, u_i) - \bar{E}_{i-1} \tag{10-5-1}$$

Once the i-th firm has selected its optimal location $\bar{l}_i$ and its optimal rate of discharge $\bar{u}_i$, the charge levied on it by the Control Agency is

$$\bar{\tau}_i = \tau_i(\bar{l}_i, \bar{u}_i)$$

This charge can be separated into two parts

$$\bar{\tau}_i = \bar{\tau}_i^1 + \bar{\tau}_i^2$$

where $\bar{\tau}_i^1$ is equal to the sum of the external costs borne by the predecessors of i and $\bar{\tau}_i^2$ is the surplus of damages imposed on the rest of the society (aesthetics, vegetation, goods, health, recreation, ...).

Rule 2 *The charge $\bar{\tau}_i$ cannot be varied unless a predecessor j of i leaves the region. In this case, the charge $\bar{\tau}_i$ is reset on the basis of the entering order (i.e., the firm i has to pay a tax which is equal to the surplus of damage it would have generated by entering a system composed by all its predecessors with the exception of the j-th one).* This rule implies that when the j-th firm leaves the system the taxes of all its successors $(j + 1), (j + 2), \ldots,$ are lowered by the Control Agency.

Rule 3 *If a firm i reduces its pollutant flow rate from the value $\bar{u}_i$ to a value $u'_i < \bar{u}_i$ its charge and the charges of its successors are reset as if u'_i would have been its discharge rate since the beginning of its activity.*

Rule 4 *If a firm i raises its pollutant flow rate from the value $\bar{u}_i$ to a value $u'_i > \bar{u}_i$, the increment of its tax is equal to the charge that would be levied on a new firm entering the regional system in point $\bar{l}_i$ with a discharge rate $u'_i - \bar{u}_i$.*

Rules 1–4 define a taxation scheme which satisfies our three basic principles.

Two Optimality Properties

Assume that each firm selects the location $\bar{l}_i$ and the discharge rate $\bar{u}_i$ by maximizing its profit. Then, the taxation scheme specified above gives a solution which, from the social point of view, can be called "pseudo-efficient." In fact, each time a firm enters the regional system the social benefit (total profit) is certainly improved, although it is not maximized. Moreover, suppose that the taxation scheme satisfies the following property.

Rule 5 *The part $\left(\sum_i \tau_i^1\right)$ of the money collected via the effluent charge system is used to refund the external costs supported by each firm.*

This rule implies that the profit of each firm is independent of the actions taken by its successors (of course, we are restricting our discussion to the effects induced by pollution). Thus, as long as production and treatment technologies are unchanged no firm is interested in modifying its decision (location $\bar{l}_i$ and discharge rate $\bar{u}_i$), unless some predecessor changes. This kind of equilibrium, typical of non-cooperative situations is called *Nash equilibrium* and has been investigated in the literature on game theory (Nash, 1951). It is to be expected that these studies would help to clarify the studies on regional environmental management. It is also easy to check that the solution induced by the taxation scheme is *Pareto optimal* in the sense specified in Sec. 10-2: a perturbation of the discharge rates of a subset of the firms operating in the system induces a loss of benefit to at least one firm of the subset. Indeed, the oldest firm of the subset cannot modify its discharge rate without lowering its benefit which is independent of the discharge rates of the successors. This means that in a co-operative situation modifications can only be achieved by forming coalitions in which side payments between firms are suitably calibrated.

Applicability of the Taxation Scheme

Unfortunately, the taxation scheme specified above is difficult to implement for two reasons. First, because the damages induced by a given waste discharge are strongly dependent upon the natural environmental conditions. For instance, the effect of a biodegradable discharge depends upon river flow rate and water temperature. Thus, the damage is time dependent and, as a consequence, the tax should reflect the expected (mean) value of the damage. This expected value is difficult to determine because the water quality models are nonlinear with respect to flow rate and temperature. Then the only practical alternative is to define the tax by means of the damage corresponding to a suitable steady state. Of course, under these conditions it cannot be guaranteed that the charges will cover the total damage, so that the optimality properties discussed above cannot be proved anymore.

The second (and main) weak point of this taxation scheme is that it is practically impossible to estimate the environmental damage in monetary terms. Kneese and Bower (1968) and Wyzga (1974) report some promising works constituting a first step towards the estimation of the damage function, "but it is hard to be sanguine about the availability in the foreseeable future of a comprehensive body of statistics reporting the expected net environmental damages induced by the various externality generating activities" (Baumol and Oates, 1971). As far as the external costs imposed on firms are concerned a lot of information would be required even for a rough estimate. This means that extremely high costs would be required to make the taxation scheme operative.

For this reason, one can try to estimate the damage E by means of a synthetic pollution index P, which, in turn, is a function of the discharges. In order to have a safe estimate of the damage it is worth assuming that E is infinite for values of P greater than a standard value P^*. The use of the so called *stream standards* corresponds, for example, to assuming that $E = 0$ for $P \le P^*$ and that $E = \infty$ for $P > P^*$. Better estimates are obtained by assuming that the damage is an increasing convex function of the pollution index, for example,

$$E = \alpha \frac{P}{P^* - P} \tag{10-5-2}$$

where α is a suitable conversion factor. Then, Eq. (10-5-1) becomes

$$\tau_i(l_i, u_i) = \alpha \frac{P^*(P_i - \bar{P}_{i-1})}{(P^* - P_i)(P^* - \bar{P}_{i-1})} \tag{10-5-3}$$

where

$$P_i = P(\bar{u}_i, \ldots, \bar{u}_{i-1}, \bar{l}_i, \ldots, \bar{l}_{i-1}, u_i, l_i)$$
$$\bar{P}_i = P(\bar{u}_i, \ldots, \bar{u}_{i-1}, \bar{l}_i, \ldots, \bar{l}_{i-1}, \bar{u}_i, \bar{l}_i)$$

The pollution index P can be, for example, the total (integral) oxygen deficit (see Sec. 10-3) in the basin or the maximum (pointwise) oxygen deficit in the basin. When the second index is used, it is easy to devise situations where a polluter has a zero tax up to some value of discharge. Obviously, this fact is undesirable. Thus, in this case, a slightly different approach, based on the direct estimation of the surplus of damage τ_i, is to be preferred. Again, it is reasonable to assume that this surplus of damage is increasing and convex with the discharge u_i and that it approaches infinity in a non-integrable way when u_i approaches the value u_i^* for which the pollution index P is equal to the standard P^*. For example, one can use the following expression for the tax τ_i

$$\tau_i(l_i, u_i) = \beta(l_i) \frac{u_i}{u_i^*(l_i) - u_i} \tag{10-5-4}$$

where $\beta(l_i)$ is a suitable function of space and $u_i^*(l_i)$ is the solution of the following parametric mathematical programming problem

$$\max_{u_i} [u_i]$$

subject to

$$P(\bar{u}_i, \ldots, \bar{u}_{i-1}, \bar{l}_1, \ldots, \bar{l}_{i-1}, u_i, l_i) \leq P^*$$

Both Eqs. (10-5-3) and (10-5-4) satisfy our basic principles and have the advantage of being relatively simple for the Control Agency to apply (no explicit information on the technology of the firms are required).

In the following pages, the results of a simulation study are shown in which the total oxygen deficit in the river basin (see Sec. 10-3) and the maximum (pointwise) oxygen deficit in the river basin are considered as pollution indices. In the first case a taxation scheme is used based on Eq. (10-5-3) (Scheme A), while in the second the scheme used is based on Eq. (10-5-4) (Scheme B).

A Simulation Study

In this simulation study, the Rhine river between Mannheim–Ludwigshafen and the Dutch–German border (see Fig. 5-3-1) has been considered (see Secs. 5-3, 8-4, and 10-3). The initial discharge pattern of the biodegradable pollutants is shown in Fig. 10-5-1. The discharges shown in Fig. 10-5-1a are lower than the discharges shown in Fig. 5-3-2, which correspond roughly to the situation in 1970. The reason for this is that the 1970 situation (and the present one, also) is too poor to be a realistic starting point for a dynamic taxation scheme. Therefore, it has been assumed that at the implementation of the scheme some of the polluters already present are forced to reduce their discharges. The pollutants brought in by the tributaries are included in Fig. 10-5-1a. The corresponding profiles of dissolved oxygen and bacteria concentrations are shown in Fig. 10-5-1b for mean flow rate and water temperature of 20 °C. Notice the sags downstream from the major pollution sources (see Sec. 5-3), while in the Middle Rhine, between Rüdesheim (550 km) and Köln (700 km), the oxygen content reaches relatively high values. The reasons for the latter fact are the absence of industries and the high reaeration rate. The bacterial biomass concentration profile exhibits maxima in correspondence with minima of the DO concentration. Notice in particular the low bacterial density before Köln which is due to scarcity of easily digestable food in the Middle Rhine.

The growth of wastewater production from 1970 to 1985 was estimated (see Sec. 8-4) assuming a constant growth factor equal to the present trend (3 percent

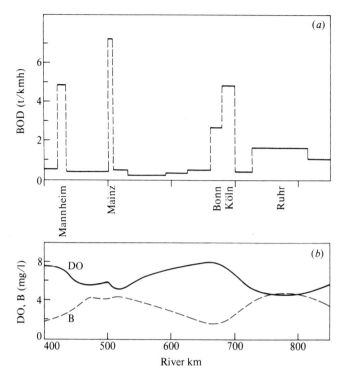

Figure 10-5-1 Assumed pollutional situation on the Rhine river in 1970:
(a) pollutional load
(b) dissolved oxygen (DO) and bacterial biomass (B).

per year) and the assumption was made that this corresponds to 120 large industrial firms, with a maximum discharge rate of 3.5 tons of BOD per hour. About 220 positions were considered as possible locations for the firms along the river course. Each time a new firm i considers the possibility of entering the system five potential locations $l_i^1, l_i^2, \ldots, l_i^5$ are randomly selected. For each one of these locations which is still available, the Control Agency computes the tax to be proposed to the firm and then the firm selects the best location $\bar{l}_i$ and the best discharge $\bar{u}_i$ by maximizing its profit (the profit functions of the firms take into account the treatment costs and were suitably generated).

From now on the results of the simulation study will be shown only for the period 1970–1980, (with the exception of Fig. 10-5-7) since in this case the effects of the different taxation schemes can be better ascertained.

Taxation scheme A It is now assumed that the damage E can be estimated by means of Eq. (10-5-2) where P is the total oxygen deficit, i.e.,

$$P = \int_{\mathscr{L}} A(l)(c_s - c(l)) \, dl$$

where $\mathscr{L}$ is the section of the Rhine river considered and $A(l)$ and $c(l)$ are the cross-sectional area of the river and the dissolved oxygen concentration respectively.

Figure 10-5-2a shows the pattern of the increase of the discharge in the period 1970–1980 obtained by the Streeter-Phelps model. It can be seen that the area of major industrial development is located in the Ruhr district (a minor industrial development is present in the Middle Rhine) and that no firm enters the region upstream from Mainz. This spatial distribution is easily understandable. The firms located near the Dutch–German border take advantage of the fact that the deficit they create mainly develops in the Netherlands, while the tax does not take into account the corresponding damages. The existence of a minor pole of attraction in the Middle Rhine may be due to the higher reaeration capacity of this area. The change in the oxygen profile between 1970 and 1980 (see Fig. 10-5-2b) is essentially what one would expect: a decrease of the oxygen level along the Rhine, which is maximal at the border.

Figure 10-5-3a shows the expected increase of the pollutional load when the ecological model is used. Again, the industrial development takes place in the Middle Rhine and near the Dutch–German border, but this time it can be seen that the major development is in the Middle Rhine area. A possible way of understanding the phenomenon may be found in an odd effect induced by the trophic chain: consider Fig. 10-5-3b where both the new (1980) and old (1970) oxygen profiles are shown. Despite the growth of wastewater production, the oxygen levels from the 760th to the 810th kilometer are higher than in 1970. This singular behavior can be explained by observing the bacterial biomass

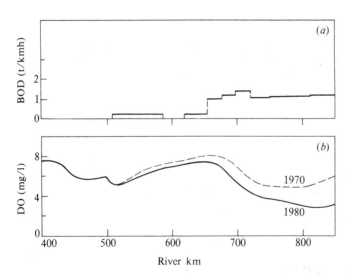

Figure 10-5-2 Effects of taxation scheme A (Streeter-Phelps model):
(a) additional pollutional load
(b) dissolved oxygen profile in 1970 and 1980.

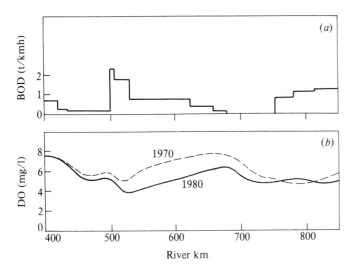

Figure 10-5-3 Effects of taxation scheme A (ecological model):
(a) additional pollutional load
(b) dissolved oxygen profile in 1970 and 1980.

profiles (see Fig. 10-5-4). In 1970 the low discharge rate in the Middle Rhine greatly reduces the bacterial biomass. Thus, downstream from Bonn (690 km), when the pollution load again increases, the bacterial biomass needs a certain time to develop and destroy the pollutants accumulated in the meantime. Therefore, the point of maximum deficit is relatively far downstream from Bonn, in the Ruhr district. On the other hand, in 1980, the continuous discharge in the Middle Rhine sustains a more uniform distribution of bacterial biomass. Thus, the high discharge in the Bonn–Köln area can be more rapidly oxidized and the oxygen content in the Ruhr district increases. The spreading of the development area in the Middle Rhine proves that the firms take advantage of the upstream shift of the point of maximum deficit (recall that the charge depends on the integral of the oxygen deficit). This effect is not shown by the Streeter-Phelps model (see Fig. 10-5-2) since this model does not describe the dynamics of the bacteria.

The differences between the geographical distribution of preferences shown

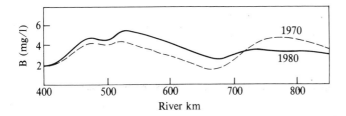

Figure 10-5-4 Bacterial biomass profile in 1970 and 1980.

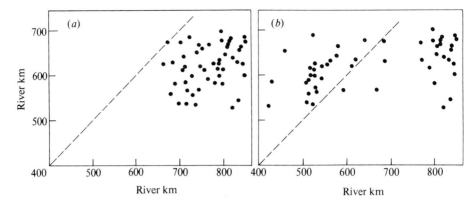

Figure 10-5-5 The barycenter of the five randomly generated locations versus the selected one (taxation scheme A):
(a) Streeter-Phelps model
(b) ecological model.

by the two models are well illustrated in Fig. 10-5-5. The mean value $l_i^* = \frac{1}{5}\sum_{j=1}^{5} l_i^j$ of the five positions considered by each firm i is quoted versus the final selected location $\bar{l}_i$ in the case of the Streeter-Phelps model (Fig. 10-5-5a) and in the case of the ecological model (Fig. 10-5-5b). Since the five potential positions l_i^j are randomly generated, their mean value l_i^* lies, in general, around the middle of the river (km 625). By analyzing Fig. 10-5-5a (Streeter-Phelps model) it can be seen that all the firms select points which are close to the Dutch–German border, while in Fig. 10-5-5b (ecological model), a greater number of firms select a location which is not very far from the mean of the possible locations. (Only the first 50 firms are shown in the figure because the lack of available positions implied by a higher number of firms would not allow this asymmetry to be detected).

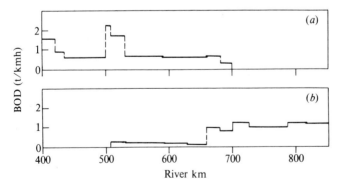

Figure 10-5-6 Additional pollutional load in 1980 (Streeter-Phelps model, damage evaluated by means of Eq. (10-5-5)):
(a) FOD* = 30 mg/l
(b) FOD* = 60 mg/l.

As we have seen, some firms select their locations as close as possible to the border, because the tax does not take into account the damages induced in the Netherlands. This is not unrealistic since no limits are set today on the amount of pollutant crossing the Dutch–German border. However, a more realistic taxation scheme has somehow to take into account the effects of the pollutants on the other side of the border. Since only biodegradable pollutants have been considered a natural measure for these effects is the total amount of oxygen necessary to restore the "clean water condition" via the natural pathway. This measure will be referred to as *final oxygen demand* (FOD), and is nothing but the sum of the oxygen deficit and the BOD load at the border.

Assume then that there is an upper bound FOD* on the final oxygen demand at the border. Then the firms will be charged on the basis of the following environmental damage

$$E = \alpha \frac{\int_{\mathscr{L}} A(l)(c_s - c(l))\, dl}{P^* - \int_{\mathscr{L}} A(l)(c_s - c(l))\, dl} + \gamma \frac{\text{FOD}}{\text{FOD}^* - \text{FOD}}$$

The results of two simulations with this damage function are shown in Fig. 10-5-6. With low values of FOD* the major development takes place in the Upper and Middle Rhine (see Fig. 10-5-6a), while with high values of FOD* the firms select more or less the same positions as in the case with no limits on the FOD (see Fig. 10-5-6b). All intermediate situations can of course be obtained by properly selecting the value of FOD*. In Fig. 10-5-7 the mean of the positions selected by the firms newly entering is shown as a function of time both for

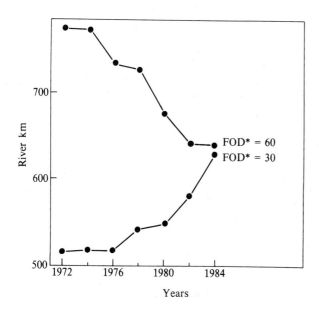

Figure 10-5-7 The barycenter of the positions selected by the newly-entering firms as a function of time for high and low values of FOD* (Streeter-Phelps model).

364 MANAGEMENT OF THE RIVER BASIN

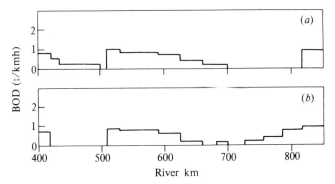

Figure 10-5-8 Additional pollutional load in 1980 induced by taxation scheme B:
(*a*) Streeter-Phelps model
(*b*) ecological model.

high and low values of FOD*. If the bound on FOD is low (30 mg/l) the industrialization develops from upstream toward downstream. In other words, only lack of space forces the firms to consider locations which are closer and closer to the border. On the other hand, if the bound on FOD is high (60 mg/l) the industrialization develops from the border upstream.

Taxation scheme B The effects of a taxation scheme based on a direct evaluation of the damage surplus (see Eq. (10-5-4)) are now shown. In this simulation study the pollution index is assumed to be the maximum (pointwise) oxygen deficit in the basin and its standard value P^* is fixed at 4 mg/l.

The patterns of the 1980 discharges obtained with the Streeter-Phelps and the ecological model are shown in Fig. 10-5-8. No essential differences in their spatial distribution can be noted. Again two major areas of development can be recognized. The first one, in the Middle Rhine is centered upstream from the point where the oxygen content in 1970 has the highest value, while the second one is just upstream from the Dutch–German border (the maximum oxygen deficit created by these firms is in The Netherlands).

The 1980 dissolved oxygen profiles obtained with the ecological model and the taxation schemes A and B are compared in Fig. 10-5-9.

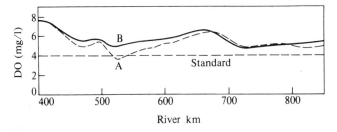

Figure 10-5-9 Dissolved oxygen in 1980 with taxation schemes A and B.

The curve obtained with the taxation scheme B is flatter than the other, as one would intuitively expect. Finally, it is worth noting that both taxation schemes induce an industrial development in the Middle Rhine area (see Figs. 10-5-2, 10-5-3, 10-5-8), due to the fact that no spatial inhomogeneity has been introduced in the taxation schemes (for example, $\beta(l)$ has been assumed to be constant in Eq. (10-5-4)). This would not be acceptable to the public, of course, because the Middle Rhine is the only part of the river section considered which still shows some natural beauties. In order to obtain acceptable industrial development, the Control Agency must levy a charge on the user dependent on its location (for example, the parameter $\beta(l)$ in Eq. (10-5-4) must be greater in the Middle Rhine area).

REFERENCES

Section 10-1

Cohon, J. L. and Marks, D. H. (1975). A Review and Evaluation of Multiobjective Programming Techniques. *Water Resour. Res.*, 11, 208–220.

David, L. and Duckstein, L. (1975). Long Range Planning of Water Resources: a Multiple Objective Approach. *UNPD/UN Interregional Seminar on River Basin and Interbasin Development*, Budapest, Hungary.

Himmelblau, D. M. (1973). *Decomposition of Large-Scale Problems*. North-Holland, Amsterdam.

Keeney, R. L., Wood, E. F., David, L., and Csontos, K. (1976). Evaluating Tisza River Basin Development Plans Using Multiattribute Utility Theory. *IIASA Collaborative Publication CP-76-3*, International Institute for Applied System Analysis, Laxenburg, Austria.

Kneese, A. V. and Bower, B. T. (1968). *Managing Water Quality: Economics, Technology, Institutions*. John Hopkins Press, Baltimore, Md.

Raiffa, H. (1968). *Decision Analysis: Introductory Lectures on Choices under Uncertainty*. Addison-Wesley, Reading, Mass.

Stehfest, H. (1978). On the Monetary Value of an Ecological River Quality Model. Report IIASA RR-78-1, International Institute for Applied Systems Analysis, Laxenburg, Austria.

Yu, T. K. and Seinfeld, J. H. (1973). Observability and Optimal Measurement Location in Linear Distributed Parameter System. *Int. J. Control*, 18, 785–799.

Section 10-2

Beeson, R. M. and Meisel, W. S. (1971). The Optimization of Complex Systems with Respect to Multiple Criteria. *Proc. of Systems, Man and Cybernetics Conference*, IEEE, Anaheim, Calif.

Cohon, J. L. and Marks, D. H. (1975). A Review and Evaluation of Multiobjective Programming Techniques. *Water Resour. Res.*, 11, 208–220.

Dorfman, R. (1972). Conceptual Model of a Regional Water Quality Authority. In *Models for Managing Regional Water Quality* (Dorfman, R., Jacoby, H. D., and Thomas, Jr., H. A. eds.). Harvard University Press, Cambridge, Mass.

Haimes, Y. Y., Hall, W. A., and Freedman, H. T. (1975). *Multiobjective Optimization in Water Resources Systems: The Surrogate Worth Trade-off Method*. Elsevier, Amsterdam.

Keeney, R. L. and Raiffa, H. (1976). *Decisions with Multiple Objectives: Preference and Value Trade-offs*. Wiley, New York.

Koopmans, T. C. (1951). Analysis of Production as an Efficient Combination of Activities. In *Activity Analysis of Production and Allocation, Cowles Comm. Monogr. 13*, (T. C. Koopmans, ed.). Wiley, New York.

Kuhn, H. W. and Tucker, A. W. (1951). Nonlinear Programming. In *Proc. of the Second Berkeley Symp. on Mathematical Statistics and Probability*, (J. Neyman, ed.). University of California Press, Berkeley, Calif., USA.
Monarchi, D. E., Kisiel, C. C., and Duckstein, L. (1973). Interactive Multiobjective Programming in Water Resources: A Case Study. *Water Resour. Res.*, **9**, 837–850.
Ostrom, A. R. and Gros, J. G. (1975). Application of Decision Analysis to Pollution Control: The Rhine River Study. Report IIASA RM-75-45, International Institute for Applied Systems Analysis, Laxenburg, Austria.
Reid, R. W. and Vemuri, V. (1971). On the Noninferior Index Approach to Large Scale Multicriteria Systems. *Journal of the Franklin Institute*, **291**, 241–254.
Roy, B. (1971). Problems and Methods with Multiple Objective Functions. *Mathematical Programming*, **1**, 239–266.
Zadeh, L. A. (1963). Optimality and Non-Scalar-Valued Performance Criteria. *IEEE Trans. on Automatic Control*, **8**, 59–60.

Section 10-3

Brown, R. M., McClelland, N. I., Deininger, R. A., and O'Connor, M. F. (1972). A Water Quality Index: Crashing the Psychological Barrier. In *Indicators of Environmental Quality* (W. A. Thomas, ed.). Plenum Press, New York.
Deininger, R. A. (1965). Water Quality Management: The Planning of Economically Optimal Pollution Control Systems, Systems Research Memorandum No. 125, Northwestern University, Illinois, USA.
Kneese, A. V. and Bower, B. T. (1971). *Managing Water Quality: Economics, Technology, Institutions*. John Hopkins Press, Baltimore, Md.
Revelle, C., Dietrich, G., and Stensel, D. (1969). The Improvement of Water Quality under a Financial Constraint: A Commentary on Linear Programming Applied to Water Quality Management. *Water Resour. Res.*, **5**, 507.
Rinaldi, S., (ed.) (1975). *Topics in Combinatorial Optimization*. Springer-Verlag, Wien, New York.

Section 10-4

Ferrar, T. A. (1973). Progressive Taxation as a Policy for Water Quality Management. *Water Resour. Res.*, **9**, 563–568.
Hass, J. E. (1970). Optional Taxing for the Abatement of Water Pollution. *Water Resour. Res.*, **6**, 353–365.
Kamien, M. S., Schwartz, N. L., and Dolbear, F. T. (1966). Asymmetry between Bribes and Charges. *Water Resour. Res.*, **2**, 147–157.
Kneese, A. V. (1971). Strategies for Environmental Management. *Public Policy*, **XIX**, Winter, 37–52.
Kneese, A. V. and Bower, B. T. (1968). *Managing Water Quality: Economics, Technology, Institutions*. John Hopkins Press, Baltimore, Md.
Mäler, K. G. (1974). *Environmental Economics: A Theoretical Inquiry*. John Hopkins Press, Baltimore, Md.
Rinaldi, S., Soncini-Sessa, R., and Whinston, A. B. (1977). Stable Taxation Schemes in Regional Environmental Management. Report IIASA RR-77-10, International Institute for Applied Systems Analysis, Laxenburg, Austria (to be published in *J. Env. Econ. Manag.*).
Shapley, L. S. (1971). Cores of Convex Games. *Int. J. of Game Theory*, **1**, 11–26.
Upton, C. (1968). Optimal Taxing of Water Pollution. *Water Resour. Res.*, **4**, 865–875.

Section 10-5

Baumol, W. J. and Oates, W. E. (1971). The Use of Standards and Prices for Protection of the Environment. *Swedish Journal of Economics*, **1**, 42–54.

Guariso, G., Rinaldi, S., and Soncini-Sessa, R. (1977). A Taxation Scheme for Environmental Quality Control. *Proc. IFAC Symposium on Environmental Systems Planning, Design and Control, Kyoto, Japan, August, 1–5*. Pergamon Press, Oxford, England.

Guariso, G. and Soncini-Sessa, R. (1976). The Control of Environmental Quality and Regional Industrial Development: A Simulation Study on the Rhine River Basin. *Proceedings of the XIV Automation and Resources Conference, Milano, Italy, 23–29 Nov. 1976*. FAST, Milano, Italy.

Kneese, A. and Bower, B. T. (1968). *Managing Water Quality: Economics, Technology, Institutions*. John Hopkins Press, Baltimore, Md.

Mäler, K. G. (1974). *Environmental Economics: A Theoretical Inquiry*. John Hopkins Press, Baltimore, Md.

Meade, J. E. (1973). *The Theory of Economic Externalities*. A. W. Sigthoff-Leiden, Institut Universitaire de Haute Etudes Internationales, Géneve, Switzerland.

Nash, J. (1951). Noncooperative Games. *Ann. Math.*, **54**, 286–295.

Pigou, A. C. (1929). *The Economics of Welfare*. Macmillan, London.

Wyzga, R. E. (1974). *A Survey of Environmental Damage Functions*. OECD, Paris.

INDEX

Activated sludge, 223
Additivity property, 327
Adenosine bi-phosphate (ADP), 28
Adenosine tri-phosphate (ATP), 28
Adjoint variable, 287
Admissible region, 263
Adsorption, 42, 223
  modeling of, 90
Aeration:
  artificial instream, 235, 255
    optimal allocation of, 255, 267–277
    real time control of, 286–292
  biogenic, 88
  physical, 74–77
(*see also* Aerator, Phototrophs, Reaeration rate)
Aerator:
  cost, 237, 279
  critical distance, 274
  differential efficiency, 236
  diffusion, 235, 238, 289–290
  economies of scale, 237
    properties induced by, 272, 276
  mechanical surface, 235, 238
  optimal allocation of, 267–277
  real time control of, 286–292, 296
  standard efficiency, 235
Aggregated variables, selection of, 77, 81–82
Air pollution, 309
Air temperature, 21–23, 70
Akaike's criterion, 200

Algae (*see* Phototrophs, Water Plant)
Algorithm (*see* Method)
Allocation:
  of artificial aeration, 267–277
  of cooling effort, 257–260
  of flow augmentation, 257–260
  of treatment effort, 246–254, 257–260, 262–264
Ammonia, 41–42
Ammonium, 24–27, 30, 36, 41, 79, 135
(*see also* Nitrification)
Anaerobic conditions, 27, 31, 223–226
  effect of phototrophs on, 36, 225
  in sediments, 41, 80
(*see also* Bacteria: anaerobic)
Approximation:
  of a convex programming problem, 250
  kinematic, 61
  linearization, 13
  low frequency, 123
  one-dimensional, 55
  parabolic, 62
(*see also* Discretization)
Arrhenius' law, 88
Assignment:
  input–output, 292
  pole, 294
Asymptotically stable system, 7

Bacteria, 26–32
  aerobic, 30

† The letter f after a page number indicates that the key word appears on a figure.

INDEX **369**

Bacteria (*continued*)
 anaerobic, 30
 autotrophic, 30
 benthic, 32
  model of biomass dynamics, 86
 community, 31
  influence on BOD_5, 78
 death rate, 83
 growth rate, 35
  maximum specific, 87
 heterotrophic, 30
 impact of: flow rate, 41
  protozoa, 33, 87
 maintenance rate, 83
 measurement of bacterial mass, 132
 metabolism, 30–32
 nitrifiers, 30–31
  growth rate, 31
 planctonic, 32
  model of biomass dynamics, 82–87
 spores, 27
Bacterial mass:
 measurement, 132
 model of the dynamics, 82–89
Balance equation, 47–56
 benthal constituents, 56
 heat, 69–72
 mass, 59–61
 momentum, 61–62
 one-dimensional, 54–55
  limitations of, 55–56
 oxygen, 73
 planctonic constituents, 47–56
 three-dimensional, 48–51
 two-dimensional, 51–53
Bayes' rule, 142, 311
Bayesian estimator, 143
Bellman's principle, 215
Benefit, 340
 economic, 334
 environmental, 334
 sharing of, 334
 social welfare, 334–341
Benthic organism, 32, 36, 129
 balance equation of, 56
Best compromise solution, 320–323
Biocenosis, 24–39
Biochemical Oxygen Demand (*see* BOD)
Biochemical phenomena, 24–41
Biochemical submodel, 43, 73–93, 261
Biological slime, 224
Biomass, 81–82
Bisection procedure, 316
Blow-down water, 233
BOD, 77–78
 benthal BOD decay, 79–80
 carbonaceous, 108
 decay: chemical models, 77–80
  ecological models, 82–87
  as a first order reaction, 79, 94–95, 107–108

BOD, decay (*continued*)
  phases, 78, 80, 108
  as a second order reaction, 79, 109
 definition, 77–78
 distributed load, 107, 113–115
 lag phase, 108–109
 measurement, 77, 132
  how to avoid 152, 188, 199, 286
 nitrogenous, 108
 sensitivity with respect to: equilibrium
   temperature, 102, 133
  heat discharge, 104
  pollutant load, 101
 standards, use of, 247–251, 257
 suspended BOD decay, models of, 79–80
Bormida river, 147–157
 Streeter–Phelps model of, 152–157
Bottom deposit (*see* Sediments)
Boundary:
 lateral, 51, 54
 surface, 51
Boundary conditions:
 in three dimensions, 51
 in two dimensions, 52
 in one dimension, 54, 56
Boundary value problems, multi-point, 159
Bowen ratio, 71
Bribes, 334
Budget constraint, 245, 246, 249
Buoyancy forces, 22
By-pass piping, 241, 257
By-product production, 238

Cam river, 195
 feedforward emergency control on, 305–307
 model of, 195–198
Camp model, 108
Carbon dioxide, 25, 36, 89
Characteristic equation, 7
Characteristic function, 338
 convexity of, 339
 superadditivity of, 339
Characteristic line, 57
Characteristic polynomial, 9
Characteristics, method of, 56–58
Charges on effluent, 317, 334, 346, 353–365
Chemical model, 59, 77–80
Chemical Oxygen Demand (*see* COD)
Chemotrophs, 27–37
Chlorinated hydrocarbons, 30
Ciliates, raptorial, 27, 33, 35
(*see also* Protozoa)
Cloudiness ratio, 70
Coalition, 317, 334–344
Coarse-fine grid method, 264
COD, 78–79
 measurement of, 78–79
Combinatorial Optimization, 325
Compensation, 300–307
Concentration, 48

Concentration (*continued*)
   average, as index, 55
Conceptualization, 14
Condenser, 232
   temperature, 234
Conditions:
   boundary: in one dimension, 54, 56
      in three dimensions, 51
      in two dimensions, 52
   initial, 56
   Kuhn-Tucker, 213, 218
   Lagrange first order necessary, 217
   for a local minimum, 216
Congestion effect, 346
Conjugate gradient method, 216–217
Conservation principle, 49
Constraint:
   BOD discharge, 247–249
   budget, 245, 246, 249
   discretization of, 250
   elimination of redundant, 251–254
   equality, 217
   inequality, 218
   linear, 208–218
   quality standard, 247
   (*see also* Standards)
Constraint method, 321
Consumers:
   growth rate, 36
   role in self-purification, 33
Continuity equation, 60
Continuous model/system, 4
Continuous stirred tank reactor model
      (*see* CSTR model)
Continuum approach, 47
Control:
   design: of real time control systems, 283–307
      of steady state control systems, 244–246
   feedback, 284
   feedforward, 284, 300–307
   local, 291
   open-loop, 283
   optimal emergency, 303
   (*see also* Control problem)
Control Agency, 309, 336
   cost function of a, 337
   as a profit corporation, 348–353
Control law, 285
Control problem, 204–208
   deterministic, 206
   optimal, 206–208
   real time, 283–285
   steady state, 207, 244–246
   stochastic, 206
   unsteady state, 207, 283–285
Control variable, 204
   input–output assignment of, 292
Controllability, 294–295
Controller:
   local, 291

Controller (*continued*)
   supervisory, 292
   (*see also* Control, Regulator)
Convex optimization problem, 216, 218, 246, 255
Convex simplex method, 218
Cooling tower, 232
   mechanical-draft, 232
   natural-draft, 232
   selection of optimal efficiency, 257–260
Core, 343
Cost:
   of aerators, 237, 279
   of low flow augmentation, 240
   minimum treatment, 246
   of reservoirs, 240
   of temperature control, 234
   of treatment network, 235
   of treatment plants, 228–230
   of waste water transmission, 241
Costate, 287
Covariance matrix, 178
Criterion:
   Akaike's, 200
   design, 205, 294
   Dobbins', 120
   myopic, 316, 326
Critical deficit, 98–99
   approximated formulae, 252–254
Critical point, 98–99, 110
   properties of, 99, 110
   upper and lower bound of, 251
Crustacea, 38
CSTR model, 123
   example application, 136, 195–198
   in real time control, 297–300
   with time delay, 124
Cubic interpolation, 217

Damage (*see* Environmental damage, External diseconomy)
Death rate, of bacteria, 83
Decision structure:
   cooperative, 246
   decentralized, 317
   multilevel, 317
   non-cooperative, 317, 356
   single decision maker, 207
Decision vector, 208
Decomposition, 251, 310
   Dantzig–Wolfe technique, 219
   nested technique, 219
Deficit:
   critical (*see* Critical deficit)
   oxygen, 74
Degradation, 27
   pathways of, 31
   rate of, 79
      temperature dependence, 80, 88
   (*see also* Bacteria, Enzyme)
Dehydrogenation, 28

Delaware estuary, 135, 242, 354
Denitrification, 30
Density, 48
Deoxygenation rate, 79
  temperature dependence, 80, 88
De Saint Venant, equations of, 59–63
  kinematic approximation, 61
  parabolic approximation, 62
Description:
  external, 3
  internal, 3
Design criteria, 205, 294
Detention tank, 298–299
Diffusion:
  eddy, 22
  molecular, 20, 50
Diffusion term, 50
Discharge:
  constraint on, 247–249
  regulation of, 241, 298–300
  (see Outlet structure, Standards)
Discrete model/system, 4
Discretization:
  of constraint, 250
  in space, 4, 206, 250–251, 269
  in time, 4, 206
Discrimination, of models, 193–198
Diseconomy, external, 309, 353
Dispersion, 20
  Dobbins' criterion, 128
  longitudinal, 54
    coefficient of, 54
  turbulent, 50–51
Dispersion model, 59
  approximated, 117–128
  Streeter-Phelps, 117
Dispersion term, 50
Dissolved Oxygen (see DO)
Distributed BOD load, 107–113
Distributed lag model, 126
  discrete, 181
    parameter estimation of, 181–187
    state estimate of, 187–191
  example of application, 180–191
    in real time control, 301–303
Distributed parameter model/system, 4, 59
Disturbance, 204
  counteraction against, 283–285
  impulsive, 300
  measurable, 284
  selection of nominal value, 207, 280
  stochastic, 286
  (see also Noise)
DO, 73–77
  balance equation, 73–83
  consumption of, 78
    by bacteria, 30–31
    by protozoa, 33
  critical deficit, 98–99, 110
    approximated formulae of, 252–254

DO (continued)
  critical point, 98–99, 110
    properties of, 99, 110
    upper and lower bound of, 251
  deficit, 74
  effects on DO of: BOD distributed load, 113
    cooling towers, 234
    reservoirs, 240
    photosynthetic oxygen production, 111–113
    surfactants, 116
  measurement: estimate of BOD from, 152, 188, 199, 286
    parameter estimation with DO measurements only, 152, 188, 199
  photosynthetic production of, 36, 89, 111, 225
    modeling of, 108, 135
    temperature dependence, 39
  reaeration rate (see Reaeration rate)
  oversaturation, 37, 117, 237
  saturation concentration, 74
    temperature dependence, 76
  sensitivity with respect to
    algal population, 115–117
    BOD load variations, 101
    discharge pattern, 241
    equilibrium temperature, 102–103, 133
    heat discharge, 104–106
    surfactants concentration, 116
    standards, 247–254, 257
    transport mechanism in the water body, 41, 74
    in the sediments, 41
Dobbins' criterion, 120
Dobbins' model, 108
  examples of application, 147, 255
  parameter estimation, 147–152
    from BOD, DO data, 151
    from DO data only, 152
  sensitivity analysis of, 115–117
Dual problem:
  in convex programming, 250
  in linear programming, 211
  in geometric programming, 213–214
Dual variable, 211
Dynamic programming, 214–215
  coarse-fine grid method, 264
  functional equation, 215
  relaxation method, 264
  in water quality control problems, 260–280
Dynamical system, 2

Ecological model, 59, 80–82
  comparison with Streeter-Phelps model, 267, 330–333, 358–365
  an example of, 129–133
    equilibria, 132
    sensitivity, 133
  examples of application, 158–173, 260–280
  initial guess of parameters, 167–169
  parameter estimation of, 158–166
  survey of, 134–137

372 INDEX

Economies of scale:
  of aerator, 237, 269
    properties induced by, 272, 276
  of a treatment network, 345
    implications on taxation schemes, 347–353
  of treatment plants, 229–230
    advantages of, 334
  in wastewater transmission, 242
Eddy diffusion, 22
Efficiency:
  of aerators, 235
  of an estimate, 142
  of power plants, 231
  of taxation schemes, 342
  of treatment plants, 222–230
    in control problems, 251
  in a river basin, 326
Effluent:
  charges on, 317, 334, 346, 353–365
  real time control of, 292–294, 294–300
  redistribution of, 241
  reuse of, 238
  standard, 205, 244–245, 317, 334, 346
  tax, 317, 334, 340, 346, 353–365
  (see also Outlet structure)
Effluent flow rate, real time control of, 298–300
Eigenvalue, 7, 294
Electre method, 323
Emergency control, 300–307
Emissivity, 70
Endoenzyme, 29
Endogenous respiration (see Respiration: endogenous)
Environmental damage, 337, 356
  congestion effect, 346
    implications of, 348–353
  as an element of control problems, 244–245
Enzyme, 28–30
  allosteric inhibition, 32
    modeling of, 84–86
  competitive inhibition, 32
    modeling of, 84–86
  constituent, 30
  endoenzyme, 29
  exoenzyme, 29
    impact of turbulence on, 41
  inducable, 30
  permease, 29
  reaction controlled by, 82
  repression, 32
    modeling of, 86
Equation:
  balance, 47–56
  characteristic, 7
  continuity, 60
  De Saint Venant, 59–63
  Fredholm integral, 200
  functional, 215
  Michaelis-Menten, 82–84
  measurement, 193

Equation (*continued*)
  Riccati, 179, 287, 289
Equilibrium, 14
  Nash, 356
  temperature, 23, 69, 259
Error:
  estimation, 177, 315
  measurement, 14, 141, 178
Estimate (*see* Estimation techniques)
Estimation:
  accumulative, 145
  of drifting parameters, 187
  off-line, 145
  on-line, 145
  parameter, 140–200
    survey of applications, 198–200
  of pollution sources, 200
  recursive, 145
    example of application, 174–193
  state, 140–141, 174–179
    example of application, 187–191
    survey of applications, 198–200
  (*see also* Estimation techniques)
Estimation error, 177, 315
Estimation techniques:
  Bayesian, 143
  generalized least-squares, 144
  instrumental variable, 199
  Kalman filtering, 178–179
    extended, 193–194
    variance perturbation method, 187
  least-squares, 145
    examples of applications, 147–157, 158–173
  Markov, 144, 146
  maximum likelihood, 143
  property of, 142
  quasilinearization, 159–161
  (*see also* Estimation)
Estimator, 142
  property of, 142
  (*see also* Estimation, Estimation techniques)
Eutrophication, 36, 135, 222, 225, 239
Evaporation, 17, 22, 71, 232, 239
Exoenzyme, 29, 31
  impact of turbulence on, 41
External description, 3
External diseconomy, 309, 353

Feasibility set, 206, 208
Feasible direction method, 218
Fermentation, alcoholic, 30
Fibonacci search, 217
Filter:
  comparison between different types of, 191–193
  extended Kalman, 193–194
  gain matrix of, 179
  Kalman, 146, 178–179
  low pass, 11, 12f
  nonlinear, 194
  recursive in time and space (RFTS), 190–191

INDEX 373

Filter (*continued*)
  suboptimal recursive (SMART), 189–190
Firm, 335–353, 355, 365
Firm yield, 240
Fish:
  abundance, 135
  raptorial fish, 36
Flagellates, saprozoic, 26, 31
Fletcher-Powell method, 217
Fletcher-Reeves conjugate gradient method, 217
Flocculation, 42
  modeling of, 90, 107
Flood wave, 19
  peak flow rate, 17
  spreading of, 61, 63
Flow, 50
  overland, 17
Flow rate, 60
  influence on: self purification, 40–41, 171, 239
    reaeration rate, 41, 77
    stream velocity, 20
  peak, 17
Flow time, 58
Flow time model, 58–59
Flux, 50
Food chain, 27, 38f
Food preferences, 87
Food web, 27, 38f
Forecast, 179
Fourier series, 113
Fredholm integral equation, 200
Free energy, 81
Frequency domain, 9
Frequency response, 11
  of a river to the BOD load, 114
Friction slope, 62
Frictional heat, 22
Function:
  characteristic, 339
  cost (*see* Cost, Objective function)
  generalized Lagrangian, 219
  Hamiltonian, 237
  indifference, 315, 322
  Lagrangian, 218
  likelihood, 143
  objective, 206, 208
  penalty, 219
  posynomial, 213
  probability density, 142
  profit, 335
    properties of, 345
  transfer, 8–11
  transition, 3
  utility, 314, 321
Fungi, 27

Gain, static, 9
Gain matrix, 289, 294
  in Kalman filter, 179
Games, theory of, 317, 339, 356

Gauss-Newton algorithm, 161
Generalized Lagrange multiplier method, 256
Generalized least-squares estimator, 144
Generalized reduced gradient method, 219
Genossenschaften, 354
Geometric programming, 213–214
Glucose, a laboratory experiment with, 83–85
Goal (*see* Attribute)
Gradient method, 217
  conjugate, 217
    Fletcher-Reeves, 217
  generalized reduced, 219
Groundwater:
  flow in, 17
  recharge of, 239
Growth limiting factors, modeling of, 89
Growth rate, 35–39
  of bacteria, 35
  of nitrifiers, 31
  dependence on temperature, 39, 89
  maximum specific, 87
  of phototrophs, 37
  of protozoa, 35

Hamiltonian function, 287
Heat:
  balance equation, 69–72
  frictional, 22
  latent, 22, 71
  sensible, 22, 71
  transfer, 21, 69
  waste, 22, 72
    impact on BOD and DO, 104
Heavy metals, 32, 42–43
Hierarchical optimization, 304
Humidity, 70–71
Hydrogen sulfide, 26–27, 31
  (*see also* Anaerobic conditions)
Hydrograph, 19
  unit, 65
Hydrologic phenomena, 17–20
Hydrologic submodel, 43, 59–68
Hypothesis separation, 16, 143

Identifiability, 145, 160
Implementation, 316
Impulse response, 7
Impulsive disturbance, 300
Index:
  performance, 206, 208
  pollution, 246, 249, 324–325, 357
    additivity property of, 327–330
Indicators, water quality, 39, 324
Indifference curves, 315, 322
  (*see also* Multiobjective Programming)
Infiltration, 17
Information matrix, 143
  in quasilinearization technique, 161
Inhibition:
  allosteric, 32, 131

Inhibition: allosteric (*continued*)
    modeling of, 84–86
  competitive, 32, 131
    modeling of, 84–86
Input, 1
Input–output:
  assignment, 292
  model, 3
  relationship, 3, 141
    of linear systems, 7
Insensitivity, as a design criterion, 205
(*see also* Sensitivity analysis)
Instrumental variable method, 199
Integration schemes, 56–58, 63–64
Interception, 17
Interflow, 17
Internal description, 2f, 3
Interpolation:
  cubic, 217
  quadratic, 217
Ion exchange, 42

Jacobian matrix, 160

Kalman filter, 146, 178–179
  examples of application, 179–193
  extended, 193–194
    example of application, 195–198
  gain matrix, 179
Kinematic approximation, 61
Kuhn–Tucker conditions, 213, 218

Lagrange, first order necessary condition of, 218
Lagrange duality, 303
Lagrange multipliers, 218, 246
Lagrangian, 218
  generalized, 219
Land disposal, 238
Land use management, 309
Laplace transform, 8
Large-scale mathematical programming, 219
(*see also* Decomposition)
Law:
  Arrhenius', 88
  of mass action, 82
  Stefan-Boltzmann, 70
Lexicographic method, 319, 324
Lexicographic preference structure, 319, 324
Light, 21
  incident, 21
  intensity of, 37, 39, 89
Likelihood function, 143
Linear model/system, 5, 59
Linear programming problem, 208–212
  dual problem, 211
  feasibility set, 209
  primal problem, 211
  standard form, 209
  in water quality control problem, 246–254

Linear regulator, 299
Linearization, 13
Linearized model/system, 14
Longitudinal dispersion, 54
  coefficient of, 54
  prediction, 54
Low-flow augmentation, 239, 257
  cost of, 240
Low flow condition, reference:
  on the Rhine river, 267, 277
  selection of, 280
Low frequency approximation, 123
Low-pass filter, 11, 12f
  the river as a, 113–114
Luenberger state reconstructor, 176–178
Lumped parameter model/system, 4, 59

Maintenance rate, 83
Markov estimator, 144
Mass:
  balance equation, 59–61
  bulk density, 48
  law of mass action, 82
Mathematical programming, 206, 208–219
  constrained nonlinear, 217–219
  convex, 216, 246–250, 255
  dynamic, 214–215
    (*see also* Dynamic programming)
  geometric, 213–214
  large scale, 219
    (*see also* Decomposition)
  linear, 208–212
    (*see also* Linear programming problem)
  multiobjective, 310
    (*see also* Multiobjective programming)
  multistage, 214, 219, 260
  linear, 251
  nonconvex, 256
  nonlinear, 215–219
    in water quality control problem, 254–260
  quadratic, 212–213, 256
  unconstrained nonlinear, 216–217
Matrix:
  controllability, 295
  covariance, 178
  gain, 289, 294
    of Kalman filter, 179
  Hessian, 216, 253
  information, 143
    in quasilinearization technique, 161
  Jacobian, 160
  observability, 176
  positive definite, 178
  positive semidefinite, 178
  transition, 6
Measurement:
  of bacterial mass, 132
  of BOD, 77, 132
    how to avoid, 152, 188, 199, 286
  of COD, 78, 79

INDEX **375**

Measurement (*continued*)
　of pollutant, 132
　of protozoan mass, 132
Measurement equation, 193
Measurement network, design of, 315
Measurement noise, 14, 141, 178
Michaelis-Menten expression, 82–84
　growth limiting factors, 89
Migration term, 50
Migration velocity, 50
Minimum:
　global, 215
　local, 215
Model, 1
　approximated dispersion, 117–128
　attributes of, 59
　Camp, 108
　chemical, 59, 77–80, 94–117
　Continuously Stirred Tank Reactor (CSTR), 123
　　with time delay, 124
　　(*see also* CSTR model)
　discrete distributed lag, 181
　　(*see also* Distributed lag model)
　dispersion, 59
　distributed parameter, 4, 59
　distributed lag, 126
　　(*see also* Distributed lag model)
　Dobbins', (*see* Dobbins' model)
　ecological, 59, 80–88
　　(*see also* Ecological model)
　flow time, 59
　input–output, 3
　linear, 5, 59
　lumped parameter, 4, 59
　mechanistic, 3
　Nash, 64–66
　nonlinear, 5, 59
　O'Connor's, 108
　plug-flow, 58–59
　rainfall–runoff, 17
　state, 3
　steady state, 59
　Streeter-Phelps (*see* Streeter-Phelps model)
　structure of, 14
　Thomas, 107
　time invariant, 4, 59
　time varying, 4, 59
　unsteady state, 59
　(*see also* System)
Model discrimination, 193–198, 262–267
　selection of the "best" model, 310–311
Modeling process, 14–16
Molecular approach, 47
Molecular diffusion, 20
Momentum:
　balance equation, 61–62
　density of, 48
Motion, 5
　forced, 5

Motion (*continued*)
　free, 5
Multilevel decision structure, 317
Multiobjective programming, 310, 318–323
　best compromise solution, 310, 320–323
　method of solution: constraint, 321
　　Electre, 323
　　lexicographic, 319, 324
　　Semops, 323
　　surrogate worth trade-off, 323
　　weighting, 320
　noninferior solution, 319
　Pareto optimal solution, 319
　preference structure (*see* Preference: structure)
Multipoint boundary value problem, 159
Multistage decision problem, 214, 219, 251, 260
Multivariable model/system, 5
Myopic optimization criterion, 316, 326

Nash equilibrium, 356
Nash model, 64–66
　transfer function, 66
Network:
　measurement, design of, 315
　of treatment plant, 241, 334–335
　　cost of, 335
　　economies of scale in a, 345
Newton's method, 217
Nitrate, 24, 89
Nitrification, 26, 135
　in trickling filters, 225
Nitrifiers (*see* Bacteria: nitrifiers)
Nitrite, 24
Nitrogen, 222, 239
Noise:
　measurement, 14, 141, 178
　process, 14, 141, 178
　output, 14
　(*see also* Disturbance)
Nominal value, 206, 288
　selection of, 207, 280
Noninferior solution, 319
Nonlinear filter, 179, 194
Nonlinear model/system, 5, 59
Nonlinear programming, 215
　constrained, 217
　unconstrained, 216

Objective function, 206, 208, 318
　(*see also* Attribute)
Observability, 145
　complete, 176
Observability matrix, 176
O'Connor's model, 108
Once-through cooling, 232
Open-loop control, 283
Operating rule, 241
Optimal control problem, 206
　real time, 283–285
　steady state, 207, 244–246

Optimal solution, local, 260
Optimality:
 Bellman's principle of, 215
 as a design criterion, 205
Optimization problem:
 primary, 323
 secondary, 310
 (*see also* Control problem, Mathematical programming, Multiobjective programming, Second variation control theory)
Optimization methods:
 bisection, 316
 coarse-fine grid, 264
 conjugate gradient, 216–217
 constraint, 321
 convex simplex, 218
 cubic interpolation, 217
 Dantzig-Wolfe decomposition, 219
 direct search, 260
 Electre, 323
 feasible direction, 218
 Fibonacci search, 217
 Fletcher-Powell, 217
 Fletcher-Reeves conjugate gradient, 217
 Gauss-Newton, 161
 generalized Lagrange multiplier, 256
 generalized reduced gradient, 219
 gradient, 217
 Lagrange multiplier, 217
 nested decomposition, 219
 Newton's, 217
 penality function, 219
 quadratic interpolation, 217
 relaxation, 264
 Rosen's gradient projection, 255
 Semops, 323
 sequential unconstrained minimization, 255
 simplex, 210
 steepest descent, 216–217
 surrogate worth trade-off, 323
 variable metric, 216–217
 variance perturbation, 187
 weighting, 320
 Wolfe and Beale, 213
Outlet structure, influence on:
 use of one-dimensional models, 55
 water temperature, 23
Output, 1
 noise, 14
Output prediction, 179
Output transformation, 3
Overland flow, 17
Oversaturation (*see* DO)

Parabolic approximation, 62
Parallel connection, 10f
Parameter, 15
 dependence on flow rate and temperature, 87–88

Parameter estimation (*see* Estimation, Estimation techniques)
Pareto optimal solution, 319
Particulate pollutants, 42–43, 81
 modeling the decay of, 36
Penalty function method, 219
Peptone, an experiment with, 24–27
Percolation, 17
Performance index, 206, 208
Permease, 29
Perturbation (*see* Disturbance)
PH, 41–43
Phosphates, 36, 89
Phosphorous, 222, 238
Phototrophs, 36–38
 algal bloom, 36
 dependence on light intensity, 37, 89
 endogenous respiration, 39, 89
 growth limiting factors, 89
 growth rate, 37
  dependence on temperature, 39, 89
 impact: on self-purification, 36, 225
  on DO concentration, 111–113, 115–117
 models of:
  photosynthetic oxygen production, 89
  biomass dynamic, 89
P.I.D. regulator, 299–300
Planctonic constituents, balance equation of, 47–56
Plug flow model, 58–59
Pole, 7, 294
Pole assignment, 294–296
Pollutant:
 degradation by bacteria, 27–32
 dissolved, 81
  modeling the decay of, 77–86
 nondegradable, 132, 222, 239
 particulate, 42–43, 81
  modeling the decay of, 86
 sediment, 87
 suspended, 41
Pollution index (*see* Index)
Polynomial bounded algorithm, 325
Posynomial function, 213
Potassium dichromate, 78
Potassium permanganate, 78
Power plant:
 efficiency of, 231
 throttling of, 232
 waste heat production of, 230
Precipitation, 17, 42, 222
 modeling of, 90
Prediction, 179
Preference structure, 316, 319
 indifference curves, 322
 lexicographic, 319, 324
 optimal weights, 322
 utility functions, 314, 321
(*see also* Multiobjective programming)
Prey-predator relationship, 133

Prices, as dual variables in linear
    programming, 212, 254
Primal problem, 211
Primal-dual pair, 211
Primary optimization problem, 323
(see also Control problem, Mathematical
    programming, Multiobjective
    programming)
Principle:
  basic, of taxation schemes, 354–355
  general conservation, 49
  of optimality, 215
  superposition, 5
Probability:
  conditional, 142
  a posteriori, 142
  a priori, 143, 312
Probability density function, 142
Process noise, 14, 141, 178
Producers, 335
Profit, as an element of control problems, 244
Profit function, 335
  aggregated, 338
  properties of, 345
Programming (see Mathematical programming)
Protozoa, 24–27, 33–36
  ciliata, 27, 33, 35
  growth rate of, 35
  interactions with bacteria, 33, 87
  measurement of protozoan mass, 132

Quadratic interpolation, 217
Quadratic programming, 212, 256
Quantities:
  extensive, 48
  intensive, 48
Quasilinearization technique, 158–161

Radiation:
  atmospheric, 21, 70
  effect on water temperature, 21
  solar, 21, 70
  of water, 21, 70
Rainfall–runoff model, 17
Rate:
  death, of bacteria, 83
  degradation, 79
    temperature dependence, 80, 88
  deoxygenation, 79, 95
    temperature dependence, 80, 88
  growth (see Growth rate)
  maintenance, 83
  reaeration (see Reaeration rate)
  respiration, endogenous, 88
  self-purification, 98
Ratio:
  Bowen, 71
  cloudines, 70
Reactions:
  energy yielding, 28–30

Reactions (continued)
  enzyme catalyzed, 28
Reaeration process (see Aeration, Reaeration
    rate)
Reaeration rate, 73–74
  dependence: on flow rate, 41, 77
    on temperature, 76
  approximated formulae, 75–77
Real time control, 207, 283–285
  of artificial instream aeration, 289–292, 296
  of effluent discharge, 292–293, 297–300
Realization problem, 3
  of linear systems, 11
Reconstructor, Luenberger's state, 176
Recovery, material, 238
Recursive estimation, 145
  example of application, 174–193
Regional profit, 337–338
Regional treatment facilities, 241, 335
Regulation of discharge, 241, 298–300
Regulator, 285
  linear, 299
  industrial, 299–300
  P.I.D., 299
(see also Control)
Relaxation method, 264
Release time, 95
Repression, 32
  modeling of, 86
Reservoir:
  impact on water temperature, 23
  firm yield, 240
  operating rule, 241
  time constant of, 65
Respiration, endogenous, 27, 33
  dependence on temperature, 39, 88–89
  modeling of, 108, 131
  in treatment plants, 223
  rate of, 88
Resuspension, 41
  modeling of, 89–90, 107
Rhine river, 158
  bacterial biomass concentration, 35f
  control of the industrial development on the,
    358–365
  hydrology, 163
  models of, 161–173
    sensitivity analysis, 169–173
  optimal allocation on the:
    of treatment effort, 264–267
    of artificial instream aeration, 26–29, 267
  optimal sequencing in treatment plant
    construction, 330–333
  pollution sources, 163
Riccati equation, 179, 287, 289
Rivers:
  particular effects in: deep, 37, 129
    fast flowing, 129
    highly polluted, 89, 129
    impounded, 31

Rivers: particular effects in (*continued*)
    shallow, 37
    slow flowing, 31, 41
    as a sequence of: reservoirs, 64–66
    reservoirs and channels, 66–67
Rosen's gradient projection method, 255
Rule:
    Bayes', 142
    operating, 241
Runoff, surface, 17
Runoff model, 17

Saddle point, 119
Sag curve, 98
Saint Venant (*see* De Saint Venant)
Sampling time, 4
Saprobial system, 39
Saturation (*see* DO)
Scour, 107
Second variation control theory, 286–289
Second variation problem, 286
Secondary optimization problem, 310, 323
Sediment, 41
    decay of BOD in the, 79–80, 87
Sedimentation, 41, 239
    modeling of, 89–90, 107
Sedimentation tank, 222, 227
Self-purification:
    biochemical, 27
    dependence on: flow rate, 40–41, 171, 239
        temperature, 39
    physical, 41
    role in self-purification of: bacteria, 27–32
        chemical agents, 41
        higher order consumer, 33–36
        physical agents, 39–41
        phototrophs, 36–38
Self-purification phenomena, 24–41
Self-purification rate, 98
Semops method, 323
Sensitivity analysis, 99–101
    of Dobbins' model, 115–116
    of an ecological model, 133
    in linear programming problems, 211–212
    of Streeter-Phelps model, 101–106
Sensitivity system, 101
Sensitivity theory, 99–101
Sensitivity vector, 100
Separation hypothesis, 16, 143
Sequencing problem, 317, 323–330
Sequential unconstrained minimization method, 255
Serial connection, 10f
Set:
    feasibility, 206, 209
    input, 2
    output, 2
    state, 2
Simplex method, 210

Slack variable, 209
Slope:
    bottom, 61
    friction, 62
Sludge recycle, 223
Social welfare, 334–336
Solid, suspended, 132, 222, 239
Solid pollution, 309
Solution:
    best compromise, 310, 320–323
    feasible, 206
    local optimal, 260
    noninferior, 319
    Pareto optimal, 319
Sorbitol, an experiment with, 32, 85
Sorption, 42
Sources:
    estimation of pollution sources, 200
    type of, in the balance equation, 54–55
(*see also* Outlet structure)
Stability:
    as a design criterion, 205, 294
    of a system, 7
Stage discharge relationship, 60–62, 155
Stagnant zone, 55
Standard:
    BOD, 247–252, 257
    as a component of control problems, 244–246
    DO, 247–254, 257, 266–267
    effluent, 205, 227, 244–245, 317, 334, 346
    non-degradable pollutants, 266
    selection of, 246, 254
    stream, 205, 227, 244–245, 246–247, 255, 357
    water temperature, 72, 257
State, 2
    augmented, 189
State estimation (*see* Estimation, Estimation techniques)
State model, 3
State prediction, 179
State reconstructor, 174–179
    Luenberger's, 176
    in a regulator, 285
Static gain, 9
Steady state control, 207
    design of, 244–426
Steady state model, 59
Steepest descent method, 216–217
Stefan-Boltzmann law, 70
Stochastic control, 206
Stochastic system, 14
Stream standard, 205, 227, 244–245, 246–247, 255, 357
Streeter-Phelps model, 79–80, 94–106
    approximated dispersion model, 123
    comparison: with an ecological model, 267, 330–333, 358–365
    with measured data, 81f, 153
    with dispersion, 117

Streeter-Phelps model (*continued*)
  modified, 107–117
    examples of applications, 147–157, 163–173, 255–260, 261–267
    parameter estimation of:
      from BOD, DO data, 151
      from DO data only, 152–155
    sensitivity analysis, 101–106
Structure:
  decision (*see* Decision structure)
  preference (*see* Preference: structure)
  outlet (*see* Outlet structure)
Submodel:
  biochemical, 43, 73–93, 261
  hydrologic, 43, 59–68
  interrelationship between the submodels, 43–44
  thermal, 43, 68–73
Suboptimal recursive filter, 189
Subsidy, 334
Substrate degradation, model of, 82–87
(*see also* Bacteria, BOD, Pollutants)
Succession, 24
Sulfate, 30
Sulfur, 30
Superposition principle, 5
Surface:
  boundary, 51
  layer, 74–75
Surface runoff, 17
Surfactants, impact on DO, 116
Surrogate worth trade-off method, 323
Synthesis stage, 27–31, 223
System, 1
  asymptotically stable, 7
  completely controllable, 295
  completely observable, 176
  continuous, 4, 5–14
  discrete, 4, 12–13
  distributed parameter, 4, 59
  dynamical, 2
  linear, 5–14
    block diagram representation, 6f
  linearized, 14
  lumped parameter, 4, 5–14, 59
  multivariable, 5, 12
  nonlinear, 5, 13–14
  selection of the order of the, 200
  sensitivity, 101
  single input–single output, 5
  stability, 7
  stochastic, 14
  time-invariant, 4, 59
  time-varying, 4, 59
(*see also* Model)
System identification, 15
(*see also* Estimation, Estimation techniques)

Tax, 340
(*see also* Charge)

Taxation schemes, 340–344, 354–358
  acceptable, 142
  analysis of: dynamic, 353–365
    static, 334–353
  attributes of, 342–344
  basic principles, 354–355
  efficient, 342
  lexicographic, 341
  properties of, 344
  pseudo-efficient, 356
  stable, 342
Temperature (*see* Air temperature, Water temperature)
Temperature dependence of:
  degradation rate, 80, 88
  DO saturation concentration, 76
  endogenous respiration rate, 39, 88–89
  maximum specific growth rate, 39, 89
  phototrophs activity, 39
  reaeration rate, 76
  self-purification, 39
Thermal phenomena, 20–24
Thermal submodel, 43, 68–73
Time, 1
  discrete, 12
  flow, 58
  release, 95
Time constant, 9
  of a reservoir, 67
Time invariant model/system, 4, 59
Time varying model/system, 4, 59
TOC, 81
Total Organic Carbon (TOC), 81
Transfer function, 8
  of the approximated dispersion model, 123
  of Nash model, 66
  of Streeter-Phelps model, 114–115
Transform:
  Laplace, 8
    property of, 9
  Z, 13
Transition function, 3
Transition matrix, 6
Treatment network, 241, 334–335
  cost of a, 335
  economies of scale in a, 345
  implications on taxation schemes, 347–353
Treatment plant, 225–230
  advanced, 222, 226, 235, 239, 240, 268
  biological unit processes, 222, 226
    activated sludge, 223, 235
    aerobic, 222
    aerobic stabilization pond, 222, 225, 235
    aerobic–anaerobic, 222, 226
    anaerobic, 222, 226
    digestor, 226
    facultative pond, 226
    trickling filter, 223–224
  chemical unit processes, 221–222, 227

Treatment plant (*continued*)
    cost function of, 227–230
        determination of, 228–230
        properties of, 230
    design of, 226–228
    economies of scale, 229–230
    efficiency, 222–230
        real time control of, 297–298
        with respect to a river basin, 326
    modeling of, 228
    primary, 226
    regional network of, 241, 334–335
        (*see also* Treatment network)
    secondary, 226
    selection of optimal efficiency in river basin control problems, 246–254, 257–260
    tertiary, 222, 227, 235, 239, 240, 268
    unit operation, 221, 226
    unit processes, 221–226
Turbulence:
    impact on exoenzyme, 41
    role in self-purification, 41

Unbiasedness, of an estimate, 142
Unit hydrograph, 65
Unsteady state control, 207, 283–285
    design of, 283–307
Unsteady state model, 59
Utility function, 314, 321
(*see also* Multiobjective programming)

Validation, 15
Value:
    mean, 207
    nominal, 206, 288
        selection of, 207, 280
    worst possible, 207
Vapor pressure, 71
Variable:
    adjoint, 287
    aggregated, selection of, 77, 81–82
    control, 204, 292
    decision, 208
    dual, 211–212
    input, 1
    instrumental, method of, 99
    output, 1
    slack, 209
    state, 2
Variable metric method, 216–217

Variance perturbation method, 187
Vector optimization (*see* Multiobjective programming)
Velocity:
    bulk, 48
    diffusive, 50
    effects of non-uniform velocity profile, 53, 54
    equivalent flow, 120
    migration, 50
    water, 48, 60
    of a wave, 61
Waste heat, 22, 72
    impact on BOD and DO, 104
Waste treatment (*see* Treatment plant)
Wastewater transmission, 241
Water cycle, 17, 18f
Water inflow, 55
Water plants, role of, in nitrification, 31–32
Water quality:
    definition of, 38–39
    indicators, 39, 324
    management problems, definition of, 309
Water resources development, 309
Water surface layer, 74–75
Water temperature:
    control of, 230–234
    equilibrium, 23, 69, 259
        sensitivity of DO with respect to, 102, 133
    influence on:
        DO concentration, 116–117
        self-purification, 39–40
    modeling the dynamic of, 68–73, 259
    natural, 23, 72
    processes affecting the, 21–24
    as quality indicator, 21
    standards, 72, 257
Weighting factors, 246
    optimal, 322
    selection of, 246
    (*see also* Lagrange, Prices)
Weighting method, 320
Wind velocity, 22, 71
Worms, 36

Yield factor, 83
Yomo river, 180
    model of, 184–191

Z-transform, 13